An Introduction to Hopf Algebras

Robert G. Underwood

An Introduction
to Hopf Algebras

 Springer

Robert G. Underwood
Department of Mathematics
Auburn University
Montgomery, AL 36124
USA
runderwo@aum.edu

ISBN 978-1-4899-9784-5   ISBN 978-0-387-72766-0 (eBook)
DOI 10.1007/978-0-387-72766-0
Springer New York Dordrecht Heidelberg London

Mathematics Subject Classification (2010): 13AXX, 16TXX, 20AXX

Printed on acid-free paper

Springer is part of Springer Science+Business Media (www.springer.com)

To my wife, Rebecca

Preface

The purpose of this book is to provide an introduction to Hopf algebras. This book differs from other texts in that Hopf algebras are developed from notions of topological spaces, sheaves, and representable functors. This approach has certain pedagogical advantages, the foremost being that algebraic geometry and category theory provide a smooth transition from modern algebra to Hopf algebras. For example, the motivation for the definition of an exact sequence of Hopf algebras is best established by first defining exactness for a sequence of representable group functors.

Hopf algebras are attributed to the German mathematician Heinz Hopf (1894–1971). The study of Hopf algebras spans many fields in mathematics, including topology, algebraic geometry, algebraic number theory, Galois module theory, cohomology of groups, and formal groups. In this work, we focus on applications of Hopf algebras to algebraic number theory and Galois module theory. By the end of the book, readers will be familiar with established results in the field and should be poised to phrase research questions of their own.

An effort has been made to make this book as self-contained as possible. That said, readers should have an understanding of the material on groups, rings, and fields normally covered in a basic course in modern algebra. Also, most of the groups given here are Abelian. All of the rings are non-zero commutative rings with unity, and we only consider commutative algebras.

This work contains 12 chapters. Each chapter contains an exercise set with questions ranging in level of difficulty from elementary to advanced. In the last chapter, we list some open problems and research questions.

In Chapter 1, we begin by introducing Spec A, the spectrum of a commutative ring with unity A. Included is a section on the role played by nilpotent elements as the zero functions on Spec A. In Chapter 2, we endow Spec A with the Zariski topology and study its properties as a topological space. We next introduce presheaves and sheaves on Spec A. The structure sheaf on Spec A is then generalized to the functor $\mathrm{Hom}_{R\text{-alg}}(A, -)$, represented by the R-algebra A. In Chapter 3, we endow $\mathrm{Hom}_{R\text{-alg}}(A, -)$ with the structure of a group functor, which leads in Chapter 4 to the definition of an R-Hopf algebra H. In this chapter, we prove in detail the fundamental result of R. Larson and M. Sweedler that states that if H is a cocommutative Hopf algebra of finite rank over a Dedekind domain R, then the linear dual H^* is isomorphic to $H \otimes_R \int_{H^*}$ as H-modules, where $\int_{H^*}$ is the ideal of integrals of H^*.

We specialize to Hopf orders in group rings in Chapters 5 through 9. In Chapter 5, we review R. Larson's construction of Hopf orders in KG, where G is an arbitrary finite Abelian group, and in Chapter 6 we employ formal groups to construct a collection of R-Hopf orders in KC_{p^n}, where C_{p^n} denotes the cyclic group of order p^n. The material in this chapter is joint work with Lindsay Childs of the University at Albany, SUNY. In Chapter 7, we review the classification of Hopf orders in KC_p due to J. Tate and F. Oort, and in Chapter 8 we employ C. Greither's cohomological argument to give a complete classification of Hopf orders in KC_{p^2} in the case where K is a finite extension of $\mathbb{Q}_p$ containing a primitive p^2nd root of unity. In Chapter 9, we review several constructions of Hopf orders in KC_{p^3} (also joint work with L. Childs).

We next turn to some important applications of Hopf algebras. In Chapter 10, we use results of Lindsay Childs and Nigel Byott to describe the ring of integers in an algebraic number field as a Hopf-Galois module. In Chapter 11, we give a topological proof of the finiteness of the class group $\mathcal{C}(R)$, following J. W. S. Cassels. We then generalize the class group to Hopf orders H and give A. Fröhlich's hom-description of the class group $\mathcal{C}(H)$. Next, we investigate the structure of the Hopf-Swan subgroup of $\mathcal{C}(H)$. In Chapter 12, we discuss some open questions and research problems.

This book grew out of lectures given during a year-long course in modern algebra at Auburn University, Montgomery Campus, over the period 2006–2007. I would like to thank Professors Lindsay Childs, Jorg Feldvoss, James Carter, Tim Kohl, and the late Bettina Richmond, who kindly agreed to read a draft of the book. I am especially grateful to Lindsay Childs, who introduced me to Hopf algebras and whose body of work has informed and influenced my study of them. I would also like to thank former and current students who read and commented on early drafts of the class notes.

I wish to thank the Department of Mathematics at Auburn University, Montgomery, for its support during my sabbatical year, 2008–2009, and the Department of Mathematics at Florida State University for its hospitality and support during my sabbatical year while I worked on this book. In particular, I thank Professor Warren Nichols for our many discussions about coalgebras, bialgebras, Hopf algebras, and data mining.

I wish to thank my editors at Springer, Ann Kostant and Elizabeth Loew, for their patience with my earlier drafts and their support and guidance throughout this project.

Finally, I thank my wife, Rebecca Brower, my son, Andre, and my parents, especially my father, the late Kenneth L. Underwood (1933–2010), who have been a great inspiration to me and who have given me the strength to persevere and complete this work.

Figures 1.1 and 1.2 originally appeared as Figures 13 and 14, respectively, in *Basic Algebraic Geometry* by I. R. Shafarevich, © Springer-Verlag, 1974, and are used with kind permission of Springer Science and Business Media.

February 2011   Robert G. Underwood

Contents

Notation

$\mathbb{Z}$	ring of integers
$\mathbb{Z}_{>0}$	positive integers
$\mathbb{Q}$	field of rational numbers
$\mathbb{Q}^\times$	non-zero rationals
$\mathbb{R}$	field of real numbers
$\mathbb{R}_{>0}$	positive real numbers
$\mathbb{R}^\times$	non-zero reals
$\mathbb{C}$	complex numbers
1_A	unity in the ring A
λ_A	structure map of the R-algebra A
ζ_m	primitive mth root of unity
$A \subseteq B$	A is a subset of B
$A \subset B$	A is a proper subset of B
$H \le G$	H is a subgroup of G
$H < G$	H is a proper subgroup of G
$H \lhd G$	H is a normal subgroup of G
$\lvert G \rvert$	order of the finite group G
$\bar{a}$	image of a under canonical surjection $A \to A/B$
U_n	group of the nth roots of unity
C_m	cyclic group of order m
$\hat{G}$	character group of G
$[G : H]$	index of H in G
$[L : K]$	degree of L over K
R_P	localization of R at prime ideal P
$\hat{R}_P$	completion of R at prime ideal P
π	uniformizing parameter
$U_j(R)$	group of units in R of the form $1 + R\pi^j$
$\mathbb{F}_p$	finite field with p elements
$\mathbb{Z}_p$	ring of p-adic integers
$\mathbb{Q}_p$	field of p-adic rationals
$\lvert\ \rvert_v$	normalized absolute value on K

K_ν	completion of K with respect to the $\vert\;\vert_\nu$-topology
$\mathrm{disc}(S/R)$	discriminant of S over R
Spec A	spectrum of the ring A
$R[F]$	representing algebra of the representable functor F
$\mathbf{G}_m$	multiplicative group scheme
$\mathbf{G}_a$	additive group scheme
μ_n	multiplicative group of the nth roots of unity
$\mathrm{Hom}_{R\text{-alg}}(A, B)$	R-algebra homomorphisms from A into B
H	Hopf algebra, Hopf order
Δ_H	comultiplication map of H
ϵ_H	counit map of H
σ_H	coinverse of H
H^D	dual module of H
H^*	linear dual of H
$\langle\,,\,\rangle$	duality map $H^* \times H \to R$
ξ	order-bounded group valuation
$A(\xi)$	Larson order given by ξ
$\Xi(H)$	order-bounded group valuation determined by H
$H(i), H(i, j)$	Larson order in KC_p, KC_{p^2}, respectively
$F(\overline{x}, \overline{y})$	n-dimensional degree 2 formal group
$E(G, C_m)$	equivalence classes of extensions of G by C_m
$A(i, j, u)$	Greither order in KC_{p^2}
$A(i, j, k, u, v, w)$	duality Hopf order in KC_{p^3}
$\mathrm{Gal}(L/K)$	Galois group of L over K
$GE(H, R)$	Galois H-extensions of R
$SGE(H, R)$	semilocal Galois H-extensions of R
V_K	adele ring of K
J_K	group of ideles of K
I_K	ideal group of K
$\mathcal{F}(R)$	fractional ideals of K
$\mathcal{C}(R)$	class group of R
$\mathcal{C}(H)$	class group of the Hopf order H
$\mathcal{R}(H)$	realizable classes in $\mathcal{C}(H)$
$\mathcal{T}(H)$	Hopf-Swan subgroup of $\mathcal{C}(H)$

Chapter 1
The Spectrum of a Ring

1.1 Introduction to the Spectrum

Throughout this chapter, by "ring" we mean a non-zero commutative ring with unity. Our first proposition employs Zorn's Lemma [Rot02, Appendix].

Let S be a non-empty set, and let $\preceq$ be a relation on S. We say that S is **partially ordered** if $\preceq$ is reflexive, antisymmetric, and transitive. A subset T of a partially ordered set S is a **chain** if for all $x, y \in T$ either $x \preceq y$ or $y \preceq x$.

Zorn's Lemma. *Let X be a non-empty partially ordered set in which each chain has an upper bound. Then X has a maximal element.*

Proposition 1.1.1. *Let A be a ring. Then every proper ideal of A is contained in a prime ideal.*

Proof. Let J be a proper ideal of A, and let $\mathcal{P}$ denote the collection of all proper ideals of A that contain J. Since $J \in \mathcal{P}$, $\mathcal{P}$ is a non-empty set that is partially ordered under set inclusion. Let $\mathcal{C}$ be a chain in $\mathcal{P}$. Then the ideal $\cup_{I \in \mathcal{C}} I$ is an upper bound for $\mathcal{C}$. Thus, by Zorn's Lemma, $\mathcal{P}$ contains a maximal element M. We show that M is a maximal ideal by showing that A/M is a field.

Since A is a commutative ring with unity, then so is A/M. Let b be an element of A with $(b) + M \neq M$. Since $M \subseteq (b) + M$, $(b) + M$ cannot be in $\mathcal{P}$, and so $M \subseteq (b) + M = A$. Since $1 \in A$, there exist elements $r \in A$ and $m \in M$ such that $rb + m = 1$ and hence $rb = 1 - m$. Now

$$(r + M)(b + M) = rb + M$$
$$= (1 - m) + M$$
$$= (1 + M) + (-m + M)$$
$$= 1 + M,$$

and thus $b + M$ is a unit of A/M. Consequently, A/M is a field such that M is a maximal ideal of A. Since every maximal ideal is a prime ideal, the result follows.

R.G. Underwood, *An Introduction to Hopf Algebras*, DOI 10.1007/978-0-387-72766-0_1,   1
© Springer Science+Business Media, LLC 2011

Definition 1.1.1. Let A be a ring. The collection of prime ideals of A is the **spectrum of A** and will be denoted by Spec A. The elements of the spectrum are the **points** of Spec A.

Proposition 1.1.1 shows that Spec A is non-empty.

When the ideal (0) is a prime ideal of A, and hence is a point of Spec A, we write $\omega = (0)$. For example, Spec $\mathbb{Z} = \{(2), (3), (5), (7), (11), \dots, \omega\}$.

Proposition 1.1.2. *Let F be a field. Then* Spec $F = \{\omega\}$.

Proof. The only ideals of F are $\{0\}$ and F, and of these only $\{0\}$ is prime.   □

Let A be a ring and let $x \in$ Spec A. Then $S = A \backslash x$ is a multiplicative set and we can form the **localization of A at x** defined as

$$S^{-1}A = \{a/s : a \in A, s \in S\}.$$

We denote the localization of A at x by A_x. In the case where $A = \mathbb{Z}$ and p is a prime of $\mathbb{Z}$, we have the following.

Proposition 1.1.3. *Let p be a prime of $\mathbb{Z}$. Then* Spec $\mathbb{Z}_{(p)} = \{\omega, (p)\}$.

Proof. Since $\mathbb{Q}$ is an integral domain and $\mathbb{Z}_{(p)} \subseteq \mathbb{Q}$, $\mathbb{Z}_{(p)}$ is an integral domain. Therefore, $\omega \in$ Spec $\mathbb{Z}_{(p)}$. Moreover, $\mathbb{Z}_{(p)}$ is a local ring with unique maximal ideal (p), and so $(p) \in$ Spec $\mathbb{Z}_{(p)}$. We show that ω and (p) are the only points of Spec $\mathbb{Z}_{(p)}$. Let I be a proper ideal of $\mathbb{Z}_{(p)}$ other than ω or (p). By Proposition 1.1.1, I is contained in a prime ideal; in fact, it is contained in a maximal ideal and consequently $I \subset (p)$.

Now $I \cap \mathbb{Z} \subset (p) \cap \mathbb{Z} = p\mathbb{Z}$, and so $I \cap \mathbb{Z} = p^l \mathbb{Z}$ with $l > 1$. Thus, in $\mathbb{Z}_{(p)}/I$, $p^l = 0$, and so $\mathbb{Z}_{(p)}/I$ is not an integral domain and hence I is not a prime ideal. It follows that ω and (p) are the only points in Spec $\mathbb{Z}_{(p)}$.   □

More generally, let A be a ring and let $x \in$ Spec A. Then

$$\text{Spec } A_x = \{yA_x : y \in \text{Spec } A, y \subseteq x\}.$$

An element $a \in A$ is **nilpotent** if $a^n = 0$ for some integer $n > 0$. If f is a non-nilpotent element of A, then $S = \{1, f, f^2, f^3 \dots\}$ is a multiplicative set that determines the localization $S^{-1}A$.

Proposition 1.1.4. *Let f be a non-nilpotent element of A with $S = \{1, f, f^2, \dots\}$, and let A_f denote the localization $S^{-1}A$. Then*

$$\text{Spec } A_f = \{xA_f : x \in \text{Spec } A, f \notin x\}.$$

Proof. Exercise.   □

For example, Spec $\mathbb{Z}_f$, $f = 6$, consists of $\{(5), (7), (11), \dots, \omega\}$.

To compute the spectrum of a quotient ring, we need the following proposition.

Proposition 1.1.5. *Let $\phi : A \to B$ be a ring homomorphism. Suppose x is a prime ideal of B. Then $\phi^{-1}(x)$ is a prime ideal of A.*

Proof. We first show that $\phi^{-1}(x)$ is an ideal of A. Since ϕ is an additive group homomorphism, $\phi^{-1}(x)$ is an additive subgroup of A. Let $a \in A$ and let $b \in \phi^{-1}(x)$. Then $\phi(ab) = \phi(a)\phi(b) = \phi(a)c$, where $c \in x$. Thus $\phi(ab) \in x$, so that $ab \in \phi^{-1}(x)$.

We next show that $\phi^{-1}(x)$ is prime. Suppose $ab \in \phi^{-1}(x)$. Then $\phi(ab) = \phi(a)\phi(b) \in x$, so that either $\phi(a) \in x$ or $\phi(b) \in x$. This shows that either $a \in \phi^{-1}(x)$ or $b \in \phi^{-1}(x)$, and so $\phi^{-1}(x)$ is prime. □

Proposition 1.1.6. *Let I be an ideal of A. Then $\mathrm{Spec}\,(A/I)$ is in a one-to-one correspondence with the collection of prime ideals x of A for which $I \subseteq x$; that is,*

$$\mathrm{Spec}\,(A/I) = \{\overline{x} : x \in \mathrm{Spec}\,A, I \subseteq x\},$$

where $\overline{x}$ denotes the image of x under the canonical surjection $A \to A/I$.

Proof. Let x be a prime ideal of A containing I. Then $\overline{x}$ is an additive subgroup of A/I. For $\overline{a} \in A/I$, $\overline{a} \cdot \overline{x} = \overline{ax} \subseteq \overline{x}$, and so $\overline{x}$ is an ideal of A/I.

We show that $\overline{x}$ is prime. Suppose $\overline{f} \cdot \overline{g} \in \overline{x}$. Then $\overline{fg} \in \overline{x}$; that is, $\overline{fg} = \overline{s}$ for some $s \in x$. Hence, $\overline{fg - s} = \overline{0}$, and so $fg \in s + I \subseteq x$. Thus either $f \in x$ or $g \in x$, and so either $\overline{f} \in \overline{x}$ or $\overline{g} \in \overline{x}$. Thus $\overline{x}$ is in $\mathrm{Spec}\,A/I$.

Now suppose $\overline{x}$ is a prime ideal of A/I. Then, by Proposition 1.1.5, x is a prime ideal of A. □

For example, let $A = \mathbb{Z}$, $I = (6)$. Then $\mathrm{Spec}\,\mathbb{Z}/(6) = \{(2), (3)\}$, where $(2) = \{0, 2, 4\}$ and $(3) = \{0, 3\}$.

One may wonder why we don't consider the spectrum of all *maximal* ideals of A instead of just prime ideals. Proposition 1.1.5 helps explain why: Proposition 1.1.5 fails if we replace "prime ideal" with "maximal ideal." For example, the inclusion of $\iota : \mathbb{Z} \to \mathbb{Q}$ is so that $\omega \subseteq \mathbb{Q}$ is maximal while $\iota^{-1}(\omega) = \omega \subseteq \mathbb{Z}$ is not.

1.2 The Associated Map of Spectra

In this section, we introduce the basic map between two spectra. Let $\phi : A \to B$ be a ring homomorphism with $\phi(1_A) = 1_B$, and let $x \in \mathrm{Spec}\,B$. By Proposition 1.1.5, $\phi^{-1}(x) \in \mathrm{Spec}\,A$, and thus there exists a map

$$\mathrm{Spec}\,B \to \mathrm{Spec}\,A$$

defined by $x \mapsto \phi^{-1}(x)$ for $x \in \mathrm{Spec}\,B$. We call this map the **associated map of spectra** and denote it by ${}^a\phi$.

Example 1.2.1. Let (p) be a non-zero prime ideal of $\mathbb{Z}$, and consider the inclusion $\iota : \mathbb{Z} \to \mathbb{Z}_{(p)}$ given by $n \mapsto n/1$. By Proposition 1.1.3, Spec $\mathbb{Z}_{(p)} = \{\omega, (p)\}$. The associated map $^a\iota : \text{Spec } \mathbb{Z}_{(p)} \to \mathbb{Z}$ is given as $^a\iota(\omega) = (\omega); {}^a\iota((p)) = (p)$.

Example 1.2.2. Let $A = \mathbb{Z}$, and consider the canonical surjection $\jmath : \mathbb{Z} \to \mathbb{Z}/(15)$. Then $^a\jmath : \text{Spec } \mathbb{Z}/(15) \to \text{Spec } \mathbb{Z}$ is given as $(3) \mapsto (3); (5) \mapsto (5)$.

Let $\mathbb{Z}[i]$ denote the Gaussian integers. We know that the integers can be embedded into the Gaussian integers by the map $\iota : \mathbb{Z} \to \mathbb{Z}[i]$ defined as $\iota(a) = a, a \in \mathbb{Z}$. We compute the spectrum of $\mathbb{Z}[i]$ and the associated map $^a\iota : \text{Spec } \mathbb{Z}[i] \to \text{Spec } \mathbb{Z}$. Since $\mathbb{Z}[i]$ is an integral domain, $\omega \in \text{Spec } \mathbb{Z}[i]$, and since $\iota^{-1}(\omega) = \omega, {}^a\iota(\omega) = \omega$.

Candidates for the other elements of Spec $\mathbb{Z}[i]$ include the principal ideals $(2), (3), (5), \ldots$ of $\mathbb{Z}[i]$. However, we must check whether these ideals are prime in $\mathbb{Z}[i]$. For example, one has $2 = (1 + i)(1 - i)$, and so (2) is not a prime ideal of $\mathbb{Z}[i]$. Moreover, $(1 + i) = (1 - i)$ since $-i(1 + i) = (1 - i)$.

We claim that $(1 + i)$ is prime. In the quotient $\mathbb{Z}[i]/(1 + i)$, $i = -1$; thus $\mathbb{Z}[i]/(1 + i) \cong \mathbb{Z}$, which is an integral domain, and so $(1 + i)$ is prime. Moreover, $\iota^{-1}(1 + i) = (2)$, and so $^a\iota(1 + i) = (2)$.

But what about the ideals generated by the odd primes of $\mathbb{Z}$? The following proposition, known as *Fermat's Two-Squares Theorem*, tells us precisely when (p), $p > 2$, factors in $\mathbb{Z}[i]$.

Proposition 1.2.1. *Suppose $p > 2$. Then $p = (a + bi)(a - bi)$ for some integers a, b if and only if $p \equiv 1 \pmod 4$.*

Proof. We first suppose that p factors as $(a + bi)(a - bi)$ and consider cases determined by whether a and b are even or odd.

Case 1. a and b are both even or a and b are both odd. In this case, p is even, which is impossible.

Case 2. a is even and b is odd (or vice versa). Without loss of generality, we suppose that $a = 2k, b = 2l + 1$, for some integers k, l. Then

$$p = (2k)^2 + (2l + 1)^2 = 4k^2 + 4l^2 + 4l + 1$$

$$\equiv 1 \pmod 4.$$

This proves the "only if" part of the theorem.

For the converse, we suppose $p \equiv 1 \pmod 4$ and show that there exist integers a, b, such that $p = a^2 + b^2$. Let m be such that $p - 1 = 4m$. Then $m = (p - 1)/4$. Since $\mathbb{Z}/(p)$ is a finite field, its multiplicative group of units, $U(\mathbb{Z}/(p))$, is cyclic and isomorphic to $\mathbb{Z}/(p - 1)$. Let r be a generator for $U(\mathbb{Z}/(p))$. Since the order of r in $\mathbb{Z}/(p)$ is $p - 1$, $r^{2m} \equiv -1 \pmod p$.

Put $s = r^m$. Then $s^2 + 1 \equiv 0 \pmod p$, so that $s^2 + 1 = (s + i)(s - i) = pn$ for some $n \in \mathbb{Z}$. Now, if (p) is a prime ideal in $\mathbb{Z}[i]$, then either $s + i \in (p)$ or $s - i \in (p)$, or, in other words, either p divides $s + i$ or p divides $s - i$. We may assume without losing generality that $p(c + di) = s + i$ for integers c, d. But this says that $pd = 1$, which is a contradiction. So (p) is not prime and thus, since $\mathbb{Z}[i]$ is a PID, p is reducible. Hence $p = (a + bi)(c + di)$ for some integers a, b, c, d. But this implies that $p = (a - bi)(c - di)$, and so

$$p^2 = (a^2 + b^2)(c^2 + d^2).$$

Now, since p is prime, $p = a^2 + b^2$, and hence $p = (a + bi)(a - bi)$.   □

Now we know precisely when (p) is prime in $\mathbb{Z}[i]$.

Proposition 1.2.2. *Suppose $p > 2$. The ideal (p) is prime in $\mathbb{Z}[i]$ if and only if $p \not\equiv 1 \pmod 4$.*

Proof. By Proposition 1.2.1, p is a reducible element of $\mathbb{Z}[i]$ if and only if $p \equiv 1 \pmod 4$. Thus (p) is a prime ideal in $\mathbb{Z}[i]$ if and only if $p \not\equiv 1 \pmod 4$.   □

When p, $p > 2$, does factor, it turns out that the ideals $(a + bi)$ and $(a - bi)$ are distinct prime ideals of $\mathbb{Z}[i]$. In fact, these ideals are maximal. In what follows, we put $\mathbb{F}_p = \mathbb{Z}/(p)$.

Proposition 1.2.3. *Suppose $p > 2$, $p = (a + bi)(a - bi)$. Then $(a \pm bi)$ are distinct maximal ideals of $\mathbb{Z}[i]$.*

Proof. Put $I = (a + bi)$. We show that $\mathbb{Z}[i]/I$ is a field. We have $x + yi \equiv z + yi$ (mod I) if and only if $x \equiv z \pmod p$. Likewise, $x + yi \equiv x + wi \pmod I$ if and only if $y \equiv w \pmod p$. Thus $\mathbb{Z}[i]/I \cong \mathbb{F}_p[i]/J$, where $J = (\bar{a} + \bar{b}i)$. Here, $\bar{a}, \bar{b}$ are the images of a, b under the canonical surjection $\mathbb{Z} \to \mathbb{F}_p$.

Now $\bar{b}$ is a unit in $\mathbb{F}_p$, and hence $J = (\bar{c} + i)$ for some $\bar{c} \in \mathbb{F}_p$. Thus, in the quotient $\mathbb{F}_p[i]/J$, $i \equiv -\bar{c}$, and so $\mathbb{F}_p[i]/J \cong \mathbb{F}_p$. Thus, $\mathbb{Z}[i]/I \cong \mathbb{F}_p$, a field.

A similar argument shows that $\mathbb{Z}[i]/(a - bi) \cong \mathbb{F}_p$. We leave it as an exercise to show that $(a + bi) \neq (a - bi)$.   □

If (p) is prime in $\mathbb{Z}[i]$, then $\mathbb{Z}[i]/(p)$ is an integral domain with p^2 elements, and hence $\mathbb{Z}[i]/(p)$ is a field. In fact, every finite field has q^k elements for some prime q and integer $k \geq 1$ [La84, Chapter 7, §5].

We now have Spec $\mathbb{Z}[i]$ computed:

$$\text{Spec } \mathbb{Z}[i] = \{(1 + i), (3), (2 + i), (2 - i), (7), (11), \ldots, \omega\}.$$

The spectral diagram that illustrates the associated map $^a\iota : \text{Spec } \mathbb{Z}[i] \to \text{Spec } \mathbb{Z}$ appears as Fig. 1.1 below.

Let $C_2 = \langle \tau \rangle$ denote the cyclic group of order 2. Another important computation of the spectral diagram involves the group ring

$$\mathbb{Z}C_2 = \{a + b\tau : a, b \in \mathbb{Z}\}.$$

There is a natural inclusion $\iota : \mathbb{Z} \to \mathbb{Z}C_2$, defined as $a \mapsto a1 + 0\tau$. As we did with the Gaussian integers, we compute the associated map $^a\iota : \text{Spec } \mathbb{Z}C_2 \to \text{Spec } \mathbb{Z}$.

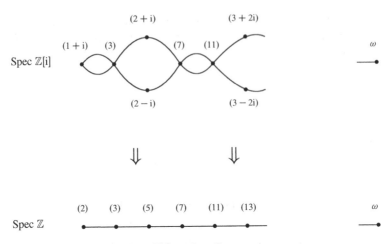

Spec $\mathbb{Z}[i]$

Spec $\mathbb{Z}$

Fig. 1.1 The map of spectra $^a\iota$: Spec $\mathbb{Z}[i] \rightarrow$ Spec $\mathbb{Z}$

Proposition 1.2.4. *Suppose* $p > 2$. *Then*

$$p = \left(\frac{p+1}{2} + \frac{p-1}{2}\tau\right)\left(\frac{p+1}{2} - \frac{p-1}{2}\tau\right).$$

Proof. Observe that

$$p = \left(\frac{p+1}{2}\right)^2 - \left(\frac{p-1}{2}\right)^2 = \left(\frac{p+1}{2}\right)^2 - \left(\frac{p-1}{2}\tau\right)^2,$$

and factor as the difference of two squares.   □

Let us try to recover this factorization using the method of Proposition 1.2.1. Since p is odd, $m = (p-1)/2$ is an integer. Let r be a generator for the multiplicative group of $\mathbb{F}_p$, and put $s = r^m$. Then $s^2 - 1 \equiv 0 \pmod{p}$, so that $s^2 - 1 = (s + \tau)(s - \tau) = pn$ for some $n \in \mathbb{Z}$.

Now, if (p) is a prime ideal in $\mathbb{Z}C_2$, then p divides either $s + \tau$ or $s - \tau$. We may assume without losing generality that $p(c + d\tau) = s + \tau$ for integers c, d. But this says that $pd = 1$, which is a contradiction.

So (p) is not a prime ideal. But does p factor in $\mathbb{Z}C_2$? The answer is "yes," but not because (p) fails to be prime in $\mathbb{Z}C_2$ ($\mathbb{Z}C_2$ is not a PID, so we do not automatically know that p is reducible).

We seek integers a, b, c, d such that

$$p = (a + b\tau)(c + d\tau).$$

If such integers exist, then $p = ac + bd$ and $0 = ad + bc$, so that $pa = a^2c + abd$ and $0 = abd + b^2c$. Thus $b^2c = -abd$, which yields $pa = a^2c - b^2c$. So let $c = a$.

Then $p = a^2 - b^2 = (a + b)(a - b)$. Now, since p is prime in $\mathbb{Z}$, we have either $p = a + b$, $p = -a - b$, $p = a - b$, or $p = -a + b$. Without loss of generality, we can assume that $p = a + b$. Then $1 = a - b$, so that $2a = p + 1$, and so $a = c = \frac{p+1}{2}$ and $b = \frac{p-1}{2}$, which yields $p = (\frac{p+1}{2} + \frac{p-1}{2}\tau)(\frac{p+1}{2} - \frac{p-1}{2}\tau)$.

Put $I^+ = \left(\frac{p+1}{2} + \frac{p-1}{2}\tau\right)$ and $I^- = \left(\frac{p+1}{2} - \frac{p-1}{2}\tau\right)$.

Proposition 1.2.5. *For $p > 2$, the ideals I^+ and I^- are distinct prime ideals of $\mathbb{Z}C_2$.*

Proof. We first show that the ideals $I^\pm$ are maximal. Arguing as in Proposition 1.2.3, one concludes that $\mathbb{Z}C_2/I^+ \cong \mathbb{F}_pC_2/\overline{I^+}$, where

$$\overline{I^+} = \left(\overline{\frac{p+1}{2} + \frac{p-1}{2}\tau}\right) = (\overline{1} + \tau).$$

Thus $\mathbb{Z}C_2/I^+ \cong \mathbb{F}_pC_2/(\overline{1} + \tau) \cong \mathbb{F}_p$, which is a field. Consequently, $I^\pm$ are maximal ideals.

Now suppose $I^+ = I^-$. Then

$$\frac{p+1}{2} + \frac{p-1}{2}\tau + \frac{p+1}{2} - \frac{p-1}{2}\tau = p + 1 \in I^+,$$

which is impossible since $p \in I^+$ and $(p, p + 1) = 1$. $\qquad\square$

Next, we consider the prime ideal $(2) \subseteq \mathbb{Z}$. Since $(1 + \tau)^2 \equiv 0 \pmod 2$, the ideal $(2) \subseteq \mathbb{Z}C_2$ is not prime. It is irreducible, however: there is no factorization of 2 in $\mathbb{Z}C_2$. So, which prime ideals $x \in \operatorname{Spec} \mathbb{Z}C_2$ satisfy $\iota^{-1}(x) = (2)$? Any such prime ideal x contains (2), and since 2 is irreducible, x is an ideal of the form $(2, a + b\tau)$ for some $a + b\tau \in \mathbb{Z}C_2$. Since $\mathbb{F}_2C_2/(\overline{a} + \overline{b}\tau)$ is an integral domain if and only if $\overline{a} = \overline{b} = 1$, $x = (2, 1 + \tau)$.

Finally, the prime ideal ω of $\mathbb{Z}$ factors into two distinct prime ideals $(1 + \tau)$ and $(1 - \tau)$ of $\mathbb{Z}C_2$. We conclude that

$$\operatorname{Spec} \mathbb{Z}C_2 = \{(2, 1 + \tau), (2 - \tau), (2 + \tau), (3 - 2\tau), (3 + 2\tau), \ldots, (1 + \tau), (1 - \tau)\},$$

and the spectral diagram appears as Fig. 1.2 below.

For another example, let C_3 denote the cyclic group of order 3, generated by τ. Using the ideas employed in the computation of $\operatorname{Spec} \mathbb{Z}C_2$, one can compute $\operatorname{Spec} \mathbb{Z}C_3$ as well as the map of spectra ${}^a\iota : \operatorname{Spec} \mathbb{Z}C_3 \to \operatorname{Spec} \mathbb{Z}$ induced by the embedding $\iota : \mathbb{Z} \to \mathbb{Z}C_3$.

Note that $\mathbb{Z}C_3$ has zero divisors: $(\tau - 1)(\tau^2 + \tau + 1) = 0$. To compute the preimage of $(2) \subseteq \mathbb{Z}$ in $\operatorname{Spec} \mathbb{Z}C_3$, observe that $x^2 + x + 1$ is irreducible over $\mathbb{F}_2$; consequently, $(2, \tau^2 + \tau + 1)$ is a maximal ideal of $\mathbb{Z}C_3$. The complete diagram appears as Fig. 1.3 below.

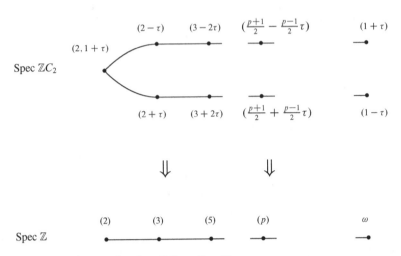

Spec $\mathbb{Z}C_2$

$(2, 1 + \tau)$

$(2 - \tau)$ $(3 - 2\tau)$ $(\frac{p+1}{2} - \frac{p-1}{2}\tau)$ $(1 + \tau)$

$(2 + \tau)$ $(3 + 2\tau)$ $(\frac{p+1}{2} + \frac{p-1}{2}\tau)$ $(1 - \tau)$

(2) (3) (5) (p) ω

Spec $\mathbb{Z}$

Fig. 1.2 The map of spectra $^a\iota$: Spec $\mathbb{Z}C_2 \to$ Spec $\mathbb{Z}$

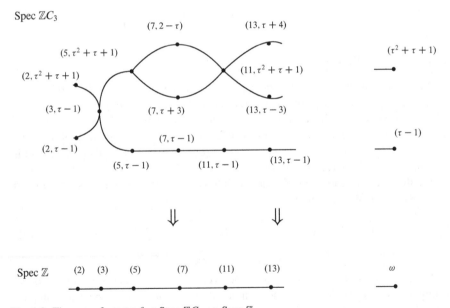

Spec $\mathbb{Z}C_3$

$(5, \tau^2 + \tau + 1)$

$(2, \tau^2 + \tau + 1)$

$(3, \tau - 1)$

$(2, \tau - 1)$

$(7, 2 - \tau)$ $(13, \tau + 4)$

$(11, \tau^2 + \tau + 1)$ $(\tau^2 + \tau + 1)$

$(7, \tau + 3)$ $(13, \tau - 3)$

$(7, \tau - 1)$ $(\tau - 1)$

$(5, \tau - 1)$ $(11, \tau - 1)$ $(13, \tau - 1)$

Spec $\mathbb{Z}$ (2) (3) (5) (7) (11) (13) ω

Fig. 1.3 The map of spectra $^a\iota$: Spec $\mathbb{Z}C_3 \to$ Spec $\mathbb{Z}$

1.3   Nilpotent Elements

Let A be a ring. Recall that an element $f \in A$ is nilpotent if there exists a positive integer n for which $f^n = 0$.

Proposition 1.3.1. *The collection of nilpotent elements forms an ideal of A.*

Proof. We first show that the set S of nilpotent elements is a subgroup. Let $a, b \in S$ with $a^n = 0$, $b^m = 0$, for positive integers m, n. Then

$$(a + b)^{m+n-1} = \sum_{i=0}^{m+n-1} \binom{m+n-1}{i} a^{m+n-1-i} b^i.$$

If $m + n - 1 - i \geq n$, then the term in the sum on the right-hand side is 0. Otherwise, if $n > m + n - 1 - i$, then $i > m - 1$ and so the sum on the right is 0. Thus $a + b \in S$. Clearly, $0 \in S$. Moreover, $(-a)^n = a^n = 0$ if n is even, and $(-a)^n = -a^n = 0$ if n is odd, so that $-a \in S$ if $a \in S$. Thus S is a subgroup of A.

Let $r \in A$, $a \in S$. Then $(ra)^n = r^n a^n = 0$, and so $ra \in S$. Thus S is an ideal of A. □

The ideal of nilpotent elements is called the **nilradical of** A and is denoted by $\mathrm{nil}(A)$. The nilpotent elements play an important role in the study of Spec A.

Let $x \in \mathrm{Spec}\, A$. Then A/x is an integral domain and as such has a field of fractions $\mathrm{Frac}(A/x)$. There is a natural inclusion $\iota_x : A/x \to \mathrm{Frac}(A/x)$, depending on x, and the composition of this map with the canonical surjection $J_x : A \to A/x$ yields the ring homomorphism

$$\ell_x : A \to \mathrm{Frac}(A/x), \quad \ell_x = \iota_x J_x.$$

Let $f \in A$. Then f determines a function on Spec A,

$$f : \mathrm{Spec}\, A \to \bigcup_{x \in \mathrm{Spec}\, A} \mathrm{Frac}(A/x),$$

where $f(x) = \ell_x(f)$ for $x \in \mathrm{Spec}\, A$. In this manner, A can be viewed as a collection of functions on Spec A.

Example 1.3.1. Let $A = \mathbb{Z}$, and let $f = 12$. Then 12 determines a function on Spec $\mathbb{Z}$,

$$12 : \mathrm{Spec}\, \mathbb{Z} \to \bigcup_{x \in \mathrm{Spec}\, \mathbb{Z}} \mathrm{Frac}(\mathbb{Z}/x) = \left(\bigcup_{p \in \mathbb{Z}} \mathbb{F}_p \right) \cup \mathbb{Q},$$

with $12((2)) = 0$, $12((3)) = 0$, $12((5)) = 2$, $12((7)) = 5$, $12((11)) = 1$, and $12(x) = 12$ for all other points $x \in \mathrm{Spec}\, \mathbb{Z}$.

In Example 1.3.1, the first and second zeros are taken to be equal. This is technically incorrect, however, since the first zero is the residue class of 0 in $\mathbb{F}_2$, while the second 0 is the residue class in $\mathbb{F}_3$. But for our purposes we shall assume they are equal.

In the ring A, the element $0 \in A$ is a zero function on Spec A since $0(x) = 0$ for all $x \in$ Spec A. Moreover, $f \in A$ is a zero function on Spec A if and only if $f(x) = 0$ for all $x \in$ Spec A; that is, if and only if $f \in x$ for all $x \in$ Spec A. These zero functions have an elegant characterization.

Proposition 1.3.2. *The following statements are equivalent:*

(i) The element $f \in A$ is a zero function on Spec A.
(ii) The element f belongs to every prime ideal of A.
(iii) The element f is nilpotent.

Proof. $(ii) \Leftrightarrow (iii)$. Suppose f is nilpotent. Then $f^n = 0$ for some integer $n > 0$, so $f^n \in x$ for each $x \in$ Spec A. Thus either $f \in x$ or $f^{n-1} \in x$. Repeating this process (if necessary) shows that $f \in x$.

For the converse, we suppose f is not nilpotent and show that there exists a prime ideal x for which $f \notin x$; that is, $f \notin \cap_{x \in \text{Spec } A} x$. Consider the collection $\mathcal{I}$ of ideals of A that contain no power of f. This collection is partially ordered by inclusion $\subseteq$ and is non-empty since it contains ω. Each chain $\{I_\alpha\}_{\alpha \in J}$ has an upper bound $\cup_{\alpha \in J} I_\alpha$. Thus, by Zorn's Lemma, $\mathcal{I}$ has a maximal element M.

We claim that M is a prime ideal of A. To this end, we show that A/M is an integral domain. Suppose $b_1 + M$ and $b_2 + M$ are two elements of A/M with $b_1, b_2 \notin M$. We have $M \subseteq (b_1) + M$, so if $(b_1) + M$ contains no power of f, then $M = M + (b_1)$, which says that $b_1 \in M$, a contradiction. Thus there exists an integer $n_1 > 0$ with $f^{n_1} \in (b_1) + M$. By similar reasoning, there exists an integer $n_2 > 0$ with $f^{n_2} \in (b_2) + M$. Now $f^{n_1+n_2} \in (b_1 b_2) + M$, so if $b_1 b_2 \in M$, then $f^{n_1+n_2} \in M$, a contradiction. It follows that $b_1 b_2 \notin M$, and hence $b_1 b_2 \neq 0$ in A/M. Thus $M \in$ Spec A with $f \notin M$ since M contains no power of f.

It is immediate that $(i) \Leftrightarrow (ii)$. $\square$

The presence of zero functions other than $0 \in A$ is not desirable, and we would like to remove them from consideration. This is done by a simple process.

Proposition 1.3.3. *Let A be a ring. Then the quotient ring $A/\text{nil}(A)$ is a set of functions on* Spec $(A/\text{nil}(A))$ *in which there is a unique zero function, namely $0 \in A/\text{nil}(A)$.*

Proof. In view of Proposition 1.3.2, this amounts to showing that the only nilpotent element in $A/\text{nil}(A)$ is $\text{nil}(A)$. Let $a + \text{nil}(A) \in A/\text{nil}(A)$, and suppose that $(a + \text{nil}(A))^n = \text{nil}(A)$ for some $n > 0$. Then $a^n \in \text{nil}(A)$, and thus there exists an integer $m > 0$ for which $(a^n)^m = a^{mn} = 0 \in A$. Thus $a \in \text{nil}(A)$, which says that $a + \text{nil}(A) = \text{nil}(A)$. $\square$

As a consequence of Proposition 1.3.3, we can replace A with the ring $A/\text{nil}(A)$, which has no non-trivial nilpotent elements. In practice, we shall assume that A satisfies $\text{nil}(A) = 0$. Suppose this is the case. Let $S \subseteq$ Spec A and let $f \in A$. Then the restriction of f to S denoted by f_S is a function on S. Moreover, given

$f, g \in A$, with $g_S(x) \neq 0$ for all $x \in S$, the function $q = f_S/g_S$ is defined, with its value at $x \in S$ given as

$$q(x) = f_S(x)/g_S(x).$$

The function q is a **rational function on** S. Note that q may not be defined on all of Spec A.

1.4   Chapter Exercises

Exercises for §1.1

1. Compute Spec $(\mathbb{Z}/(60))$.
2. Compute Spec $(\mathbb{Z} \times \mathbb{Z})$.
3. (from the text) Let A be a ring, and let $x \in$ Spec A. Show that

$$\text{Spec } A_x = \{yA_x : y \in \text{Spec } A, y \subseteq x\}.$$

4. Compute Spec $\mathbb{Q}(\sqrt{2})$.
5. Suppose $\phi : A \to B$ is a ring homomorphism with B an integral domain. Show that $\ker(\phi) \in$ Spec A.
6. Prove that $\omega \in$ Spec A if and only if A is an integral domain.
7. Compute Spec $\mathbb{Q}[t]$, with t indeterminate.
8. Compute Spec $\mathbb{Z}V$, where V denotes the Klein 4-group.

Exercises for §1.2

9. Construct the spectral diagram associated to the inclusion $\mathbb{Z} \to \mathbb{Z}_{(3)}$.
10. Construct the spectral diagram associated to the canonical surjection $j : \mathbb{Z} \to \mathbb{Z}/(10)$.
11. Construct the spectral diagram associated to the inclusion $\mathbb{R} \to \mathbb{R}[t]$.
12. Construct the spectral diagram associated to the evaluation homomorphism $\phi_{\sqrt{2}} : \mathbb{Q}[t] \to \mathbb{R}$.
13. Let $\phi : R \to R'$ be a surjection of rings. Prove that $^a\phi$ is an injection.
14. Assume that $\iota : R \to R'$ is an injection of rings. Is $^a\iota$ necessarily an injection?
15. Consider the group ring $\mathbb{Z}C_3$ with $\langle \tau \rangle = C_3$.

    (a) Show that the ideal $(19, 3 - 2\tau)$ is principal.
    (b) Prove that (p) factors in $\mathbb{Z}C_3$ if $\frac{4p-1}{3}$ is the square of an integer.
    (c) Does the converse of (b) hold?

16. Verify that the spectral diagram in Figure 1.3 is correct.
17. Let $R = \mathbb{Z}/(20)$ and let $S = \{1, 2, 4, 8, 12, 16\}$. Construct the spectral diagram associated to the homomorphism $\mathbb{Z}/(20) \to S^{-1}(\mathbb{Z}/(20))$ defined by $n \mapsto n/1$.

18. Let $\psi : \text{Spec } \mathbb{Z}C_5 \to \text{Spec } \mathbb{Z}$ be the map associated to the inclusion $\mathbb{Z} \to \mathbb{Z}C_5$. Compute $\psi^{-1}((11))$.

19. Let $\phi : \mathbb{Z}C_4 \to \mathbb{Z}C_2$ denote the ring homomorphism defined as $g \mapsto g^2$, $\langle g \rangle = C_4$. Construct the map of spectra $\text{Spec } \mathbb{Z}C_2 \to \text{Spec } \mathbb{Z}C_4$.

Exercises for §1.3

20. Compute the ideal of nilpotent elements in $\mathbb{Z}/(100)$.

21. Show that the set of nilpotent elements in the non-commutative ring $\text{nil}(\text{Mat}_2(\mathbb{Z}))$ does not form an ideal.

22. Give an example of a ring A that has zero divisors and satisfies $\text{nil}(A) = \{0\}$.

23. Let $S = \{(3), (5), (7), \dots, \omega\} \subseteq \text{Spec } \mathbb{Z}$. Compute the collection of all rational functions on S.

24. Let $S = \{\omega\} \subseteq \text{Spec } \mathbb{Z}$. Compute the collection of all rational functions on S.

25. Let $S = \{(2791)\} \subseteq \text{Spec } \mathbb{Z}$. Compute the collection of all rational functions on S.

Chapter 2
The Zariski Topology on the Spectrum

The goal of this chapter is to introduce the Zariski topology on Spec A. Throughout this chapter, by "ring" we mean a non-zero commutative ring with unity.

2.1 Some Topology

Definition 2.1.1. Let X be an arbitrary set. **A topology** on X is a collection $\mathcal{T}$ of subsets of X for which:

(i) $\emptyset$ and X are in $\mathcal{T}$.
(ii) The union of the elements of a subcollection of $\mathcal{T}$ is in $\mathcal{T}$. ($\mathcal{T}$ is closed under arbitrary unions.)
(iii) The intersection of the elements of a finite subcollection of $\mathcal{T}$ is in $\mathcal{T}$. ($\mathcal{T}$ is closed under finite intersections.)

The set X together with a topology $\mathcal{T}$ is a **topological space**. A subset U of X is **open** if $U \in \mathcal{T}$. A subset V of X is **closed** if $X \backslash V$ is open. For example, $\emptyset$ is open and X is closed. At the same time, X is open and $\emptyset$ is closed.

Example 2.1.1. Let $X = \mathbb{R}$, endowed with the usual absolute value $|\ |$, and define $\mathcal{T}$ to be the collection of all arbitrary unions of intervals

$$\{x \in \mathbb{R} : |x - a| < \varepsilon\},$$

where $a \in \mathbb{R}$ and $\varepsilon \in \mathbb{R}_{>0}$. Then $\mathcal{T}$ is the **standard topology on $\mathbb{R}$ induced by** $|\ |$.

In the standard topology on $\mathbb{R}$, $(2, 4) = \{x \in \mathbb{R} : |x - 3| < 1\}$ is open and $(-\infty, 2] \cup [3, \infty) = \{x \in \mathbb{R} : |x - 3| \geq 1\}$ is closed.

R.G. Underwood, *An Introduction to Hopf Algebras*, DOI 10.1007/978-0-387-72766-0_2,   13
© Springer Science+Business Media, LLC 2011

Example 2.1.2. Let $X = \mathbb{Q}$, and let p be a prime number. Let $|\;|_p$ denote the p-adic absolute value defined on $\mathbb{Q}$. Then the $|\;|_p$-**topology on** $\mathbb{Q}$ is the collection of all arbitrary unions of intervals

$$\{x \in \mathbb{Q} : |x - a|_p < \varepsilon\},$$

where $a \in \mathbb{Q}$ and $\varepsilon \in \mathbb{R}_{>0}$.

The sequence $\{a_n\}$ in $\mathbb{Q}$ is a $|\;|_p$-Cauchy sequence in $\mathbb{Q}$ if and only if for each $\epsilon > 0$ there exists a positive integer N for which a_n is in the open set $|a_n - a_m|_p < \epsilon$ whenever $m, n > N$.

We are interested in defining a topology on our set of prime ideals Spec A. Let E be a subset of A, and put

$$V(E) = \{x \in \text{Spec } A : E \subseteq x\}.$$

Let $\mathcal{T}$ be the collection of all subsets U of Spec A such that $U = \text{Spec } A \backslash V(E)$ for some $E \subseteq A$. To show that $\mathcal{T}$ is a topology on Spec A, we need two lemmas.

Lemma 2.1.1. *Let $\{E_\alpha\}$ be an arbitrary collection of subsets of A indexed by the set J. Then $\cap_{\alpha \in J} V(E_\alpha) = V(\cup_{\alpha \in J} E_\alpha)$.*

Proof. Let $x \in \cap_{\alpha \in J} V(E_\alpha)$. Then $x \in V(E_\alpha)$ for all $\alpha \in J$, and consequently $E_\alpha \subseteq x$ for all α, which says that $\cup_{\alpha \in J} E_\alpha \subseteq x$ and hence $x \in V(\cup_{\alpha \in J} E_\alpha)$.

Now suppose that $x \in V(\cup_{\alpha \in J} E_\alpha)$. Then $E_\alpha \subseteq \cup_{\alpha \in J} E_\alpha \subseteq x$ for all α, which says that $x \in V(E_\alpha)$ for all α. The lemma follows.   □

Lemma 2.1.2. *Let $\{E_j\}$ be a finite collection of subsets of A indexed by the set $J = \{1, 2, \ldots, k\}$. For each j, $1 \le j \le k$, let (E_j) denote the ideal of A generated by the elements of E_j. Then $\cup_{j=1}^k V((E_j)) = V(\cap_{j=1}^k (E_j))$.*

Proof. Exercise.   □

Proposition 2.1.1. *$\mathcal{T}$ is a topology on Spec A.*

Proof. We show that conditions (i), (ii), and (iii) of Definition 2.1.1 are satisfied. For (i) we have $\emptyset = \text{Spec } A \backslash V(\{0\})$ and Spec $A = \text{Spec } A \backslash V(A)$, so $\emptyset$ and Spec A are in $\mathcal{T}$.

For (ii) consider an arbitrary subcollection $\mathcal{S} \subseteq \mathcal{T}$ indexed by the set J. We have $\mathcal{S} = \{\text{Spec } A \backslash V(E_\alpha)\}_{\alpha \in J}$ for some collection of subsets $\{E_\alpha\}_{\alpha \in J}$ of A. Now,

$$\bigcup_{\alpha \in J} \text{Spec } A \backslash V(E_\alpha) = \text{Spec } A \backslash \bigcap_{\alpha \in J} V(E_\alpha) \quad \text{by DeMorgan's laws}$$

$$= \text{Spec } A \backslash V \left(\bigcup_{\alpha \in J} E_\alpha \right) \quad \text{by Lemma 2.1.1,}$$

and so (ii) holds.

To prove (iii), consider the finite subcollection $\mathcal{S} = \{\text{Spec } A \backslash V(E_i)\}_{i=1}^{k}$ of subsets of $\mathcal{T}$. We have

$$\bigcap_{j=1}^{k} \text{Spec } A \backslash V(E_j) = \text{Spec } A \backslash \bigcup_{j=1}^{k} V(E_j) \quad \text{by DeMorgan's laws}$$

$$= \text{Spec } A \backslash V \left(\bigcap_{j=1}^{k} (E_j) \right) \quad \text{by Lemma 2.1.2.}$$

Thus (iii) holds. □

The topology given by Proposition 2.1.1 is the **Zariski topology** on Spec A.

A subset of a topological space can be endowed with a topology in a natural way.

Definition 2.1.2. Let X be a topological space with topology $\mathcal{T}$, and let Y be a subset of X. Let $\mathcal{T}_Y$ denote the collection of all subsets of Y of the form

$$\{U \cap Y : U \in \mathcal{T}\}.$$

Then $\mathcal{T}_Y$ is a topology on Y called the **subspace topology**.

A surjective map from a topological space to a set can induce a topology on the set.

Definition 2.1.3. Let $\phi : X \to Y$ be a surjective map, where X is a topological space and Y is an arbitrary set. Let $\mathcal{T}$ denote the collection of subsets of Y defined as

$$\{U \subseteq Y : \phi^{-1}(U) \text{ is open in } X\}.$$

Then $\mathcal{T}$ is a topology on Y called the **quotient topology**.

A topology given a collection of topological spaces can be defined as follows.

Definition 2.1.4. Let $\{X_\alpha\}_{\alpha \in J}$ be a family of topological spaces, and consider the Cartesian product $P = \prod_{\alpha \in J} X_\alpha$. Let

$$\prod_{\alpha \in J} U_\alpha \tag{2.1}$$

be a subset of P, where U_α is open in X_α for $\alpha \in J$ and $U_\alpha = X_\alpha$ for all but a finite number of α. Let $\mathcal{T}$ be the collection of all arbitrary unions of subsets of the form (2.1). Then $\mathcal{T}$ is a topology on P called the **product topology on P**.

One can generalize the product topology as follows. Let $\{X_\alpha\}_{\alpha \in J}$ be a family of topological spaces, and for $\alpha \in J$ let Y_α be an open subset of X_α Let Y denote the family $\{Y_\alpha\}_{\alpha \in J}$.

Let X_Y be the subset of $\prod_{\alpha \in J} X_\alpha$ consisting of all $(x_\alpha)_{\alpha \in J}$ for which $x_\alpha \in Y_\alpha$ for all but a finite number of α. On X_Y define a topology by defining the open sets to be the collection of all unions of subsets of the form

$$\prod_{\alpha \in J} U_\alpha,$$

where U_α is open in X_α, for all α, and $U_\alpha = Y_\alpha$ for all but a finite number of α. We call this topology the **restricted product topology on** $X = \prod_{\alpha \in J} X_\alpha$ **with respect to the family** $Y = \{Y_\alpha\}$. Clearly, the restricted product topology on X with respect to the family $\{X_\alpha\}_{\alpha \in J}$ is the ordinary product topology on X.

Perhaps the simplest topology we can endow on a set X is the **discrete topology**. Here, every subset U of X is an open subset of X. In other words, $\mathcal{T}$ is the power set of X.

Let X and Y be topological spaces. Let $\psi : X \to Y$ be a function from X to Y. Then ψ is **continuous** if $\psi^{-1}(U)$ is open in X whenever U is open in Y. For example, if $\mathbb{R}$ is endowed with the standard topology, then the map $f : \mathbb{R} \to \mathbb{R}$ given by $x \mapsto x^2$ is continuous. Moreover, if Y is given the quotient topology, then the surjective map of topological spaces $\phi : X \to Y$ is continuous.

Continuous maps can be defined in terms of closed subsets.

Proposition 2.1.2. *A map $\psi : X \to Y$ is continuous if and only if $\phi^{-1}(V)$ is closed in X whenever V is closed in Y.*

Proof. Exercise. $\qquad\square$

The next proposition shows that the associated map is continuous in the Zariski topology.

Proposition 2.1.3. *Let $\phi : A \to B$ be a homomorphism of rings. Endow* Spec B *and* Spec A *with the Zariski topology. Then $^a\phi :$ Spec $B \to$ Spec A is continuous.*

Proof. Put $\psi = {}^a\phi$, and let $V(E)$ be a closed set in Spec A. We show that $\psi^{-1}(V(E))$ is closed in Spec B. Let $x \in \psi^{-1}(V(E))$. Then $\psi(x) \in V(E)$, so that $E \subseteq \psi(x) = \phi^{-1}(x)$, and thus $\phi(E) \subseteq x$, which says that $x \in V(\phi(E))$ and therefore $\psi^{-1}(V(E)) \subseteq V(\phi(E))$.

Now suppose $x \in V(\phi(E))$. Then $\phi(E) \subseteq x$, so that $E \subseteq \phi^{-1}(x) = \psi(x)$, which yields $\psi(x) \in V(E)$. Thus $x \in \psi^{-1}(V(E))$, which gives $V(\phi(E)) \subseteq \psi^{-1}(V(E))$. It follows that $\psi^{-1}(V(E)) = V(\phi(E))$; that is, the pullback of a closed set is closed. $\qquad\square$

Under what conditions are two topological spaces essentially the same? We next study the analog of the notion of a group or ring isomorphism.

Definition 2.1.5. A map $\gamma : X \to Y$ is a **homeomorphism of topological spaces** if

(i) γ is a bijection and
(ii) γ and γ^{-1} are continuous.

Two spaces are **homeomorphic** if there exists a homeomorphism $\gamma : X \to Y$. We then write $X \cong Y$.

For example, if we consider $\mathbb{R}$ with the standard topology and endow the subset $(-1, 1) \subseteq \mathbb{R}$ with the subspace topology, then the map $f : (-1, 1) \to \mathbb{R}$ defined as $f(x) = \frac{x}{1-x^2}$ is a homeomorphism of topological spaces.

Here is an important example of a homeomorphism.

Proposition 2.1.4. *Let $\phi : A \to A/\mathrm{nil}(A)$ be the canonical surjection of rings, and let $^a\phi : \mathrm{Spec}\,(A/\mathrm{nil}(A)) \to \mathrm{Spec}\,A$ be the associated map of topological spaces. Then $^a\phi$ is a homeomorphism of topological spaces.*

Proof. We first show that $^a\phi$ is injective. Suppose that $x \neq y$ in $\mathrm{Spec}\,(A/\mathrm{nil}(A))$. Then $\phi^{-1}(x) \neq \phi^{-1}(y)$, and so $^a\phi(x) \neq {}^a\phi(y)$, which says that $^a\phi$ is an injection. Now let $y \in \mathrm{Spec}\,A$. Then $x = y + \mathrm{nil}(A)$ is a prime ideal of $\mathrm{Spec}\,(A/\mathrm{nil}(A))$ with $\phi^{-1}(x) = y$. Thus $^a\phi(x) = y$, and so $^a\phi$ is a bijection.

We already know that $^a\phi$ is continuous (Proposition 2.1.3), so it remains to show that $(^a\phi)^{-1} : \mathrm{Spec}\,A \to \mathrm{Spec}\,(A/\mathrm{nil}(A))$ is continuous. But this is equivalent to showing that $^a\phi(V(E))$ is closed in $\mathrm{Spec}\,A$ for any subset $E \subseteq A/\mathrm{nil}(A)$, and this holds since $^a\phi(V(E)) = V(F)$, where F is the subset of A for which $E = F + \mathrm{nil}(A)$. $\qquad\square$

2.2 Basis for a Topological Space

Another way to describe a topology on a set is to give a set of "building blocks" for the open sets of the topology.

Definition 2.2.1. Let X be a topological space. A **basis** for X is a collection $\mathcal{B}$ of open subsets of X that satisfies:

(i) For each $x \in X$, there exists at least one subset $B \in \mathcal{B}$ for which $x \in B$.
(ii) For each $x \in B_1 \cap B_2$ with $B_1, B_2 \in \mathcal{B}$, there exists a subset $B_3 \in \mathcal{B}$ for which $x \in B_3$ and $B_3 \subseteq B_1 \cap B_2$.

If $\mathcal{B}$ is a basis for the topological space X, then the collection of all unions of elements in $\mathcal{B}$ is a topology on X called the **topology on X generated by $\mathcal{B}$**. Of course, the topology on X generated by $\mathcal{B}$ coincides with the given topology on X.

We are interested in obtaining a basis for the Zariski topology on $\mathrm{Spec}\,A$. Let f be a non-nilpotent element of A. Then

$$V(\{f\}) = \{x \in \mathrm{Spec}\,A : f \in x\}$$

is a closed set in $\mathrm{Spec}\,A$ (since $\mathrm{Spec}\,A \backslash V(\{f\})$ is open by definition). In what follows, we shall write $V(f)$ in place of $V(\{f\})$, and we will set $D(f) = \mathrm{Spec}\,A \backslash V(f)$. For an ideal I of A, we put $D(I) = \mathrm{Spec}\,A \backslash V(I)$.

The open set $D(f)$ has some important properties.

Lemma 2.2.1. *Let f, g be non-nilpotent elements of A. Then*

(i) $D(f) \cap D(g) = D(fg)$;
(ii) $D(f) \cup D(g) = D((f, g))$.

Proof. We prove (i) and leave (ii) as an exercise.

Let $x \in D(f) \cap D(g)$. Then

$$x \in \text{Spec } A \backslash V(f) \cap \text{Spec } A \backslash V(g) = \text{Spec } A \backslash (V(f) \cup V(g))$$

by DeMorgan's laws. Hence, $f \notin x$ and $g \notin x$. Suppose that $fg \in x$. Then, since x is prime, either $f \in x$ or $g \in x$, which is a contradiction. Thus $fg \notin x$, and so $x \notin V(fg)$. Hence, $x \in D(fg)$.

For the converse, suppose that $x \in D(fg)$. Then $x \notin V(fg)$, and so $fg \notin x$. Now $f \notin x$, for otherwise $gf = fg \in x$ since x is an ideal. Moreover, by the same reasoning, $g \notin x$. It follows that $x \in D(f) \cap D(g)$. $\qquad\square$

Lemma 2.2.2. *Let f, g be non-nilpotent elements. Then $D(f) \subseteq D(g)$ if and only if $f^n = ga$ for some $n > 0$ and some $a \in A$.*

Proof. Suppose $D(f) \subseteq D(g)$. Then, by DeMorgan's laws, $V(g) \subseteq V(f)$, which says that if g is contained in the prime ideal $x \in \text{Spec } A$, then so is f. Now, by Proposition 1.1.6, $\text{Spec } (A/(g)) = \{\overline{x} : x \in \text{Spec } A, g \in x\}$; hence f is contained in every prime ideal of $A/(g)$. By Proposition 1.3.2, $f + (g)$ is nilpotent in $A/(g)$. Thus $f^n \in (g)$ for some integer $n > 0$; that is, $f^n = ga, a \in A$.

For the converse, suppose $f^n \in (g), n > 0$, and let $g \in x$. Then $f^n \in x$, so that $f \in x$. Thus $V(g) \subseteq V(f)$, which gives $D(f) \subseteq D(g)$. $\qquad\square$

Proposition 2.2.1. *The collection*

$$\mathcal{B} = \{D(f) : f \text{ is a non-nilpotent element of } A\}$$

is a basis for the Zariski topology on $\text{Spec } A$.

Proof. We show that conditions (i) and (ii) of Definition 2.2.1 hold.

For (i), let $x \in \text{Spec } A$. Since $x \neq A$, $A \backslash x$ is non-empty, and thus there exists an element $f \in A \backslash x$ that satisfies $f \notin x$. Thus $x \notin V(f)$. Now, by Proposition 1.3.2, f is non-nilpotent. It follows that $x \in D(f)$, where $D(f) \in \mathcal{B}$.

For condition (ii), we suppose that $x \in D(f) \cap D(g)$ for f, g non-nilpotent. By Lemma 2.2.1(i), $D(f) \cap D(g) = D(fg)$. We claim that fg is non-nilpotent for otherwise $V(fg) = \text{Spec } A$ by Proposition 1.3.2, and thus $\text{Spec } A \backslash V(fg) = \emptyset$, which is a contradiction. Thus $D(fg)$ is an element of $\mathcal{B}$ with $x \in D(fg)$ and $D(fg) = D(f) \cap D(g)$. $\qquad\square$

Example 2.2.1. A basis for $\text{Spec } \mathbb{Z}$ is $\mathcal{B} = \{D(f) : f \in \mathbb{Z}, f \neq 0\}$. As an example of a basis element, we have

$$D(10) = \text{Spec } \mathbb{Z} \backslash V(10) = \{(3), (7), (11), \ldots, \omega\}.$$

Every open set $D(f)$ is topologically equivalent to the spectrum of a localized ring.

Proposition 2.2.2. *The open set $D(f)$ is homeomorphic to* Spec A_f, *where A_f is the localization $S^{-1}A$, where S is the multiplicative set $\{1, f, f^2, \ldots\}$.*

Proof. We equip $D(f)$ with the subspace topology induced by Spec A. Spec A_f, which is the collection $\{xA_f : x \in \text{Spec } A, f \notin x\}$, is given the Zariski topology. To prove the proposition, we define a map $\varphi : \text{Spec } A_f \to D(f)$ by the rule $\varphi(xA_f) = x$, noting that $x \in D(f)$. We show that φ is bijective and bicontinuous.

Suppose $\varphi(xA_f) = \varphi(yA_f)$. Then $x = y$ and so $xA_f = yA_f$, and thus φ is an injection. Next, let $y \in D(f)$. Then $yA_f \in \text{Spec } A_f$, and so $\varphi(yA_f) = y$, which shows that φ is a bijection. Now let $D(fg) = D(f) \cap D(g)$ be a basic open set in $D(f)$. To show that φ is continuous, it is enough to show that $\varphi^{-1}(D(fg))$ is open in Spec A_f. We have

$$\varphi^{-1}(D(fg)) = \{xA_f | x \in \text{Spec } A, f \notin x, g \notin x\}$$
$$= \text{Spec } A_f \backslash V(g/1),$$

and so φ is continuous.

It remains to show that φ^{-1} is continuous. Let g/f^n be a non-nilpotent element of A_f, so that $D(g/f^n)$ is a basis element of Spec A_f. We show that $\varphi(D(g/f^n))$ is open in $D(f)$. For any $xA_f \in D(g/f^n)$, one has $g/f^n \notin xA_f$, which says that $g \notin x$ and $f \notin x$. Thus

$$\varphi(D(g/f^n)) = D(fg) = D(f) \cap D(g),$$

which is open in $D(f)$.   □

Definition 2.2.2. A topological space X is **Hausdorff** if for each pair of distinct points $x_1, x_2 \in X$ there exist open sets U_1, U_2 such that $x_1 \in U_1$, $x_2 \in U_2$, and $U_1 \cap U_2 = \emptyset$.

For example, $X = \mathbb{R}$ is Hausdorff in the standard topology.

Proposition 2.2.3. Spec $\mathbb{Z}$ *is not Hausdorff.*

Proof. Let (p) and (q) be two non-trivial points of Spec $\mathbb{Z}$. Suppose that U and W are open sets of Spec $\mathbb{Z}$ with $(p) \in U$ and $(q) \in W$. Since $\{D(f)\}$, $f \neq 0$, is a basis for Spec $\mathbb{Z}$, we can assume without loss of generality that $U = D(m)$ and $W = D(n)$ for some non-zero integers m, n.

We claim that $D(m) \cap D(n) \neq \emptyset$. Since only a finite number of primes divide a non-zero integer and there are an infinite number of primes, there exists a prime s such that $s \nmid m$ and $s \nmid n$. Thus $(s) \notin V(m) \cup V(n)$. Now, by DeMorgan's laws,

$$(s) \in \mathrm{Spec}\ \mathbb{Z}\backslash(V(m) \cup V(n))$$

$$= \mathrm{Spec}\ \mathbb{Z}\backslash V(m) \cap \mathrm{Spec}\ A\backslash V(n)$$

$$= D(m) \cap D(n),$$

and so points in $\mathrm{Spec}\ \mathbb{Z}$ cannot be separated by open sets.   □

Definition 2.2.3. Let X be a topological space. An **open covering** of X is a collection $\mathcal{A}$ of open subsets of X whose union is X. That is, an open covering is a collection $\mathcal{A} = \{U_\alpha\}_{\alpha \in J}$, U_α open in X, that satisfies $X = \cup_{\alpha \in J} U_\alpha$.

Definition 2.2.4. X is **compact** if every open covering $\mathcal{A}$ of X admits a finite subcover; that is, X is compact if for every open covering $\mathcal{A} = \{U_\alpha\}_{\alpha \in J}$ of X there is a finite subcollection $\mathcal{C} = \{U_i\}_{i=1}^{k} \subseteq \mathcal{A}$ with $X = \cup_{i=1}^{k} U_i$.

Definition 2.2.5. A subset Y of a topological space X is **compact** if it is compact in the subspace topology induced by X.

Proposition 2.2.4. *Let W be a closed subset of the compact topological space X. Then W is compact in the subspace topology induced by X.*

Proof. Let $\mathcal{A} = \{U_\alpha \cap W\}$ be an open covering of W with U_α open in X. Now $\mathcal{B} = \{U_\alpha\} \cup \{X\backslash W\}$ is an open covering of X. Since X is compact, we can choose a finite subcollection $\{U_i\}_{i=1}^{k} \cup \{X\backslash W\}$ of $\mathcal{B}$ that covers X. Now $\{U_i \cap W\}_{i=1}^{k}$ is a subcollection of $\mathcal{A}$ that covers W, and so W is compact in the subspace topology.   □

Proposition 2.2.5. *Let $\phi : X \to Y$ be a continuous map of topological spaces with X compact. Then $\phi(X) \subseteq Y$ is compact in the subspace topology induced by Y.*

Proof. Let $\{U_\alpha \cap \phi(X)\}_{\alpha \in J}$ be an open covering of $\phi(X)$ in the subspace topology. Then $\{\phi^{-1}(U_\alpha)\}_{\alpha \in J}$ is an open covering of X, and since X is compact, one can extract a finite subcover $\{\phi^{-1}(U_i)\}_{i=1}^{k}$ of X. Now $\{U_i \cap \phi(X)\}_{i=1}^{k} \subseteq \{U_\alpha \cap \phi(X)\}_{\alpha \in J}$ is a finite subcover of $\phi(X)$.   □

The space $\mathrm{Spec}\ A$ with the Zariski topology enjoys the property of compactness.

Proposition 2.2.6. $\mathrm{Spec}\ A$ *is compact.*

Proof. Since $\mathcal{B} = \{D(f), f$ non-nilpotent$\}$ is a basis for $\mathrm{Spec}\ A$, we can assume that an open covering of $\mathrm{Spec}\ A$ is of the form $\mathcal{A} = \{D(f_\alpha)\}$, where $\{f_\alpha\}_{\alpha \in J}$ is a set of non-nilpotent elements of A. We have $\mathrm{Spec}\ A = \cup_{\alpha \in J} D(f_\alpha)$ with

$$\bigcup_{\alpha \in J} D(f_\alpha) = \bigcup_{\alpha \in J} \mathrm{Spec}\ A\backslash V(f_\alpha) = \mathrm{Spec}\ A\backslash \bigcap_{\alpha \in J} V(f_\alpha)$$

and thus $\displaystyle\bigcap_{\alpha \in J} V(f_\alpha) = \emptyset$. Now

$$\bigcap_{\alpha \in J} V(f_\alpha) = \bigcap_{\alpha \in J} \operatorname{Spec} A \backslash D(f_\alpha)$$

$$= \operatorname{Spec} A \backslash \bigcup_{\alpha \in J} D(f_\alpha)$$

$$= \operatorname{Spec} A \backslash D(I) = V(I),$$

where I is the ideal of A generated by the set of elements $\{f_\alpha\}_{\alpha \in J}$. Therefore, $V(I) = \emptyset$.

Since $V(I) = \emptyset$, there is no point $x \in \operatorname{Spec} A$ with $I \subseteq x$. Hence, by Proposition 1.1.1, $I = A$. Specifically, $1 \in I$, and so there is a finite subset $\{f_1, f_2, \ldots, f_k\}$ of $\{f_\alpha\}$ and ring elements $g_1, g_2, \ldots, g_k$ such that

$$g_1 f_1 + g_2 f_2 + \cdots + g_k f_k = 1.$$

It follows that $I = (f_1, f_2, \ldots, f_k) = A$, and thus

$$V(I) = \operatorname{Spec} A \backslash D(I)$$

$$= \operatorname{Spec} A \backslash \bigcup_{i=1}^{k} D(f_i)$$

$$= \bigcap_{i=1}^{k} (\operatorname{Spec} A \backslash D(f_i))$$

$$= \bigcap_{i=1}^{k} V(f_i) = \emptyset,$$

and so

$$\operatorname{Spec} A = \operatorname{Spec} A \backslash \bigcap_{i=1}^{k} V(f_i) = \bigcup_{i=1}^{k} (\operatorname{Spec} A \backslash V(f_i)) = \bigcup_{i=1}^{k} D(f_i).$$

Therefore, $\{D(f_i)\}_{i=1}^{k}$ is a finite subcollection of $\{D(f_\alpha)\}$ that covers $\operatorname{Spec} A$, and thus $\operatorname{Spec} A$ is compact. $\qquad \square$

Definition 2.2.6. Let S be a subset of a topological space X. The **closure** of S, denoted by S^c, is the intersection of all of the closed sets V that contain S; that is, $S^c = \cap_{S \subseteq V} V$.

For example, if $(-1, 1) = \{x : |x| < 1\} \subseteq \mathbb{R}$, then $(-1, 1)^c = [-1, 1] = \{x : |x| \leq 1\}$, and if $\{(p)\} \subseteq \operatorname{Spec} \mathbb{Z}$, then $\{(p)\}^c = \{(p)\}$.

Observe that a subset $S \subseteq X$ is closed if and only if $S = S^c$.

We can determine precisely when singleton subsets $\{x\}$ of $\operatorname{Spec} A$ are closed.

Proposition 2.2.7. *The subset $\{x\} \subseteq \mathrm{Spec}\ A$ is closed if and only if x is a maximal ideal of A.*

Proof. Suppose $\{x\} \subseteq \mathrm{Spec}\ A$ is closed. Then $x = V(E)$ for some subset $E \subseteq A$. Let I be an ideal of A for which $x \subseteq I \subset A$. By Proposition 1.1.1, there exists a prime ideal y for which $x \subseteq I \subseteq y \subset A$. Now, since $\{x\}^c = \{x\}$, $x = y = I$. Thus x is maximal.

We leave the converse as an exercise. $\square$

At the other extreme, we have $\{\omega\} \subseteq \mathrm{Spec}\ \mathbb{Z}$, whose closure is all of $\mathrm{Spec}\ \mathbb{Z}$. These non-closed singleton subsets are of interest.

Definition 2.2.7. A point $x \in \mathrm{Spec}\ A$ for which $\{x\}^c = \mathrm{Spec}\ A$ is a **generic point** of $\mathrm{Spec}\ A$.

Proposition 2.2.8. *$\mathrm{Spec}\ A$ has a generic point if and only if $\mathrm{nil}(A)$ is a prime ideal.*

Proof. Suppose $x \in \mathrm{Spec}\ A$ is generic. Then $\{x\}^c = \mathrm{Spec}\ A$, so that $\mathrm{Spec}\ A = \cap_{x \in V}\ V$. Hence $\mathrm{Spec}\ A = V$ for all closed V with $x \in V$. Thus $\mathrm{Spec}\ A = V(x)$, which says that x is the nilradical of A.

For the converse, suppose $\mathrm{nil}(A)$ is prime. We show that $\mathrm{nil}(A)$ is a generic point of $\mathrm{Spec}\ A$. Note $\{\mathrm{nil}(A)\}^c = \cap_{\mathrm{nil}(A) \subseteq V} V$. Now $\mathrm{Spec}\ A = V(\{\mathrm{nil}(A)\}) \subseteq V$ for each V containing $\mathrm{nil}(A)$, and thus $\{\mathrm{nil}(A)\}^c = \mathrm{Spec}\ A$. $\square$

For example, ω is a generic point of $\mathrm{Spec}\ \mathbb{Z}$, and ω is a generic point of $\mathrm{Spec}\ \mathbb{F}_p$.

Definition 2.2.8. A space X is **reducible** if there exist proper closed subsets $X_1 \subseteq X$ and $X_2 \subseteq X$ for which $X = X_1 \cup X_2$. A non-empty space X is **irreducible** if it is not reducible.

Proposition 2.2.9. *The space $\mathrm{Spec}\ A$ is irreducible if and only if it has a generic point.*

Proof. Suppose x is a generic point of $\mathrm{Spec}\ A$, and let $\mathrm{Spec}\ A = X_1 \cup X_2$ be a decomposition. The point x is in either X_1 or X_2, so let's assume $x \in X_1$. Then $\mathrm{Spec}\ A = \{x\}^c \subseteq X_1^c = X_1$. Thus $X_1 = \mathrm{Spec}\ A$ and $\mathrm{Spec}\ A$ is irreducible.

For the converse, we show that $\mathrm{Spec}\ A$ being irreducible implies the existence of a generic point. We use the contrapositive: we assume that $\mathrm{Spec}\ A$ has no generic point and show that $\mathrm{Spec}\ A$ is reducible. To this end, we assume that $\mathrm{nil}(A)$ is not prime. Then there exist f, g such that $fg \in \mathrm{nil}(A)$ with both $f \notin \mathrm{nil}(A)$ and $g \notin \mathrm{nil}(A)$. Thus $V(fg) = V(f) \cup V(g) = \mathrm{Spec}\ A$ with $V(f) \subset \mathrm{Spec}\ A$ and $V(g) \subset \mathrm{Spec}\ A$, so that $\mathrm{Spec}\ A$ is reducible. $\square$

Proposition 2.2.10. *$\mathrm{Spec}\ A$ is irreducible if and only if $\mathrm{nil}(A)$ is prime.*

Proof. Exercise. $\square$

For example, since $\mathrm{nil}(\mathbb{Z} \times \mathbb{Z})$ is not prime, $X = \mathrm{Spec}\ (\mathbb{Z} \times \mathbb{Z})$ is reducible. In fact, $X = X_1 \cup X_2$ with $X_1 = V(E_1)$, $E_1 = \{(n, 0) : n \in \mathbb{Z}\}$, and $X_2 = V(E_2)$ with $E_2 = \{(0, n) : n \in \mathbb{Z}\}$.

2.3 Sheaves

We begin this section with some category theory.

Definition 2.3.1. A **category** $\Im$ is a construction consisting of the following three components:

  (i) a collection of objects $\mathrm{Ob}(\Im)$;
 (ii) for each $A, B \in \mathrm{Ob}(\Im)$, a collection of morphisms $\Im(A, B)$;
(iii) for each $A, B, C \in \mathrm{Ob}(\Im)$, a law of composition $\Im(A, B) \times \Im(B, C) \rightarrow \Im(A, C)$, where $(f, g) \mapsto gf$, for $f \in \Im(A, B), g \in \Im(B, C)$.

Categories exist throughout mathematics. Perhaps the primary example is the category of sets. Here the objects are sets, the morphisms between objects are the functions on the sets, and the law of composition is ordinary function composition. Another important category is the category whose objects are the Abelian groups and whose morphisms are the homomorphisms between them.

We assume that a category $\Im$ satisfies the following axioms.

Axiom 1. The collections $\Im(A, B)$ and $\Im(C, D)$ are disjoint unless $A = C$ and $B = D$.

Axiom 2. If $f \in \Im(A, B), g \in \Im(B, C), h \in \Im(C, D)$, then $h(gf) = (hg)f$.

Axiom 3. For each $A \in \mathrm{Ob}(\Im)$, there exists an element $I_A \in \Im(A, A)$ for which $fI_A = f$ for $f \in \Im(A, B)$ and $I_A g = g$ for $g \in \Im(C, A)$.

Let X be a topological space, and let $\Im$ be a category.

Definition 2.3.2. A **presheaf** $\mathcal{F}$ on X with values in $\Im$ is defined by the following data: for each open set U of X, there exists an object $\mathcal{F}(U) \in \mathrm{Ob}(\Im)$, and for each inclusion $U \subseteq V$ of open sets, there exists a morphism

$$\varrho_U^V : \mathcal{F}(V) \rightarrow \mathcal{F}(U)$$

in $\Im(\mathcal{F}(V), \mathcal{F}(U))$ For which

  (i) ϱ_U^U is the identity map and
 (ii) $\varrho_U^W = \varrho_U^V \varrho_V^W$ whenever $U \subseteq V \subseteq W$.

We give the following example. Let X be a topological space, and consider X a set of points, forgetting for the moment its topological structure. Let Y be a non-empty set. Let $\Im$ be the category whose objects are given as $\mathrm{Ob}(\Im) = \{\mathrm{Map}(W, Y) : W \subseteq X\}$, where $\mathrm{Map}(W, Y)$ denotes the set of functions $f : W \rightarrow Y$. For a non-empty open subset U of X, define $\mathcal{F}(U)$ to be the collection of functions $f : U \rightarrow Y$, and for $U = \emptyset$ set $\mathcal{F}(U)$ to be the unique empty function $f : \emptyset \rightarrow Y$. Moreover, whenever $U \subseteq V$, define a morphism $\varrho_U^V : \mathcal{F}(V) \rightarrow \mathcal{F}(U)$ by $\varrho_U^V(f) = f_U$, where f_U denotes the restriction of f to U. Then $\mathcal{F}$, together with the restrictions $\{\varrho_U^V\}$, is a presheaf on X that we call the **presheaf of ordinary functions** on X.

For the case $X = \text{Spec } A$, the functions in the presheaf of ordinary functions on X, however, are different from the functions we have already defined on Spec A in §1.3. For example, if $\mathcal{F}$ is the presheaf of ordinary functions on $X = \text{Spec } \mathbb{Z}$, then $\mathcal{F}(\text{Spec } \mathbb{Z})$ is certainly not in a 1-1 correspondence with $\mathbb{Z}$, yet $\mathbb{Z}$ is identified with the functions on Spec $\mathbb{Z}$ as defined in §1.3.

Can we define a presheaf of functions on Spec A that recovers the functions of § 1.3?

We begin by setting $\mathcal{F}(\emptyset) = \{0\} \subseteq A$. We next define $\mathcal{F}(U)$ for a non-empty open set U. Let $\{f_\alpha\}_{\alpha \in J}$ denote the collection of non-nilpotent elements of A for which $D(f_\alpha) \subseteq U$. If there exist indices $\alpha, \beta \in J$ with $D(f_\alpha) \subseteq D(f_\beta)$, then by Lemma 2.2.2 there exists an integer $n_\alpha > 0$ and an element $a_\beta \in A$ for which

$$f_\alpha^{n_\alpha} = f_\beta a_\beta,$$

and thus

$$\frac{1}{f_\beta} = \frac{a_\beta}{f_\alpha^{n_\alpha}}.$$

Thus, when $D(f_\alpha) \subseteq D(f_\beta)$, there exists a homomorphism $A_{f_\beta} \to A_{f_\alpha}$ defined by

$$\frac{r}{f_\beta^l} \mapsto \frac{r a_\beta^l}{f_\alpha^{n_\alpha l}},$$

which we denote as

$$\varrho_{D(f_\alpha)}^{D(f_\beta)} : A_{f_\beta} \to A_{f_\alpha}.$$

We define $\mathcal{F}(U)$ to be the subset of $\prod_{\alpha \in J} A_{f_\alpha}$ consisting of all families $\{u_\alpha\}_{\alpha \in J}$ for which

$$\varrho_{D(f_\alpha)}^{D(f_\beta)}(u_\beta) = u_\alpha$$

whenever $D(f_\alpha) \subseteq D(f_\beta)$.

In the case where U is the basic open set $D(f)$, we have $\mathcal{F}(D(f)) = A_f$. To see this, let $\{u_\alpha\}$ be a family in $\mathcal{F}(D(f))$. Then there exists an element $u_0 \in A_f$ for which

$$\varrho_{D(f_\alpha)}^{D(f)}(u_0) = u_\alpha$$

for all α. Thus $\mathcal{F}(D(f))$ is identified with A_f. By Proposition 2.2.2, Spec A_f is homeomorphic to $D(f)$, and so we write $\mathcal{F}(\text{Spec } A_f) = A_f$.

Each element of A_f can be written as the rational expression

$$\frac{r}{f^n}$$

for $r \in A$ and integer $n \geq 0$. Moreover, $f^n \notin x$ for each $x \in D(f)$. Thus r/f^n can be viewed as a rational function on $D(f)$ of the type described in §1.3; that is, for each $x \in D(f)$, we define

$$\frac{r}{f^n}(x) = \frac{r(x)}{f^n(x)},$$

where $r(x) = r \pmod{x}$ and $f^n(x) = f^n \pmod{x}$. Thus $\mathcal{F}(\operatorname{Spec} A_f) = A_f$ is the collection of rational functions on the open set $\operatorname{Spec} A_f$.

We next show that $\mathcal{F}(\operatorname{Spec} A) = A$. By Proposition 2.2.6, $\operatorname{Spec} A$ is compact, and thus there is an open covering $\{D(f_i)\}_{i=1}^k$ of $\operatorname{Spec} A$ with $A = (f_1, f_2, \cdots, f_k)$. We assume that no element f_i is a unit. Now, for each i, $1 \le i \le k$, there exists a prime ideal $x_i \in \operatorname{Spec} A$ for which $f_i \in x_i$. (This is a consequence of Proposition 1.1.1.) Thus there are no rational functions on $\operatorname{Spec} A$ of the form a/b unless $a \in A$ and $b = 1$, and so $\mathcal{F}(\operatorname{Spec} A)$ is identified with A.

We now see that $\mathcal{F}$ defined as above recovers A as the collection of functions on $\operatorname{Spec} A$ in the manner described in §1.3. (The statement $\mathcal{F}(\emptyset) = \{0\}$ can now be interpreted as: 0 is the unique function whose domain is $\emptyset$.)

Now that we have an agreeable definition for $\mathcal{F}(U)$, $U \ne \emptyset$, we suppose $U \subseteq V$ and let $\{u_\alpha\} \in \mathcal{F}(V)$. Then $D(f_\alpha) \subseteq V$. Let $\{u_\beta\}$ denote those components of $\{u_\alpha\}$ for which each index β is such that $D(f_\beta) \subseteq U$. Since $U \subseteq V$, the family $\{u_\beta\}$ satisfies $\varrho_{D(f_\beta)}^{D(f_\gamma)}(u_\gamma) = u_\beta$ whenever $D(f_\beta) \subseteq D(f_\gamma) \subseteq U$. Thus $\{u_\beta\} \in \mathcal{F}(U)$. In this manner, we define a homomorphism

$$\varrho_U^V : \mathcal{F}(V) \to \mathcal{F}(U)$$

by $\varrho_U^V(\{u_\alpha\}) = \{u_\beta\}$.

Proposition 2.3.1. $\mathcal{F}$, together with the homomorphisms ϱ_U^V defined above, is a presheaf on $\operatorname{Spec} A$.

Proof. We show that $\mathcal{F}$ satisfies conditions (i) and (ii) in Definition 2.3.1. By construction, $\varrho_U^U(\{u_\alpha\}) = \{u_\alpha\}$ for all $\{u_\alpha\} \in \mathcal{F}(U)$, so (i) is satisfied.

For (ii), suppose there exist open sets $U \subseteq V \subseteq W$. Let $\{u_\beta\}$ be the components of $\{u_\alpha\} \in \mathcal{F}(W)$ with index β for which $D(f_\beta) \subseteq V$; let $\{u_\gamma\}$ be the components of $\{u_\beta\} \in \mathcal{F}(V)$ with index γ for which $D(f_\gamma) \subseteq U$. Then $\varrho_U^V(\varrho_V^W(\{u_\alpha\})) = \varrho_U^V(\{u_\beta\}) = \{u_\gamma\} = \varrho_U^W(\{u_\alpha\})$, so (ii) holds. $\qquad\square$

The presheaf given in Proposition 2.3.1 is the **structure presheaf on** $\operatorname{Spec} A$ and is denoted by $\mathcal{O}$.

A presheaf on X should exhibit some natural properties, which we formalize in the following definition.

Definition 2.3.3. The presheaf $\mathcal{F}$ on X is a **sheaf** if, for each open set $U \subseteq X$ and each open covering $\{U_\alpha\}_{\alpha \in J}$ of U, the following are satisfied:

(i) For $r, s \in \mathcal{F}(U)$, if $\varrho_{U_\alpha}^U(r) = \varrho_{U_\alpha}^U(s)$ for all U_α in the covering, then $r = s$.

(ii) Suppose $\{r_\alpha\}$, $r_\alpha \in \mathcal{F}(U_\alpha)$, is a family of elements indexed over the open sets in the covering of U. Suppose $\varrho_{U_\alpha \cap U_\beta}^{U_\alpha}(r_\alpha) = \varrho_{U_\alpha \cap U_\beta}^{U_\beta}(r_\beta)$ for all α, β. Then there exists an element $r \in \mathcal{F}(U)$ for which $r_\alpha = \varrho_{U_\alpha}^U(r)$ for all U_α.

For example, the presheaf of ordinary functions on X is a sheaf. Indeed, suppose that $\{U_\alpha\}_{\alpha \in J}$ is a covering of the open set $U \subseteq X$. Suppose that $f, g \in \mathcal{F}(U)$ are such that $f_{U_\alpha}(x) = g_{U_\alpha}(x)$ for all $x \in U_\alpha$ and all U_α. Then one has $f(x) = g(x)$ for all $x \in U$. This shows that condition (i) of Definition 2.3.2 holds. Moreover, suppose $\{f_\alpha\}$, $f_\alpha \in \mathcal{F}(U_\alpha)$, is a family of functions indexed over all U_α in the covering. Suppose $(f_\alpha)_{U_\alpha \cap U_\beta}(x) = (f_\beta)_{U_\alpha \cap U_\beta}(x)$ for all $x \in U_\alpha \cap U_\beta$ and all U_α. Then there exists a function $f : U \rightarrow Y$ for which $f_\alpha = f_{U_\alpha}$ for all U_α. Thus the presheaf of ordinary functions is a sheaf.

Happily, the structure presheaf on Spec A is also a sheaf.

Proposition 2.3.2. *The structure presheaf $\mathcal{O}$ on* Spec A *is a sheaf of functions.*

Proof. We restrict our proof to verifying that conditions (i) and (ii) of Definition 2.3.2 hold in the case where $U = $ Spec A. The complete proof can be found in [Sh74, Chapter V, §3].

By compactness, we can assume that an open covering of Spec A is of the form $\{D(f_i)\}_{i=1}^{k}$ for non-nilpotent elements $\{f_i\}_{i=1}^{k}$. Suppose $r, s \in \mathcal{O}($Spec $A) = A$ with

$$\varrho_{D(f_i)}^{\text{Spec } A}(r) = \varrho_{D(f_i)}^{\text{Spec } A}(s)$$

for all i. Since $D(1) = $ Spec A, this is equivalent to $\varrho_{D(f_i)}^{D(1)}(r) = \varrho_{D(f_i)}^{D(1)}(s)$ for all i. But then $r/1 = s/1$, and so condition (i) holds.

For condition (ii), Let $\{q_i\}$, $q_i = v_i/f_i^n \in \mathcal{O}(D(f_i))$, be a family of rational functions with $\varrho_{D(f_i) \cap D(f_j)}^{D(f_i)}(q_i) = \varrho_{D(f_i) \cap D(f_j)}^{D(f_j)}(q_j)$ for all i, j. Since $D(f_i) \cap D(f_j) = D(f_i f_j)$,

$$\frac{v_i f_j^n}{(f_i f_j)^n} = \frac{v_j f_i^n}{(f_i f_j)^n},$$

and so

$$v_i f_j^n f_i^n f_j^n = f_i^n f_j^n v_j f_i^n. \tag{2.2}$$

Since $D(f_j) = D(f_j^{2n})$, for $1 \leq j \leq k$, the finite covering of Spec A can be written $\{D(f_j^{2n})\}_{j=1}^{k}$, and thus there exist elements $g_1, g_2, \ldots, g_k$ of A for which $\sum_{j=1}^{k} f_j^{2n} g_j = 1$. Set $q = \sum_{j=1}^{n} v_j f_j^n g_j$. Then, for all i, $1 \leq i \leq k$,

$$f_i^{2n} q = \sum_{j=1}^{k} f_i^{2n} v_j f_j^n g_j$$

$$= \sum_{j=1}^{k} v_i f_j^n f_i^n f_j^n g_j \quad \text{by (2.2)}$$

$$= v_i f_i^n \sum_{j=1}^{k} f_j^{2n} g_j$$

$$= v_i f_i^n,$$

and so $\varrho_{D(f_i)}^{D(1)}(q) = q_i$ for all i, and hence condition (ii) holds. $\square$

2.4 Representable Functors

In the last section, we introduced the notion of a category. We now ask: What kind of mappings do we have between two categories?

Definition 2.4.1. Let $\Im$ and $\Im'$ be categories. A **(covariant) functor** $F : \Im \to \Im'$ is a rule that assigns to each object $X \in \text{Ob}(\Im)$ exactly one object $F(X) \in \text{Ob}(\Im')$ and to each morphism $f \in \Im(X, Y)$ exactly one morphism $F(f) \in \Im'(F(X), F(Y))$ and that satisfies

(i) $F(gf) = F(g)F(f)$ for all $g \in \Im(Y, Z)$ and
(ii) $F(I_X) = I_{F(X)}$.

We can also define mappings between two functors.

Definition 2.4.2. Let F and G be functors from the category $\Im$ to the category $\Im'$. Then a **natural transformation** $\psi : F \to G$ is a rule that assigns to each object $X \in \text{Ob}(\Im)$ a morphism ψ_X in $\Im'(F(X), G(X))$ such that for a given morphism $f \in \Im(X, Y)$ the diagram

$$
\begin{array}{ccc}
 & \psi_X & \\
F(X) & \to & G(X) \\
 & & \\
F(f) \downarrow & & \downarrow \; G(f) \\
 & & \\
F(Y) & \to & G(Y) \\
 & \psi_Y &
\end{array}
$$

commutes.

Here is an example of a functor that will be of great importance to us. Let R be a ring and let A be a commutative R-algebra with unity, 1_A. Let $\lambda : R \to A$ be the R-algebra structure map of A. We assume that $\lambda(1_R) = 1_A$. For a commutative R-algebra B with unity 1_B, let $\text{Hom}_{R\text{-alg}}(A, B)$ denote the collection of R-algebra homomorphisms $f : A \to B$; we assume that $f(1_A) = 1_B$.

Let $\Im_{R\text{-alg}}$ denote the category of commutative R-algebras, where the collection of morphisms $\Im_{R\text{-alg}}(A, B)$ is given as $\text{Hom}_{R\text{-alg}}(A, B)$, for $A, B \in \text{Ob}(\Im_{R\text{-alg}})$. The law of composition is ordinary function composition.

Let $\Im_{\text{sets}}$ denote the category of sets.

Fix an object $A \in \text{Ob}(\Im_{R\text{-alg}})$, and define a map $F : \Im_{R\text{-alg}} \to \Im_{\text{sets}}$ by the rule

$$F(B) = \text{Hom}_{R\text{-alg}}(A, B) \tag{2.3}$$

for all $B \in \text{Ob}(\Im_{R\text{-alg}})$.

Proposition 2.4.1. *F defined as above is a functor from the category of commutative R-algebras to the category of sets.*

Proof. Let $\varrho \in \Im_{R\text{-alg}}(S, T)$ for objects $S, T \in \text{Ob}(\Im_{R\text{-alg}})$. Then there exists a unique element $F(\varrho) \in \Im_{\text{sets}}(F(S), F(T))$, defined as

$$(F(\varrho)(h))(s) = \varrho(h(s))$$

for $h \in F(S)$, $s \in A$. So it remains to check that conditions (i) and (ii) of Definition 2.4.1 hold. We prove (i) here and leave (ii) as an exercise. Let $\tau \in \Im_{R\text{-alg}}(T, W)$ for $W \in \text{Ob}(\Im_{R\text{-alg}})$. Then

$$\begin{aligned}
(F(\tau\varrho)(h))(s) &= (\tau\varrho)(h(s)) \\
&= \tau(\varrho(h(s))) \\
&= (F(\tau)(\varrho h))(s) \\
&= F(\tau)(F(\varrho)(h))(s) \\
&= ((F(\tau)F(\varrho))(h))(s),
\end{aligned}$$

and so (i) holds. $\square$

But how does the functor F defined in (2.3) relate to the sheaf structure of Spec A? The restriction map $\varrho_{D(g)}^{D(f)} : \mathcal{O}(D(f)) \to \mathcal{O}(D(g))$ of Spec A is a homomorphism of R-algebras $\varrho_{D(g)}^{D(f)} : A_f \to A_g$, and thus by the functorality of F there exists a map

$$F\left(\varrho_{D(g)}^{D(f)}\right) : F(A_f) \to F(A_g)$$

defined as

$$\left(F\left(\varrho_{D(g)}^{D(f)}\right)(h)\right)(s) = \varrho_{D(g)}^{D(f)}(h(s))$$

for $h \in F(A_f)$, $s \in A$. Put

$$\varrho_{A_g}^{A_f}(h)(s) = \left(F\left(\varrho_{D(g)}^{D(f)}\right)(h)\right)(s).$$

Then F is a kind of sheaf on Spec A with A_f playing the role of the basic open set $D(f)$ and the maps $\varrho^{A_f}_{A_g}$ playing the role of the restrictions $\varrho^{D(f)}_{D(g)}$.

To see this for the case $U = \text{Spec } A$, let $\{D(f_j)\}^m_{j=1}$ be a finite open covering of Spec A. To show that property (i) of Definition 2.3.2 holds, let $h, k \in F(A)$, and suppose that

$$\varrho^A_{A_{f_j}}(h) = \varrho^A_{A_{f_j}}(k)$$

for all $D(f_j)$ in the covering. Then

$$\varrho^{\text{Spec } A}_{D(f_j)}(h(s)) = \varrho^{\text{Spec } A}_{D(f_j)}(k(s))$$

for all $s \in A$, $1 \le j \le m$. Thus, by the sheaf property of $\mathcal{O}$, $h(s) = k(s)$, and so $h = k$, which is property (i).

Next, let $\{h_j\}$, $h_j \in F(A_{f_j})$, be a collection of homomorphisms where j ranges over all of the open sets in the covering. Suppose that

$$\varrho^{A_{f_i}}_{A_{f_i f_j}}(h_i) = \varrho^{A_{f_j}}_{A_{f_i f_j}}(h_j)$$

for all i, j. (Here, $A_{f_i f_j}$ plays the role of $D(f_i f_j) = D(f_i) \cap D(f_j)$ since Spec $A_{f_i f_j}$ is homeomorphic to $D(f_i) \cap D(f_j)$.) Thus,

$$\varrho^{D(f_i)}_{D(f_i) \cap D(f_j)}(h_i(s)) = \varrho^{D(f_j)}_{D(f_i) \cap D(f_j)}(h_j(s))$$

for all i, j and $s \in A$. Now $h_i(s) \in A_{f_i}$ and $h_j(s) \in A_{f_j}$, so, by the sheaf property of $\mathcal{O}$, there exists an element $h_s \in A$, dependent on s, for which

$$\varrho^{\text{Spec } A}_{D(f_i)}(h_s) = h_i(s)$$

for all i. Thus there exists an element $q \in F(A)$, defined by $s \mapsto h_s$, for which $\varrho^A_{A_{f_i}}(q) = h_i$. So property (ii) of Definition 2.3.2 holds.

So the functor F of (2.3) corresponds to the structure sheaf $\mathcal{O}$ on Spec A. We shall henceforth identify $F = \text{Hom}_{R\text{-alg}}(A, -)$ with Spec A, and we say that F is a **representable functor** that is **represented by** A. We write $R[F] = A$.

We next construct a new category. An algebra over R is an R-module A for which the scalar multiplication $R \times A \to A$ is given by a ring homomorphism $\lambda : R \to A$; that is, $(r, a) \mapsto \lambda(r)a$ for all $r \in R$, $a \in A$. The map λ is the R-algebra structure map of A. We set $\lambda_A = \lambda$. Let $\mathfrak{I}_R$ denote the category whose objects $\text{Ob}(\mathfrak{I}_R)$ consist of all R-algebra structure maps $\lambda_A : R \to A$. For objects $\lambda_A, \lambda_B \in \mathfrak{I}_R$, the elements of $\mathfrak{I}_R(\lambda_A, \lambda_B)$ consist of ring homomorphisms $\phi : A \to B$ for which $\phi\lambda_A = \lambda_B$. For such ϕ and $r \in R$, $a \in A$,

$$\phi(ra) = \phi(\lambda_A(r)a)$$
$$= \phi(\lambda_A(r))\phi(a)$$

$$= \lambda_B(r)\phi(a)$$

$$= r\phi(a).$$

It follows that $\Im_R(\lambda_A, \lambda_B)$ is in a 1-1 correspondence with $\Im_{R\text{-alg}}(A, B)$.

Let $\lambda_A, \lambda_B \in \mathrm{Ob}(\Im_R)$. Let $A \otimes B$ denote the tensor product (over R) of the R-modules A and B. Then $A \otimes B$ is an R-algebra with structure map $\lambda_{A \otimes B}$. Let $\phi_1 : A \to A \otimes B$, $\phi_2 : B \to A \otimes B$ be morphisms (R-algebra maps) defined as $a \mapsto a \otimes 1$, $b \mapsto 1 \otimes b$, respectively. As shown in [La84, I,§7],

$$(\lambda_{A \otimes B}, \phi_1, \phi_2)$$

is a **coproduct** in the category $\Im_R$. This says that for morphisms $\psi : A \to S$ and $\rho : B \to S$ there exists a unique morphism $\varrho : A \otimes B \to S$ for which the diagram

$$
\begin{array}{ccc}
 & \phi_1 & \\
\lambda_A & \to & \lambda_{A \otimes B} \\
\psi \downarrow & \varrho \nearrow \quad \uparrow & \phi_2 \\
\lambda_S & \leftarrow & \lambda_B \\
 & \rho &
\end{array}
$$

commutes. An important consequence of this is the following.

Proposition 2.4.2. *Let S be an R-algebra. Then*

$$\mathrm{Hom}(A, S) \times \mathrm{Hom}(B, S) = \mathrm{Hom}(A \otimes B, S).$$

Proof. Let $\psi \in \mathrm{Hom}_{R\text{-alg}}(A, S)$, $\rho \in \mathrm{Hom}_{R\text{-alg}}(B, S)$. There exists an element of $f_{\psi, \rho} \in \mathrm{Hom}_{R\text{-alg}}(A \otimes B, S)$ defined as

$$f_{\psi, \rho}\left(\sum a \otimes b\right) = \sum \psi(a)\rho(b).$$

Now suppose $f \in \mathrm{Hom}_{R\text{-alg}}(A \otimes B, S)$. Then $\psi = f|_{A \otimes 1} : A \to S$ and $\rho = f|_{1 \otimes B} : B \to S$ are morphisms. There is a morphism $f_{\psi, \rho} : A \otimes B \to S$ defined by $\sum a \otimes b \mapsto \sum \psi(a)\rho(b)$. Both f and $f_{\psi, \rho}$ satisfy the diagram above, and so $f = f_{\psi, \rho}$.   □

Let $\phi : A \to B$ be a homomorphism of R-algebras. Then, as we have seen, ϕ determines an associated map

$$^a\phi : \mathrm{Spec}\ B \to \mathrm{Spec}\ A,$$

with $^a\phi(D(f)) = \{\phi^{-1}(x) : x \in D(f)\}$ for a basic open set $D(f)$ in $\mathrm{Spec}\ B$.

Let G and F be the functors represented by B and A, respectively. Then ϕ also determines a map from G to F.

Proposition 2.4.3. *Let $\phi : A \to B$ be a homomorphism of R-algebras, and let F and G be the functors defined as $F = \mathrm{Hom}_{R\text{-alg}}(A, -)$ and $G = \mathrm{Hom}_{R\text{-alg}}(B, -)$. Then the rule $^a\phi$ that assigns to an object $S \in \mathrm{Ob}(\mathfrak{I}_{R\text{-alg}})$ the map*

$$^a\phi_S : G(S) \to F(S)$$

defined as $^a\phi_S(z)(x) = z(\phi(x))$ for $z \in G(S)$, $x \in A$, is a natural transformation of functors $^a\phi : G \to F$.

Proof. Clearly, the map $^a\phi_S$ is an element of $\mathfrak{I}_{\mathrm{sets}}(G(S), F(S))$. Let $\varrho : S \to T$ be an R-algebra homomorphism. Then, for R-algebra map $z : B \to S$,

$$
\begin{aligned}
F(\varrho)(^a\phi_S(z)) &= F(\varrho)(z\phi) \\
&= \varrho(z\phi) \\
&= (\varrho z)(\phi) \\
&= {}^a\phi_T(\varrho z) \\
&= {}^a\phi_T(G(\varrho)(z)),
\end{aligned}
$$

and so the diagram

$$
\begin{array}{ccc}
& {}^a\phi_S & \\
G(S) & \to & F(S) \\
G(\varrho) \downarrow & & \downarrow F(\varrho) \\
G(T) & \to & F(T) \\
& {}^a\phi_T &
\end{array}
$$

commutes. Thus $^a\phi : G \to F$ is a natural transformation. $\qquad\square$

And, with a little more work, one can prove the following.

Proposition 2.4.4. *[Yoneda's Lemma] Let $G = \mathrm{Hom}_{R\text{-alg}}(B, -)$ and $F = \mathrm{Hom}_{R\text{-alg}}(A, -)$ be representable functors from the category of commutative R-algebras to the category of sets. Then the collection of R-algebra homomorphisms $A \to B$ is in a 1-1 correspondence with the collection of natural transformations $G \to F$.*

Proof. We show that $\phi \mapsto {}^a\phi$ is a bijection between $\mathfrak{I}_{R\text{-alg}}(A, B)$ and the collection of natural transformations $G \to F$. Let $\phi : A \to B$ be an R-algebra homomorphism. Then, by Proposition 2.4.3, $^a\phi$ is a natural transformation.

We next show that a natural transformation $\psi : G \to F$ can be written in the form $\psi = {}^a\phi$ for some R-algebra homomorphism $\phi : A \to B$. Since $B \in \mathrm{Ob}(\mathfrak{I}_{R\text{-alg}})$, there is a function $\psi_B : G(B) \to F(B)$. Let $I_B : B \to B$ denote the identity map.

Put $\phi = \psi_B(I_B) \in F(B)$. Then $\phi : A \to B$ is a map of R-algebras. Now, for an R-algebra map $z : B \to S$, one has $\psi_S : G(S) \to F(S)$ with

$$\psi_S(z) = \psi_S(zI_B)$$
$$= \psi_S(G(z)(I_B))$$
$$= F(z)(\psi_B(I_B))$$
$$= F(z)(\phi)$$
$$= z\phi$$
$$= {}^a\phi_S(z).$$

Consequently, $\psi = {}^a\phi$. Thus the map $\phi \mapsto {}^a\phi$ is surjective. Suppose that ${}^a\phi = {}^a\omega$. Then ${}^a\phi_B(I_B) = I_B\phi = \phi$ and ${}^a\omega_B(I_B) = I_B\omega = \omega$. Thus $\phi = \omega$, and so the map $\phi \mapsto {}^a\phi$ is a bijection. $\qquad \square$

2.5   Chapter Exercises

Exercises for §2.1

1. Let S be a subset of a topological space X. Suppose that for each $a \in S$ there exists an open set $U \subseteq S$ for which $a \in U$. Prove that $X \backslash S$ is a closed subset of X.
2. List all of the possible topologies on the set $X = \{a, b, c, d\}$.
3. Prove Proposition 2.1.2.
4. Consider the Zariski topology on Spec $\mathbb{Z}$. Determine whether the following subsets are closed, open, neither, or both.

    (a) $\{(5), (7), (11), \ldots, \omega\}$.
    (b) $\{(11)\}$.
    (c) $\{(5), (7), (11), \ldots\}$.

5. Let p be a prime of $\mathbb{Z}$. List all of the open sets in the Zariski topology of Spec $(\mathbb{Z}/(p))$.
6. Suppose $\mathbb{Q}$ is endowed with the $|\ |_p$-topology. Show that the map $\phi : \mathbb{Q} \to \mathbb{Q}$ defined as $\phi(x) = x^2$ is continuous. *Hint: First show that the maps $f(x) = x$ and $g(x) = -x$ are continuous.*
7. Prove that $\mathbb{Q}$ in the $|\ |_p$-topology is not homeomorphic to $\mathbb{Q}$ endowed with the subspace topology induced by the standard topology on $\mathbb{R}$.
8. Let $\phi : X \to Y$ be a map of topological spaces where X has the discrete topology. Prove that ϕ is continuous.

Exercises for §2.2

9. Prove that $\mathbb{Q}_p$ is Hausdorff in the $|\ |_p$-topology.
10. Show that $\mathbb{R}$ is Hausdorff in the discrete topology.

11. Prove that Spec $\mathbb{Z}[i]$ is irreducible. What features in the spectral diagram indicate irreducibility?

12. Prove that Spec $\mathbb{Z}C_2$ is reducible. What features in the spectral diagram indicate reducibility?

13. Prove that Spec $\mathbb{Z}C_3$ is reducible. What features in the spectral diagram indicate reducibility?

14. Prove Lemma 2.2.1(ii).

15. Prove that Spec $(\mathbb{Z}/(8))$ is Hausdorff.

16. Compute the closure of $S = \{(2), (3), (5)\}$ in Spec $\mathbb{Z}$.

17. Compute the closure of $S = \{(3), (2 + i), (7)\}$ in Spec $\mathbb{Z}[i]$.

18. Compute the closure of $\{(1 + \tau)\}$ in Spec $\mathbb{Z}C_2$.

19. Prove the converse of Proposition 2.2.7.

20. Let $S = \{(1+i), (3), (2+i), (7), (11), (3+2i), \ldots, \} \subseteq$ Spec $\mathbb{Z}[i]$ be endowed with the subspace topology. Determine whether S is homeomorphic to Spec $\mathbb{Z}$.

21. Let $S = \{(2, 1 + \tau), (2 - \tau), (3 - \tau), \ldots, (\frac{p+1}{2} - \frac{p-1}{2}\tau), \ldots, (1 + \tau)\}$ in Spec $\mathbb{Z}C_2$ be endowed with the subspace topology. Determine whether S is homeomorphic to Spec $\mathbb{Z}$.

Exercises for §2.3

22. Let X be a topological space endowed with the discrete topology, and let A be a non-empty set with at least two elements. Let $a \in A$. Let $\mathcal{F}(\emptyset) = \{a\}$ and, for each non-empty open subset U of X, let $\mathcal{F}(U) = A$. For non-empty open subsets $U, V, U \subseteq V$, let ϱ_U^V be the identity function on A; otherwise, for V open, put $\varrho_\emptyset^V : \mathcal{F}(V) \to \{a\}$. Show that $\mathcal{F}$ is a presheaf.

23. Prove that the presheaf defined in Exercise 22 is not a sheaf.

24. Let $\mathbb{R}$ be endowed with the standard topology, and for an open subset U let $\mathcal{F}(U)$ be the collection of all continuous functions $\phi : U \to \mathbb{R}$. For $U \subseteq V$, let $\varrho_U^V : \mathcal{F}(V) \to \mathcal{F}(U)$ be the restriction of $\phi \in \mathcal{F}(V)$ to U. Prove that $\mathcal{F}$ is a sheaf.

25. Classify all of the sheaves of sets on Spec $(\mathbb{Z}/(8))$.

26. Suppose that A is a PID, and let $\mathcal{O}$ denote the structure presheaf on Spec A. Let $U = D(f) \cup D(g)$, where f, g are non-nilpotent elements of A.

    (a) Compute $\mathcal{O}(U)$.

    (b) Prove that $\mathcal{O}(\text{Spec } A) \subseteq \mathcal{O}(U) \subseteq \mathcal{O}(D(f))$.

Exercises for §2.4

27. Finish the proof of Proposition 2.4.1.

28. Let $\mathfrak{I}_{com}$ denote the category of commutative rings with unity, and let $\mathfrak{I}_{Abel}$ denote the category of Abelian groups. Define $F : \mathfrak{I}_{com} \to \mathfrak{I}_{Abel}$ by the rule $F(R) = \langle R, + \rangle$. Prove that F is a functor.

29. Let $\mathfrak{I}_{rings}$ denote the category of rings with unity, and let $\mathfrak{I}_{gps}$ denote the category of groups. Define $F : \mathfrak{I}_{rings} \to \mathfrak{I}_{gps}$ by the rule $F(R) = \langle U(R), \cdot \rangle$. Prove that F is a functor.

30. Let X be a topological space, and let $\Im_X$ be the category whose objects are the open sets of X. For $U, V \in \mathrm{Ob}(\Im_X)$ with $U \subseteq V$, the set of morphisms $\Im_X(U, V)$ consists of the inclusion map $U \subseteq V$ only. Let $\Im_{\mathrm{sets}}$ denote the category of sets. Show that $\mathcal{F}$ is a presheaf of sets on X if and only if $\mathcal{F}$ is a contravariant functor from $\Im_X$ to $\Im_{\mathrm{sets}}$.

31. Let $\Im_{\mathrm{dom}}$ denote the category of integral domains, and let $\iota : \mathbb{Z} \to \mathbb{Q}$ denote the inclusion homomorphism. Let $\alpha, \beta \in \Im_{\mathrm{dom}}(\mathbb{Q}, S)$, $S \in \mathrm{Ob}(\Im_{\mathrm{dom}})$. Prove that $\alpha\iota = \beta\iota$ implies $\alpha = \beta$.

Chapter 3
Representable Group Functors

Throughout this chapter, by "ring" we mean a non-zero commutative ring with unity.

3.1 Introduction to Representable Group Functors

Let A be a commutative R-algebra, and let F be the covariant functor defined as $F = \mathrm{Hom}_{R\text{-alg}}(A, -)$. Let S be an object in $\mathrm{Ob}(\mathfrak{I}_{R\text{-alg}})$. What structure on A do we need to endow $F(S)$ with a binary operation?

Let $\Delta : A \to A \otimes_R A$ be an R-algebra homomorphism. For $a \in A$, we shall write the image of a as

$$\Delta(a) = \sum_{(a)} a_{(1)} \otimes a_{(2)},$$

where $a_{(1)}, a_{(2)} \in A$. This is the **Sweedler notation** for $\Delta(a)$. It is important to note that "(1)" and "(2)" are not subscripts in the usual sense: $a_{(1)}$ records the left components of the tensors in the expansion of $\Delta(a)$, while $a_{(2)}$ records the right components in $\Delta(a)$.

Since Δ is an R-algebra homomorphism, $\Delta(ab) = \Delta(a)\Delta(b)$ and one writes

$$\Delta(ab) = \left(\sum_{(a)} a_{(1)} \otimes a_{(2)} \right) \left(\sum_{(b)} b_{(1)} \otimes b_{(2)} \right) = \sum_{(a,b)} a_{(1)}b_{(1)} \otimes a_{(2)}b_{(2)}.$$

Note that $\Delta(a_{(i)})$, $i \geq 1$, is written

$$\Delta(a_{(i)}) = \sum_{(a_{(i)})} a_{(i)(1)} \otimes a_{(i)(2)}.$$

R.G. Underwood, *An Introduction to Hopf Algebras*, DOI 10.1007/978-0-387-72766-0_3,   35
© Springer Science+Business Media, LLC 2011

By Proposition 2.4.3, the R-algebra homomorphism Δ determines a natural transformation of representable functors

$$^a\Delta : \mathrm{Hom}_{R\text{-alg}}(A \otimes_R A, -) \to \mathrm{Hom}_{R\text{-alg}}(A, -),$$

which by Proposition 2.4.2 can be written as

$$^a\Delta : F \times F \to F,$$

where

$$^a\Delta_S : F(S) \times F(S) \to F(S),$$

for an R-algebra S. The morphism $^a\Delta_S$ (binary operation on $F(S)$) is given by the rule

$$^a\Delta_S(f, g)(x) = (f, g)\Delta(x) = \sum_{(x)} f(x_{(1)})g(x_{(2)})$$

for $f, g \in F(S)$, $x \in A$. We set

$$f * g = {}^a\Delta_S(f, g).$$

Example 3.1.1. Let $A = RG$ with G a finite Abelian group, and let $F = \mathrm{Hom}_{R\text{-alg}}(RG, -)$ be the corresponding functor. Then the map $\Delta : RG \to RG \otimes_R RG$ defined by

$$\sum_{\tau \in G} a_\tau \tau \mapsto \sum_{\tau \in G} a_\tau (\tau \otimes \tau)$$

is an R-algebra homomorphism and determines a binary operation on $F(S)$ given by

$$(f * g)\left(\sum_{\tau \in G} a_\tau \tau\right) = \sum_{\tau \in G} a_\tau f(\tau)g(\tau).$$

Like all binary operations on sets, those on $F(S)$ can be associative or commutative, admit an identity or inverses, and so forth. Since we have defined a binary operation on $F(S)$ using an algebra map, we can also describe the properties of the binary operation by specifying conditions on this algebra map. We first look at conditions for commutativity and associativity.

The map $t : A \otimes_R A \to A \otimes_R A$ defined as $t(a \otimes b) = b \otimes a$ is the **twist map**.

Proposition 3.1.1. *Let $\Delta : A \to A \otimes_R A$ be an R-algebra map that satisfies*

$$\Delta(a) = t(\Delta(a)) \tag{3.1}$$

for all $a \in A$. Then the corresponding binary operation on $F(S) = \mathrm{Hom}_{R\text{-alg}}(A, S)$ is commutative.

Proof. We show that $(f * g)(a) = (g * f)(a)$ for all $a \in A$. We have

$$(f * g)(a) = (f, g) \sum_{(a)} a_{(1)} \otimes a_{(2)}$$

$$= (f, g) \sum_{(a)} a_{(2)} \otimes a_{(1)} \quad \text{by (3.1)}$$

$$= \sum_{(a)} f(a_{(2)}) g(a_{(1)})$$

$$= \sum_{(a)} g(a_{(1)}) f(a_{(2)}) \quad \text{since } S \text{ is commutative}$$

$$= (g, f) \sum_{(a)} a_{(1)} \otimes a_{(2)}$$

$$= (g * f)(a). \qquad \qquad \square$$

Let $I : A \to A$ denote the identity map. Let $f, g, h \in F(S)$. For $a \otimes b \otimes c \in A \otimes A \otimes A$, put $(f, g, h)(a \otimes b \otimes c) = f(a)g(b)h(c)$.

Proposition 3.1.2. *Let* $\Delta : A \to A \otimes_R A$ *be an R-algebra map that satisfies*

$$(I \otimes \Delta)\Delta(a) = (\Delta \otimes I)\Delta(a) \tag{3.2}$$

for all $a \in A$. *Then the corresponding binary operation on* $F(S)$ *is associative.*

Proof. We show that $(f * (g * h))(a) = ((f * g) * h)(a)$. Now,

$$(f * (g * h))(a) = \sum_{(a)} f(a_{(1)})(g * h)(a_{(2)})$$

$$= \sum_{(a)} f(a_{(1)}) \sum_{(a_{(2)})} g\left(a_{(2)(1)}\right) h\left(a_{(2)(2)}\right)$$

$$= \sum_{(a, a_{(2)})} f(a_{(1)}) g\left(a_{(2)(1)}\right) h\left(a_{(2)(2)}\right)$$

$$= (f, g, h) \sum_{(a, a_{(2)})} a_{(1)} \otimes a_{(2)(1)} \otimes a_{(2)(2)}$$

$$= (f, g, h) \sum_{(a, a_{(1)})} a_{(1)(1)} \otimes a_{(1)(2)} \otimes a_{(2)} \quad \text{by (3.2)}$$

$$= \sum_{(a, a_{(1)})} f\left(a_{(1)(1)}\right) g\left(a_{(1)(2)}\right) h(a_{(2)})$$

$$= \sum_{(a)} \sum_{(a_{(1)})} f\left(a_{(1)(1)}\right) g\left(a_{(1)(2)}\right) h(a_{(2)})$$

$$= \sum_{(a)} (f * g)(a_{(1)}) h(a_{(2)})$$

$$= ((f * g) * h)(a).   \square$$

What condition on Δ guarantees the existence of a multiplicative identity element? Let $m : A \otimes_R A \to A$ denote the multiplication map of A defined as $m(\sum a \otimes b) = \sum ab$, and let $\lambda : R \to S$ denote the structure map of the commutative R-algebra S. We have $\lambda(1_R) = 1_S$.

Proposition 3.1.3. *Let $\epsilon : A \to R$ be an R-algebra homomorphism for which*

$$m(I \otimes \epsilon)\Delta(a) = a = m(\epsilon \otimes I)\Delta(a) \tag{3.3}$$

for $a \in A$. Then the R-algebra homomorphism $\lambda\epsilon : A \to S$ satisfies

$$(\lambda\epsilon) * f = f = f * (\lambda\epsilon)$$

for all $f \in F(S)$. Thus $\lambda\epsilon$ is a left and right identity for the binary operation on $F(S)$.

Proof.

$$((\lambda\epsilon) * f)(a) = \sum_{(a)} \lambda(\epsilon(a_{(1)})) f(a_{(2)})$$

$$= f\left(\sum_{(a)} \epsilon(a_{(1)}) a_{(2)}\right)$$

$$= f(a) \quad \text{by (3.3).}$$

In a similar manner, one can show that $f = f * (\lambda\epsilon)$. $\square$

What condition on Δ guarantees the existence of multiplicative inverse elements?

Proposition 3.1.4. *Let $f \in F(S)$. Let $\sigma : A \to A$ be an R-algebra homomorphism for which*

$$m(I \otimes \sigma)\Delta(a) = \epsilon(a)1_A = (\sigma \otimes I)\Delta(a) \tag{3.4}$$

for $a \in A$. Then the R-algebra homomorphism $f\sigma : A \to S$ satisfies

$$(f\sigma) * f = \lambda\epsilon = f * (f\sigma).$$

Thus $f\sigma$ is a left and right inverse for f with respect to the binary operation on $F(S)$.

Proof. We have

$$((f\sigma) * f)(a) = \sum_{(a)} f(\sigma(a_{(1)})) f(a_{(2)})$$

$$= f \left(\sum_{(a)} \sigma(a_{(1)}) a_{(2)} \right)$$

$$= f(\epsilon(a) 1_A) \quad \text{by (3.4)}$$

$$= \epsilon(a) f(1_A)$$

$$= \epsilon(a) 1_S$$

$$= \lambda(\epsilon(a)).$$

Likewise, one has $\lambda\epsilon = f * (f\sigma)$. □

So we have arrived at the following.

Proposition 3.1.5. *Let* $F = \mathrm{Hom}_{R\text{-alg}}(A, -)$ *be a functor, together with additional R-algebra maps*

$$\Delta : A \to A \otimes_R A, \quad \epsilon : A \to R, \quad \sigma : A \to A,$$

that satisfy conditions (3.2), (3.3), and (3.4), respectively. Then, for each $S \in \mathrm{Ob}(\mathfrak{I}_{R\text{-alg}})$, *the set* $F(S)$ *is a group under the binary operation* $*$.

Proof. As one can easily verify, $F(S)$ together with $*$ satisfies the requirements for $F(S)$ to be a group. □

The functor F in Proposition 3.1.5 is a **representable group functor**, which is also called an **affine group scheme** or an **R-group scheme**. The R-algebra A is the **representing algebra of** F; we write $R[F] = A$. Note that F is a functor from the category of commutative R-algebras to the category of groups, where the morphisms are homomorphisms of groups. The map Δ is the **comultiplication map of** A, ϵ is the **counit map of** A, and σ is the **coinverse map of** A. When necessary to avoid confusion, we shall denote the comultiplication, counit, and coinverse maps of A by Δ_A, ϵ_A, and σ_A, respectively.

Here are some important examples of R-group schemes.

The ring R itself as an R-algebra represents the R-group scheme $F = \mathrm{Hom}_{R\text{-alg}}(R, -)$. The comultiplication on R is defined by $\Delta(1) = 1 \otimes 1$, the counit is defined by $\epsilon(1) = 1$, and the coinverse is given by $\sigma(1) = 1$. For an R-algebra S, $F(S)$ consists of a single element, the R-algebra structure map $\lambda : R \to S$, and thus F is the **trivial R-group scheme** denoted by $\mathbf{1}$, or more simply 1 when the context is clear.

For a non-trivial example, let $R[X]$ denote the algebra of polynomials in the indeterminate X. Then $F = \mathrm{Hom}_{R\text{-alg}}(R[X], -)$ is an R-group scheme, with comultiplication defined by $\Delta(X) = X \otimes 1 + 1 \otimes X$, counit defined by $\epsilon(X) = 0$, and coinverse given by $\sigma(X) = -X$.

Let us examine this group scheme more closely. Let S be an R-algebra. The group $F(S)$ consists of all R-algebra maps $\phi : R[X] \to S$. These homomorphisms are precisely the evaluation homomorphisms and are determined by sending the indeterminate X to some element a in S. We see that $F(S)$ consists of the algebra maps $\phi_a : R[X] \to S$, where $X \mapsto a, a \in S$. We have $\phi_a(1_{R[X]}) = 1_S$.

Now, how does the group product $*$ in $F(S)$ work? Let ϕ_a, ϕ_b be elements of $F(S)$, and let $m : S \otimes_R S \to S$ denote the multiplication map of S. Then

$$(\phi_a * \phi_b)(X) = m(\phi_a \otimes \phi_b)\Delta(X)$$
$$= m(\phi_a \otimes \phi_b)(X \otimes 1_{R[X]} + 1_{R[X]} \otimes X)$$
$$= m(a \otimes 1_S) + m(1_S \otimes b)$$
$$= a + b,$$

and so $\phi_a * \phi_b = \phi_{a+b}$. We identify $F(S)$ with the additive group $S, +$ of the ring S. For this reason, the group functor F is called the **additive R-group scheme**, denoted by $\mathbf{G}_a$.

For another example, let $R[X_1, X_2]$ be the R-algebra of polynomials in the indeterminates X_1, X_2. Let $I = (X_1 X_2 - 1)$, and consider the quotient ring $R[X_1, X_2]/I$. There is an isomorphism of R-algebras,

$$f : R[X_1, X_2]/I \to R[X, X^{-1}], \ X \text{ indeterminate},$$

defined by $X_1 \mapsto T$, $X_2 \mapsto X^{-1}$. The functor $F = \mathrm{Hom}_{R\text{-alg}}(R[X, X^{-1}], -)$ is an R-group scheme with comultiplication Δ defined by $\Delta(X) = X \otimes X$, counit defined as $\epsilon(X) = 1$, and coinverse given as $\sigma(X) = X^{-1}$.

Let us see how this group scheme works. Let S be an R-algebra. The group $F(S)$ consists of the R-algebra maps $\phi : R[X, X^{-1}] \to S$. These maps are determined by sending the variable X to some element $\phi(X)$ in S. But in order for ϕ to be a ring homomorphism, we must have $\phi(X^{-1}) = (\phi(X))^{-1}$, and so this element must be a unit of S. We see that $F(S)$ consists of all algebra maps $\phi_u : R[X, X^{-1}] \to S$, where $X \mapsto u, u \in U(S)$.

Now, how is the group product in $F(S)$ defined? Let ϕ_u, ϕ_v be elements of $F(S)$. Then

$$(\phi_u * \phi_v)(X) = m(\phi_u \otimes \phi_v)\Delta(X)$$
$$= m(\phi_u \otimes \phi_v)(X \otimes X)$$
$$= m(u \otimes v)$$
$$= uv.$$

Thus, $\phi_u * \phi_v = \phi_{uv}$, and we identify $F(S)$ with the multiplicative group of units in S. This is the **multiplicative R-group scheme**, which is denoted by $\mathbf{G}_m$.

Here are two more examples of R-group schemes. Let $A = R[X]/(X^n - 1)$. Then $F = \mathrm{Hom}_{R\text{-alg}}(A, -)$ is an R-group scheme with $\Delta(X) = X \otimes X$, $\epsilon(X) = 1$, and $\sigma(X) = X^{n-1} = X^{-1}$. An element $\phi \in F(S)$ is determined by sending X to an element s in S for which $s^n = 1$. For this reason, F is the **multiplicative group of the nth roots of unity**, denoted by μ_n.

Next, let $A = R[X_{1,1}, X_{1,2}, X_{2,1}, X_{2,2}]$ be the polynomial algebra in the indeterminates $X_{1,1}, X_{1,2}, X_{2,1}, X_{2,2}$. Let J be the principal ideal of A generated by $X_{1,1}X_{2,1} - X_{1,2}X_{2,2} - 1$, and consider the quotient ring $B = A/J$. Then $F = \mathrm{Hom}_{R\text{-alg}}(B, -)$ is an R-group scheme where $F(S)$ is the (multiplicative) group of 2×2 matrices M with entries in S with $\det(M) = 1$. This is the **special linear group (scheme) of order** 2, denoted by $\mathbf{SL}_2$. We leave it as an exercise to formulate the comultiplication, counit, and coinverse maps on the representing algebra B.

3.2   Homomorphisms of R-Group Schemes

What are the maps between R-group schemes?

Definition 3.2.1. Let $\psi : F \to G$ be a natural transformation of R-group schemes. Then ψ is a **homomorphism of R-group schemes** if $\psi_S : F(S) \to G(S)$ is a homomorphism of groups for all $S \in \mathrm{Ob}(\mathfrak{I}_{R\text{-alg}})$.

By Yoneda's Lemma, the homomorphism $\psi : F \to G$ corresponds to an R-algebra homomorphism $\phi : B \to A$ with $R[F] = A$, $R[G] = B$. If ψ is a homomorphism of R-group schemes, what can we infer about the map ϕ?

By Yoneda's Lemma, the comultiplication map $\Delta_A : A \to A \otimes_R A$ corresponds to the associated map $\alpha : \mathrm{Hom}_{R\text{-alg}}(A \otimes_R A, -) \to \mathrm{Hom}_{R\text{-alg}}(A, -)$. There is a map $\alpha_{A \otimes_R A} : \mathrm{Hom}_{R\text{-alg}}(A \otimes_R A, A \otimes_R A) \to \mathrm{Hom}_{R\text{-alg}}(A, A \otimes_R A)$. Observe that $\alpha_{A \otimes_R A}$ is the group product in $\mathrm{Hom}_{R\text{-alg}}(A, A \otimes_R A)$.

Let $I \in \mathrm{Hom}_{R\text{-alg}}(A, A)$ be the identity map, and let $I \otimes I \in \mathrm{Hom}_{R\text{-alg}}(A \otimes_R A, A \otimes_R A)$ be defined as $(I \otimes I)(a \otimes b) = a \otimes b$. Now, for $x \in B$,

$$\psi_{A \otimes_R A}(\alpha_{A \otimes A}(I \otimes I))(x) = \alpha_{A \otimes_R A}(I \otimes I)(\phi(x))$$

$$= (I \otimes I)(\Delta_A(\phi(x)))$$

$$= \Delta_A(\phi(x)).$$

Also by Yoneda's Lemma, the comultiplication map $\Delta_B : B \to B \otimes_R B$ corresponds to the associated map $\beta : \mathrm{Hom}_{R\text{-alg}}(B \otimes_R B, -) \to \mathrm{Hom}_{R\text{-alg}}(B, -)$. There is a map $\beta_{A \otimes_R A} : \mathrm{Hom}_{R\text{-alg}}(B \otimes_R B, A \otimes_R A) \to \mathrm{Hom}_{R\text{-alg}}(B, A \otimes_R A)$; $\beta_{A \otimes_R A}$ is the group operation in $\mathrm{Hom}_{R\text{-alg}}(B, A \otimes_R A)$.

Since ψ is a group homomorphism,

$$
\begin{aligned}
\psi_{A \otimes_R A}(\alpha_{A \otimes A}(I \otimes I))(x) &= \beta_{A \otimes_R A}(\psi_A(I) \otimes \psi_A(I))(x) \\
&= (\psi_A(I) \otimes \psi_A(I))(\Delta_B(x)) \\
&= (I \otimes I)(\phi \otimes \phi)(\Delta_B(x)) \\
&= (\phi \otimes \phi)(\Delta_B(x)),
\end{aligned}
$$

and so the R-algebra homomorphism ϕ must satisfy the property

$$
\Delta_A(\phi(x)) = (\phi \otimes \phi)\Delta_B(x) \tag{3.5}
$$

for all $x \in B$.

Moreover, since the identity element maps to the identity element under a group homomorphism,

$$
\begin{aligned}
\epsilon_A(\phi(x))1_R &= \lambda_R(\epsilon_A(\phi(x))) \\
&= (\lambda_R \epsilon_A)(\phi(x)) \\
&= \psi_R(\lambda_R \epsilon_A)(x) \\
&= (\lambda_R \epsilon_B)(x) \\
&= \epsilon_B(x)1_R,
\end{aligned}
$$

and so ϕ satisfies

$$
\epsilon_A(\phi(x)) = \epsilon_B(x) \tag{3.6}
$$

for all $x \in B$. Finally,

$$
\psi_A(I_A \sigma_A)(x) = (I_A \sigma_A)(\phi(x)) = \sigma_A(\phi(x)),
$$

which since ψ_A is a group homomorphism equals

$$
(\psi_A(I_A)\sigma_B)(x) = I_A \phi(\sigma_B(x)) = \phi(\sigma_B(x)),
$$

and so ϕ satisfies

$$
\sigma_A(\phi(x)) = \phi(\sigma_B(x)). \tag{3.7}
$$

So we have arrived at the following characterization of a homomorphism of R-group schemes.

Definition 3.2.2. Let $\psi : F \to G$ be a natural transformation of R-group schemes with $R[F] = A$, $R[G] = B$. Then ψ is a **homomorphism of R-group schemes** if the corresponding map $\phi : B \to A$ satisfies conditions (3.5), (3.6), and (3.7).

Actually, it suffices to show that condition (3.5) holds in Definition 3.2.2.

Proposition 3.2.1. *Let $\psi : F \to G$ be a natural transformation of R-group schemes with $R[F] = A$, $R[G] = B$. Then ψ is a homomorphism of R-group schemes if the corresponding map $\phi : B \to A$ satisfies the condition*

$$(\phi \otimes \phi)\Delta_B(x) = \Delta_A(\phi(x))$$

for all $x \in B$.

Proof. Exercise. $\square$

Let $\psi : F \to G$ be a homomorphism of R-group schemes with $R[F] = A$, $R[G] = B$. Let $\phi : B \to A$ denote the corresponding R-algebra homomorphism. Our goal is to arrive at a suitable definition of the kernel of ψ. One would expect $\ker(\psi)$ to be a group scheme over R. We know that $\psi_S : F(S) \to G(S)$ is a group homomorphism for each $S \in \mathrm{Ob}(\Im_{R\text{-alg}})$, and so to describe the kernel of ψ we consider $\ker(\psi_S)$. Since $\lambda\epsilon_B$ is the identity element of $G(S)$, we have

$$\ker(\psi_S) = \{f \in F(S) : \psi_S(f)(x) = (\lambda\epsilon_B)(x), \forall x \in B\}$$
$$= \{f \in F(S) : f(\phi(x)) = \lambda(\epsilon_B(x)), \forall x \in B\}.$$

Now A can be viewed as a B-algebra with scalar multiplication defined as $b \cdot a = \phi(b)a$ for $a \in A$, $b \in B$. Likewise, R can be viewed as a B-algebra with scalar multiplication defined by $b \cdot r = \epsilon_B(b)r$, and S is a B-algebra with $b \cdot s = \lambda(\epsilon_B(b))s$ for $s \in S$.

One has the tensor product $A \otimes_B R$, which is also a B-algebra, and the representable functor

$$N = \mathrm{Hom}_{B\text{-alg}}(A \otimes_B R, -),$$

which is defined on the category of commutative B-algebras.

Proposition 3.2.2. *Let S be an R-algebra and a B-algebra with scalar multiplication defined as $b \cdot s = \lambda(\epsilon_B(b))s$. Then $N(S) = \ker(\psi_S)$.*

Proof. By Proposition 2.4.2,

$$\mathrm{Hom}_{B\text{-alg}}(A \otimes_B R, S) = \mathrm{Hom}_{B\text{-alg}}(A, S) \times \mathrm{Hom}_{B\text{-alg}}(R, S).$$

Since $\mathrm{Hom}_{B\text{-alg}}(R, S)$ consists only of the map λ, $N(S)$ is identified with the elements $f \in \mathrm{Hom}_{B\text{-alg}}(A, S)$. We have

$$f(\phi(x)) = f(\phi(x)1_A)$$
$$= f(x \cdot 1_A)$$
$$= x \cdot f(1_A)$$
$$= x \cdot 1_S$$
$$= \lambda(\epsilon_B(x))1_S$$
$$= (\lambda\epsilon_B)(x).$$

Thus $N(S) = \ker(\psi_S)$.   □

Thus we have the kernel of ψ described as a representable functor on the category of B-algebras. Our next step is to translate to R-algebras.

The **augmentation ideal of** B, denoted by B^+, is the kernel of the counit map $\epsilon_B : B \to R$. We have the short exact sequence of R-modules

$$0 \to B^+ \to B \to R \to 0,$$

and so as R-modules $R \cong B/B^+$. But B/B^+ is also a B-algebra through $b \cdot \bar{c} = \overline{bc}$, and so

$$A \otimes_B R \cong A \otimes_B (B/B^+) \cong A/\phi(B^+)A$$

as R-algebras. Thus there is a representable functor on the category of commutative R-algebras defined as

$$\mathrm{Hom}_{R\text{-alg}}(A/\phi(B^+)A, -).$$

We identify $\mathrm{Hom}_{R\text{-alg}}(A/\phi(B^+)A, S)$ with $N(S)$. (Note that S is viewed simultaneously as an R-algebra and a B-algebra.)

We claim that N is an R-group scheme. To prove this, we will need some lemmas.

Lemma 3.2.1. *Let B be an R-algebra, and let J be an ideal of B. Then there is an isomorphism of R-algebras*

$$B/J \otimes_R B/J \cong (B \otimes_R B)/(J \otimes_R B + B \otimes_R J).$$

Proof. First note that there is an R-algebra map

$$\alpha : B \otimes_R B \to B/J \otimes_R B/J,$$

defined as $\alpha(a \otimes b) = \bar{a} \otimes \bar{b}$. Now $J \otimes_R B + B \otimes_R J \subseteq \ker(\alpha)$, and so there exists an R-algebra map

$$\bar{\alpha} : (B \otimes_R B)/(J \otimes_R B + B \otimes_R J) \to B/J \otimes_R B/J$$

with $\bar{\alpha}(\overline{a \otimes b}) = \bar{a} \otimes \bar{b}$.

Next, let β denote the canonical surjection of R-algebras

$$\beta : B \otimes_R B \to (B \otimes_R B)/(J \otimes_R B + B \otimes_R J)$$

defined as $\beta(a \otimes b) = \overline{a \otimes b}$. Since $\beta(J \otimes 1) = \beta(1 \otimes J) = 0$, there exists an R-algebra map

$$\overline{\beta} : B/J \otimes_R B/J \to (B \otimes_R B)/(J \otimes_R B + B \otimes_R J)$$

defined as $\overline{\beta}(\overline{a} \otimes \overline{b}) = \overline{a \otimes b}$. Clearly, $(\overline{\alpha})^{-1} = \overline{\beta}$, and thus $\overline{\beta}$ is an isomorphism. $\square$

We apply Lemma 3.2.1 to the ideal B^+ to show that $\Delta_B(B^+) \subseteq B \otimes_R B^+ + B^+ \otimes_R B$.

Lemma 3.2.2. *Let B^+ denote the augmentation ideal of B. Then $\Delta_B(B^+) \subseteq B \otimes_R B^+ + B^+ \otimes_R B$.*

Proof. Let $\Delta_R : R \to R \otimes_R R$ denote the comultiplication map of the trivial group scheme $\mathrm{Hom}_{R\text{-alg}}(R, -)$, and let $(\epsilon \otimes \epsilon) : B \otimes_R B \to R \otimes_R R$ be the R-algebra map defined by $a \otimes b \to \epsilon(a) \otimes \epsilon(b)$ for $a, b \in B$. For all $b \in B$,

$$(\epsilon \otimes \epsilon)\Delta_B(b) = \sum_{(b)} \epsilon(b_{(1)}) \otimes \epsilon(b_{(2)})$$

$$= (\epsilon \otimes 1) \sum_{(b)} b_{(1)} \otimes \epsilon(b_{(2)})$$

$$= (\epsilon \otimes 1) \sum_{(b)} b_{(1)}\epsilon(b_{(2)}) \otimes 1$$

$$= (\epsilon \otimes 1)(b \otimes 1) \quad \text{by (3.3)}$$

$$= \epsilon(b) \otimes 1$$

$$= \Delta_R(\epsilon(b)).$$

Thus, for $b \in B^+$,

$$(\epsilon \otimes \epsilon)\Delta_B(b) = \Delta_R(\epsilon(b)) = 0,$$

and so

$$\Delta_B(B^+) \subseteq \ker(\epsilon \otimes \epsilon). \tag{3.8}$$

Now, by Lemma 3.2.1 there is an R-algebra isomorphism

$$\alpha : B/B^+ \otimes_R B/B^+ \to (B \otimes_R B)/(B \otimes_R B^+ + B^+ \otimes_R B),$$

which yields a surjective homomorphism of R-algebras

$$\beta : B \otimes_R B \to (B \otimes_R B)/(B \otimes_R B^+ + B^+ \otimes_R B)$$

since $R \otimes_R R \cong B/B^+ \otimes_R B/B^+$. Consequently,

$$\ker(\epsilon \otimes \epsilon) \subseteq \ker(\beta) = B \otimes_R B^+ + B^+ \otimes_R B,$$

and so, by (3.8), $\Delta_B(B^+) \subseteq B \otimes_R B^+ + B^+ \otimes_R B$. $\square$

Lemma 3.2.3. *Let* $\psi : F \to G$ *be a homomorphism of* R-*group schemes with* $\phi : B \to A$ *the corresponding map of* R-*algebras. Then:*

(i) $\Delta_A(\phi(B^+)A) \subseteq \phi(B^+)A \otimes_R A + A \otimes_R \phi(B^+)A.$
(ii) $\epsilon_A(\phi(B^+)A) = 0.$
(iii) $\sigma_A(\phi(B^+)A) \subseteq \phi(B^+)A.$

Proof. We prove (i) and leave (ii) and (iii) as exercises. We have

$$\begin{aligned}
\Delta_A(\phi(B^+)A) &\subseteq \Delta_A(\phi(B^+))(A \otimes_R A) \\
&\subseteq ((\phi \otimes \phi)(\Delta_B(B^+)))(A \otimes_R A) \\
&\subseteq ((\phi \otimes \phi)(B \otimes_R B^+ + B^+ \otimes_R B))(A \otimes_R A) \quad \text{by Lemma 3.2.2} \\
&\subseteq \phi(B^+)A \otimes_R A + A \otimes_R \phi(B^+)A. \qquad\qquad\qquad\square
\end{aligned}$$

We are now in a position to show that N is an R-group scheme.

Proposition 3.2.3. *The representable functor* $N = \mathrm{Hom}_{R\text{-alg}}(A/\phi(B^+)A, -)$ *is an* R-*group scheme.*

Proof. Let $C = A/\phi(B^+)A$, and put $J = \phi(B^+)A \otimes_R A + A \otimes_R \phi(B^+)A$. We show that there exist R-algebra maps $\Delta : C \to C \otimes_R C$, $\epsilon : C \to R$, and $\sigma : C \to C$ that satisfy conditions (3.2), (3.3), and (3.4), respectively.

First, let $\Delta_A : A \to A \otimes_R A$ denote the comultiplication of A, and let $\beta : A \otimes_R A \to (A \otimes_R A)/J$ be the canonical surjection of R-algebras. By Lemma 3.2.3(i), $\Delta_A(\phi(B^+)A) \subseteq J$, and so there exists an R-algebra map

$$\overline{\beta \Delta_A} : C \to (A \otimes_R A)/J.$$

By Lemma 3.2.1, there is an isomorphism

$$\overline{\alpha} : (A \otimes_R A)/J \to C \otimes_R C,$$

and so the map defined as $\Delta = \overline{\alpha}\,\overline{\beta \Delta_A}$ is an R-algebra map $\Delta : C \to C \otimes_R C$, which satisfies the condition

$$(I \otimes \Delta)\Delta = (\Delta \otimes I)\Delta$$

since Δ_A satisfies condition (3.2).

Next, let $\epsilon_A : A \to R$ be the counit map. By Lemma 3.2.3(ii), there exists an R-algebra map $\epsilon : C \to R$ that satisfies, for all $c \in C$,

$$(I \otimes \epsilon)\Delta(c) = c = (\epsilon \otimes I)\Delta(c),$$

since ϵ_A satisfies (3.3).

Finally, by Lemma 3.2.3(iii), there is a map $\sigma : C \to C$, induced from σ_A, that evidently satisfies the condition

$$(I \otimes \sigma)\Delta(c) = \epsilon(c)1 = (\sigma \otimes I)\Delta(c)$$

for all $c \in C$. Thus $N = \mathrm{Hom}_{R\text{-alg}}(C, -)$ is an R-group scheme. $\qquad\square$

Now, we can make the following definition.

Definition 3.2.3. Let $\psi : F \to G$ be a homomorphism of R-group schemes, and let $\phi : B \to A$ denote the corresponding homomorphism of R-algebras. Then the **kernel of ψ** is the R-group scheme defined as

$$N = \mathrm{Hom}_{R\text{-alg}}(A/\phi(B^+)A, -).$$

Let $\psi : F \to G$ be a homomorphism of R-group schemes with $\ker(\psi) = N$. We define an **exact sequence of R-group schemes** to be the sequence

$$1 \to N \to F \xrightarrow{\psi} G$$

with $R[N] = A/\phi(B^+)A$, $R[F] = A$, and $R[G] = B$. For each $S \in \mathrm{Ob}(\mathfrak{I}_{R\text{-alg}})$, the sequence

$$1 \to N(S) \to F(S) \to G(S)$$

is an exact sequence of groups with $N(S) = \ker(F(S) \to G(S))$.

3.3 Short Exact Sequences

In the preceding section, we showed that a homomorphism of R-group schemes gives rise to an exact sequence of R-group schemes. This is analogous to the situation for ordinary abstract groups. The analogy breaks down, however, when we consider short exact sequences. For abstract groups, an exact sequence always extends to a short exact sequence, but this is not the case for R-group schemes. The problem is that the map $S \mapsto F(S)/N(S)$ is not always an R-group scheme; that is, this map does not necessarily take the form of $\mathrm{Hom}_{R\text{-alg}}(Q, -)$ for some R-algebra Q.

Let M, L be modules over a ring S, let T be a ring, and let $\varrho : S \to T$ be a ring homomorphism. We will consider T an S-module with $s \cdot t = \varrho(s)t$.

Definition 3.3.1. A ring homomorphism $\varrho : S \to T$ is **flat** if, whenever $\alpha : M \to L$ is an injection of S-modules, the map $\varphi : M \otimes_S T \to L \otimes_S T$ defined as $\varphi(m \otimes 1) = \alpha(m) \otimes 1$ is also an injection.

As an example, we prove the following.

Proposition 3.3.1. *Let f be a non-nilpotent element of the ring S. Then the localization map $S \to S_f$, $s \mapsto s/1$, is flat.*

Proof. Let $\alpha : M \to L$ be an injection of S-modules, and let $\varphi : M \otimes_S S_f \to L \otimes_S S_f$ be the map defined as $\varphi(m \otimes 1) = \alpha(m) \otimes 1$. Suppose $\varphi(m \otimes 1) = \varphi(n \otimes 1)$, so that $\alpha(m) \otimes 1 = \alpha(n) \otimes 1$. It follows that $\alpha(m - n) \otimes 1 = 0$, and so there exists an element $f^i \in \{1, f, f^2, \ldots\}$ such that $f^i \alpha(m - n) = 0$. Thus $\alpha(f^i(m-n)) = \alpha(f^i m - f^i n) = 0$, and hence $f^i m = f^i n$ since α is an injection. Consequently, $m \otimes f^i = n \otimes f^i$, and so $m \otimes 1 = n \otimes 1$ since f^i is a unit of S_f. It follows that φ is an injection. $\square$

Lemma 3.3.1. *Let $\varrho : S \to T$ be flat, and suppose that $P \cdot T \neq T$ for every maximal ideal P of S. If M is a non-zero S-module, then $M \otimes_S T$ is non-zero.*

Proof. Let $m \neq 0$ be an element of M, and let I be the annihilator ideal of m. Then $Sm \cong S/I$. Since $S/I \cong Sm \subseteq M$, the flatness of ϱ implies the existence of an injection $(S/I) \otimes_S T \to M \otimes_S T$. Note that $(S/I) \otimes_S T \cong T/(I \cdot T)$. There exists a maximal ideal P with $I \subseteq P$; hence $T/(I \cdot T) \neq 0$ since $T/(P \cdot T) \neq 0$. It follows that $M \otimes_S T \neq 0$. $\square$

Lemma 3.3.2. *Let $\varrho : S \to T$ be flat, and suppose that $P \cdot T \neq T$ for every maximal ideal P of S. Let $\alpha : M \to N$ be a map of S-modules, and let $\alpha' : M \otimes_S T \to N \otimes_S T$ be the induced map defined by $m \otimes t \mapsto \alpha(m) \otimes t$. Then, if α' is an injection, so is α.*

Proof. Suppose that $\alpha : M \to N$ has non-zero kernel L. Then, by Lemma 3.3.1, $L \otimes T \neq 0$. By the flatness of ϱ, $L \otimes T \to M \otimes_S T$ is an injection. Since $L \otimes T$ is in the kernel of α', α' is not an injection, which proves the lemma. $\square$

Definition 3.3.2. A flat map $S \to T$ is **faithfully flat** if the map $\varphi : M \to M \otimes_S T$ defined as $\varphi(m) = m \otimes 1$ is an injection for all S-modules M.

The localization map $S \to S_f$ may not be faithfully flat, though it can be used to build a faithfully flat map. Let $\{f_1, f_2, \ldots, f_n\}$ be a finite set of non-nilpotent elements of S, and suppose that the ideal generated by $\{f_1, f_2, \ldots, f_n\}$ is S. Then the map $\varrho : S \to \prod_{i=1}^{n} S_{f_i}$ defined as $s \mapsto ((s/1)_{f_i})$ is faithfully flat. (Prove this as an exercise. Hint: Use Lemma 3.3.1.)

Proposition 3.3.2. *Let $\varrho : S \to T$ be faithfully flat. Then ϱ is an injection.*

Proof. Since $S \to T$ is faithfully flat, the map $\varphi : S \to S \otimes_S T = T$ is an injection. $\square$

Proposition 3.3.3. *Let* $\alpha : S \to Q$ *be a ring homomorphism, and consider* Q *an* S-*module with* $s \cdot q = \alpha(s)q$. *Suppose* $S \to T$ *is faithfully flat. Then* $Q \to Q \otimes_S T$, $q \mapsto q \otimes 1$, *is faithfully flat.*

Proof. Let M be a Q-module (also an S-module), and let $\varrho : M \to M \otimes_Q (Q \otimes_S T)$ be the Q-module map defined by $m \mapsto m \otimes (1 \otimes 1)$. There is an isomorphism $\phi : M \otimes_Q (Q \otimes_S T) \to M \otimes_S T$ defined by $m \otimes (q \otimes t) \mapsto q \cdot m \otimes t$. Now, $\phi\varrho : M \to M \otimes_S T$ is an injection by the faithful flatness of $S \to T$. Consequently, ϱ is an injection, and so $Q \to Q \otimes_S T$ is faithfully flat. $\qquad\square$

Proposition 3.3.4. *Let* $S \to T$ *be faithfully flat, and let* $x \in \operatorname{Spec} S$. *Then the induced map* $S_x \to T \otimes_S S_x$ *is faithfully flat.*

Proof. Let M be an S_x-module, and let $\varphi : M \to M \otimes_{S_x} (T \otimes_S S_x)$ be the map defined as $m \mapsto m \otimes (1 \otimes 1)$. Note that M is also an S-module with

$$M \otimes_{S_x} (T \otimes_S S_x) \cong (M \otimes_S T) \otimes_S S_x.$$

Since $S \to T$ is faithfully flat, there is an injection $M \to M \otimes_S T$ given as $m \mapsto m \otimes 1$. Consequently, φ is also an injection. $\qquad\square$

Faithful flatness is a critical condition in view of the following.

Proposition 3.3.5. *Let* $\varrho : S \to T$ *be a flat ring homomorphism. Then* ϱ *is faithfully flat if and only if the associated map* ${}^a\varrho : \operatorname{Spec} T \to \operatorname{Spec} S$ *is surjective.*

Proof. Assume that $\varrho : S \to T$ is faithfully flat, and let $x \in \operatorname{Spec} S$. By Proposition 3.3.4, $\varrho_x : S_x \to T \otimes_S S_x$ is faithfully flat, and therefore S_x/xS_x injects into

$$(S_x/xS_x) \otimes_{S_x} (T \otimes_S S_x) \cong (T \otimes_S S_x)/(T \otimes_S xS_x).$$

It follows that $xS_x = S_x \cap (T \otimes_S xS_x)$. Thus, $T \otimes_S xS_x$ is a proper ideal of $T \otimes_S S_x$. By Proposition 1.1.1, $T \otimes_S xS_x$ is contained in a prime ideal J' of $T \otimes_S S_x$. The preimage of J' under the structure map $T \to T \otimes_S S_x$ is the prime ideal J in $\operatorname{Spec} T$. One then has ${}^a\varrho(J) = x$, and consequently ${}^a\varrho$ is surjective.

For the converse, suppose that ${}^a\varrho : \operatorname{Spec} T \to \operatorname{Spec} S$ is surjective. Let P be a maximal ideal of S. Then there exists a prime ideal $Q \in \operatorname{Spec} T$ with ${}^a\varrho(Q) = P$, and so $\varrho(P) = Q$ with $P \cdot T = \varrho(P)T = QT = Q \neq T$.

Let M be an S-module. Define a map $\varphi : M \otimes_S T \to (M \otimes_S T) \otimes_S T$ by the rule $\varphi(m \otimes t) = m \otimes 1 \otimes t$, and define a map $\omega : (M \otimes_S T) \otimes_S T \to M \otimes_S T$ by the rule $\omega(m \otimes t \otimes v) = m \otimes tv$. Then $\omega\varphi$ is the identity on $M \otimes_S T$, and so φ is an injection. An application of Lemma 3.3.2 then implies that $M \to M \otimes_S T$, $m \mapsto m \otimes 1$, is an injection. Therefore, $\varrho : S \to T$ is faithfully flat. $\qquad\square$

If $\operatorname{Spec} S$ and $\operatorname{Spec} T$ are endowed with the Zariski topology, then faithful flatness is equivalent to the notion that $\operatorname{Spec} T$ is an open covering of $\operatorname{Spec} S$.

We are now in a position to define surjectivity for homomorphisms of group schemes.

Definition 3.3.3. The homomorphism of R-group schemes $\psi : F \to G$ is an **epimorphism** if for each $g \in G(S)$ there is a faithfully flat R-algebra map $\varrho : S \to T$ for which $g' \in \psi_T(F(T))$, where $g' \in G(\varrho)(g)$.

Suppose $\psi : F \to G$ is an epimorphism. Since $\psi_T(F(T)) \cong F(T)/N(T)$, we say that G is the **quotient sheaf of F by N** and write $G = F/N$. We define a **short exact sequence of R-group schemes** to be the sequence

$$1 \to N \to F \to F/N = G \to 1.$$

If the corresponding map of a homomorphism $\psi : F \to G$ is faithfully flat, then the homomorphism is an epimorphism.

Proposition 3.3.6. *Let $\psi : F \to G$ be a homomorphism of R-group schemes with $R[F] = A$, $R[G] = B$. Suppose that the corresponding algebra map $\phi : B \to A$ is faithfully flat; that is, suppose that $\mathrm{Spec}\, A \to \mathrm{Spec}\, B$ is surjective. Then ψ is an epimorphism of group schemes.*

Proof. Let $g \in G(S)$. We consider A a B-algebra with $b \cdot a = \phi(b)a$, for $a \in A$, $b \in B$, and consider S a B-module with $b \cdot s = g(b)s$, $b \in B$, $s \in S$. Let $S' = S \otimes_B A$ denote the tensor product over B. By Proposition 3.3.3, the map $S \to S'$ defined by $s \mapsto s \otimes 1$ is faithfully flat. Let $f \in \mathrm{Hom}_{R\text{-alg}}(A, S')$ be defined by $a \mapsto 1 \otimes a$. Then

$$\begin{aligned}
f(\phi(b)) &= 1 \otimes \phi(b) \\
&= 1 \otimes \phi(b)1 \\
&= 1 \otimes b \cdot 1 \\
&= b \cdot 1 \otimes 1 \\
&= g(b) \otimes 1.
\end{aligned}$$

Define a map $g' : B \to S'$ by $b \mapsto g(b) \otimes 1$. Then $\psi_{S'}(f)(b) = g'(b)$, and ψ is an epimorphism. $\qquad\square$

Remark 3.3.1. Let $\psi : F \to G$ be an epimorphism of R-group schemes. Let $g \in G(S)$ be the trivial element of the group $G(S)$; that is, suppose $g = \lambda \epsilon_B$. Then S is a B-module with $b \cdot s = (\lambda \epsilon_B)(b)s$. For $f \in \mathrm{Hom}_{B\text{-alg}}(A, S)$,

$$\begin{aligned}
f(\phi(b)) &= f(\phi(b)1) \\
&= f(b \cdot 1) \\
&= b \cdot f(1) \\
&= b \cdot 1 \\
&= \lambda(\epsilon_B(b)).
\end{aligned}$$

In this case, there exists an element $h \in F(S) = \mathrm{Hom}_{R\text{-alg}}(A, S)$ with

$$\psi_S(h)(b) = f(\phi(b)) = \lambda(\epsilon_B(b)),$$

and so S' can be taken to be S. Indeed, we can take any

$$h \in \mathrm{Hom}_{R\text{-alg}}(A/\phi(B^+)A, S) \subseteq \mathrm{Hom}_{R\text{-alg}}(A, S).$$

Thus, the collection of all preimages of $\lambda \epsilon_B \in G(S)$ coincides with the kernel of $\psi : F \to G$ given in Definition 3.2.3.

3.4 An Example

In this section, we present an important example of a short exact sequence of group schemes.

Let K be a field, let $\mathbf{G}_m$ denote the multiplicative group scheme represented by $K[X, X^{-1}]$, and let $\mathbf{G}'_m$ denote a copy that is represented by $K[Y, Y^{-1}]$. Let m denote multiplication in the K-algebra $K[X, X^{-1}]$, and let I denote the identity map on $K[X, X^{-1}]$. For an integer $l \geq 2$, define

$$m^{(l-1)} = m(I \otimes m)(I \otimes I \otimes m) \cdots \underbrace{(I \otimes I \otimes \cdots \otimes I}_{l-2} \otimes m);$$

$$\Delta^{(l-1)} = \underbrace{(I \otimes I \otimes \cdots \otimes I}_{l-2} \otimes \Delta) \cdots (I \otimes I \otimes \Delta)(I \otimes \Delta)\Delta.$$

Then there exists a natural transformation of group schemes

$$p : \mathbf{G}_m \to \mathbf{G}'_m,$$

where $p_S : \mathbf{G}_m(S) \to \mathbf{G}'_m(S)$ is defined by

$$p_S(f)(Y) = m^{(p-1)}(\underbrace{f \otimes f \otimes \cdots \otimes f}_{p})\Delta^{(p-1)}(X).$$

Since $\Delta(X) = X \otimes X$, we have $p_S(f)(Y) = f(X)^p = f(X^p)$. The K-algebra map corresponding to p is $\phi : K[Y, Y^{-1}] \to K[X, X^{-1}]$, defined by $\phi(Y) = X^p$.

Put $\Delta' = \Delta_{K[Y,Y^{-1}]}$, $\epsilon' = \epsilon_{K[Y,Y^{-1}]}$, and $\sigma' = \sigma_{K[Y,Y^{-1}]}$. Since

$$(\phi \otimes \phi)\Delta'(Y) = \phi(Y) \otimes \phi(Y) = X^p \otimes X^p = \Delta(X^p) = \Delta(\phi(Y)),$$

$$\phi(\epsilon'(Y)) = 1 = \epsilon(\phi(Y)),$$

and

$$\phi(\sigma'(Y)) = \phi(Y^{-1}) = X^{-p} = \sigma(\phi(Y)),$$

p is a homomorphism of group schemes called the p**th power map.**

The augmentation ideal $K[Y, Y^{-1}]^+$ is $(Y - 1)$. Thus the kernel of p is the group scheme N represented by the K-algebra

$$K[X, X^{-1}]/\phi((Y - 1))K[X, X^{-1}] \cong K[X, X^{-1}]/(X^p - 1).$$

Thus there is an exact sequence of K-group schemes

$$1 \to \mu_p \to \mathbf{G}_m \xrightarrow{p} \mathbf{G}'_m.$$

In fact, we have the following.

Proposition 3.4.1. *The map* $p : \mathbf{G}_m \to \mathbf{G}'_m$ *is an epimorphism of group schemes.*

Proof. Note that $K[Y, Y^{-1}]$ is the localization $M^{-1}K[Y]$ at the multiplicative set $M = \{1, Y, Y^2, \dots\}$. Thus, by Proposition 1.1.4, Spec $K[Y, Y^{-1}]$ consists of ω together with the collection

$$\{q(Y)K[Y, Y^{-1}] : q(Y) \text{ is irreducible over } K \text{ and } Y \notin q(Y)K[Y]\}.$$

Let $(q(Y)) \in$ Spec $K[Y, Y^{-1}], q(Y) \neq 0$. We have $\phi((q(Y))) = (q(X^p))$, which is contained in some maximal ideal $(r(X))$ of Spec $K[X, X^{-1}]$. Now $\phi^{-1}((r(X)))$ is a prime ideal of $K[Y, Y^{-1}]$ containing $(q(Y))$, and hence $\phi^{-1}((r(X))) = (q(Y))$. Thus $p((r(X))) = (q(Y))$. Moreover, $p(\omega) = \omega$. Consequently, the map of spectra

$$p : \text{Spec } K[X, X^{-1}] \to \text{Spec } K[Y, Y^{-1}]$$

(which we also denote by p) is surjective, and so, by Proposition 3.3.6, p is an epimorphism. $\square$

Let S be an R-algebra and let $g \in \mathbf{G}'_m(S)$. Since $p : \mathbf{G}_m \to \mathbf{G}'_m$ is an epimorphism, there is an R-algebra S' and a faithfully flat map $S \to S'$ (a Zariski covering Spec $S' \to$ Spec S) for which g has a preimage in $\mathbf{G}_m(S')$.

We compute the structure of S'. The algebra map $g : K[Y, Y^{-1}] \to S$ is determined by sending Y to a unit $a \in S$. There exists a faithfully flat map $\varrho : S \to S'$ with $S' = S \otimes_{K[Y, Y^{-1}]} K[X, X^{-1}]$. In S',

$$(1 \otimes X)^p = 1 \otimes X^p$$
$$= 1 \otimes \phi(Y)$$
$$= g(Y) \otimes 1$$
$$= a \otimes 1.$$

And so, identifying a with $a \otimes 1$, one has $S' \cong S[T]/(T^p - a)$ for T indeterminate.
We have the short exact sequence of group schemes

$$1 \to \mu_p \to \mathbf{G}_m \xrightarrow{p} \mathbf{G}'_m \to 1 \tag{3.9}$$

with $\mathbf{G}_m/\mu_p = \mathbf{G}'_m$. Note that there are elements $S \in \mathrm{Ob}(\Im_{R\text{-alg}})$ for which

$$1 \to \mu_p(S) \to \mathbf{G}_m(S) \xrightarrow{p_S} \mathbf{G}'_m(S) \to 1$$

is not a short exact sequence of abstract groups; that is, there are R-algebras S for
which $\mathbf{G}_m(S)/\mu_p(S) \neq (\mathbf{G}_m/\mu_p)(S)$.

In Chapter 8, we shall employ short exact sequence (3.9) in the case where K is
a field containing $\mathbb{Q}_p$.

3.5   Chapter Exercises

Exercises for §3.1

1. Referring to Example 3.1.1, prove that the map $\Delta : \mathbb{Z}G \to \mathbb{Z}G \otimes_{\mathbb{Z}} \mathbb{Z}G$ defined
   by $\Delta(\sum_{\tau \in G} a_\tau) = \sum_{\tau \in G} a_\tau(\tau \otimes \tau)$ is a $\mathbb{Z}$-algebra homomorphism.
2. Let F be an R-group scheme, and let S be a commutative R-algebra. Show that
   the left/right identity element for $F(S)$ is unique.
3. Let R be a commutative ring with unity of characteristic 2. Let F be an R-
   group scheme with $R[F] = A$, and let S be a commutative R-algebra. Suppose
   that $f \in F(S)$ has order 2 in $F(S)$, and assume that $a \in A$ satisfies $\Delta(a) = a \otimes 1 + 1 \otimes a$. Prove that $\lambda\epsilon(a) = 0$.
4. Let F be an R-group scheme represented by the R-algebra A. Suppose that for
   all $a \in A$ and $\phi, \alpha, \beta \in F(S)$,

$$\sum_{(a)} \phi(a_{(1)})\alpha(a_{(2)}) = \sum_{(a)} \phi(a_{(1)})\beta(a_{(2)}).$$

Show that $\alpha = \beta$.

Exercises for §3.2

5. Prove Proposition 3.2.1.
6. Prove Lemma 3.2.3, parts (ii) and (iii).
7. Compute the augmentation ideal of $R[\mathbf{G}_a]$.
8. Compute the augmentation ideal of $R[\mathbf{G}_m]$.
9. Let $\mathbf{G}_{m,\mathbb{Z}}$ and $\mathbf{G}_{a,\mathbb{Z}}$ denote the multiplicative and additive $\mathbb{Z}$-group schemes,
   respectively. Prove that $\psi : \mathbf{G}_{m,\mathbb{Z}} \to \mathbf{G}_{a,\mathbb{Z}}$ defined as $\psi_S(x) = 0$ for all $x \in \mathbf{G}_{m,\mathbb{Z}}(S)$ is the only homomorphism of $\mathbf{G}_{m,\mathbb{Z}}$ into $\mathbf{G}_{a,\mathbb{Z}}$.

Exercises for §3.3

10. Suppose $A \to B$ and $B \to C$ are flat maps of commutative rings. Prove that $A \to C$ is flat.
11. Let A and B be commutative rings. Show that $B \to A$ is faithfully flat if and only if $M \otimes_B A = 0$ implies that $M = 0$ for all B-modules M.
12. Let A and B be commutative rings, and suppose that $B \to A$ is faithfully flat. Show that $m \cdot A \neq A$ for every maximal ideal m of B.
13. Show that the inclusion $\mathbb{Q} \to \mathbb{Q}(\sqrt{2})$ is faithfully flat.

Exercises for §3.4

14. Consider the short exact sequence of $\mathbb{Q}$-group schemes

$$1 \to \mu_p \to \mathbf{G}_m \xrightarrow{p} \mathbf{G}'_m \to 1.$$

(a) Find a $\mathbb{Q}$-algebra S for which the sequence

$$1 \to \mu_p(S) \to \mathbf{G}_m(S) \xrightarrow{p} \mathbf{G}'_m(S) \to 1$$

is a short exact sequence of abstract groups.

(b) Find a $\mathbb{Q}$-algebra S for which the sequence

$$1 \to \mu_p(S) \to \mathbf{G}_m(S) \xrightarrow{p} \mathbf{G}'_m(S) \to 1$$

fails to be short exact.

Chapter 4
Hopf Algebras

In this chapter, we focus on the structure of the representing algebra of the R-group scheme F.

4.1 Introduction to Hopf Algebras

Definition 4.1.1. An **R-Hopf algebra** is a commutative R-algebra H that is the representing algebra of an R-group scheme F. That is, an R-Hopf algebra is a commutative R-algebra H, together with R-algebra homomorphisms

$$\Delta : H \to H \otimes_R H \quad \text{comultiplication,}$$

$$\epsilon : H \to R \quad \text{counit,}$$

$$\sigma : H \to H \quad \text{coinverse,}$$

which satisfy, for all $a \in H$, the conditions

$$(I \otimes \Delta)\Delta(a) = (\Delta \otimes I)\Delta(a) \quad \text{coassociativity property,}$$

$$m(I \otimes \epsilon)\Delta(a) = a = m(\epsilon \otimes I)\Delta(a) \quad \text{counit property,}$$

$$m(I \otimes \sigma)\Delta(a) = \epsilon(a)1_H = m(\sigma \otimes I)\Delta(a) \quad \text{coinverse property,}$$

where $m : H \otimes_R H \to H$ is multiplication in H defined as $m(\sum(a \otimes b)) = \sum ab$.

We will usually denote an R-Hopf algebra by H, but when convenient we will use the notation A, B, or C.

Proposition 4.1.1. *There is a 1-1 correspondence between the collection of R-group schemes and the collection of Hopf algebras over R.*

R.G. Underwood, *An Introduction to Hopf Algebras*, DOI 10.1007/978-0-387-72766-0_4,   55
© Springer Science+Business Media, LLC 2011

Proof. The map $F \mapsto R[F]$ is a bijection between the collection of R-group schemes and the collection of R-Hopf algebras, with inverse $H \mapsto F = \mathrm{Hom}_{R\text{-alg}}(H, -)$. $\square$

Remark 4.1.1. We should note that one can perfectly well define Hopf algebras that are non-commutative R-algebras. In addition, the requirement that the coinverse be an algebra map can be omitted. A simple example, due to M. Sweedler, is the $\mathbb{Q}$-algebra $H = \mathbb{Q}\langle 1, g, x, gx \rangle$ modulo the relations $g^2 = 1$, $x^2 = 0$, $xg = -gx$. The comultiplication map $\Delta : H \to H \otimes_{\mathbb{Q}} H$ is given by $g \mapsto g \otimes g$, $x \mapsto x \otimes 1 + g \otimes x$, the counit map $\epsilon : H \to \mathbb{Q}$ is defined as $g \mapsto 1$, $x \mapsto 0$, and the coinverse map $\sigma : H \to H$ is defined by $g \mapsto g$, $x \mapsto -gx$ [Mo93, 1.5.6].

Numerous other non-commutative R-Hopf algebras can be constructed (see, for example, [Mo93, Appendix], where quantum groups are considered).

For the purposes of this book, however, we assume that our Hopf algebras are commutative. We do this for three basic reasons. The first is that our notion of Hopf algebra arises as the representing algebra of the structure sheaf Spec A, which is only defined for commutative rings A; in fact, without the assumption of commutativity, we lose the correspondence of Proposition 4.1.1. Second, assuming commutativity facilitates the theory and the computations that follow below. Third, in assuming commutativity, we lose none of the richness of our applications of Hopf algebras that follow.

We give some examples of Hopf algebras over R. The ring R itself is an R-Hopf algebra corresponding to the trivial R-group scheme 1; $R[X]$ is an R-Hopf algebra corresponding to the additive group scheme $\mathbf{G}_a$; and $R[X, X^{-1}]$ is an R-Hopf algebra that corresponds to the multiplicative group scheme $\mathbf{G}_m$.

As we have seen, if $\psi : F \to G$ is a group scheme homomorphism with algebra map $\phi : B \to A$, then $\ker(\psi) = N = \mathrm{Hom}_{R\text{-alg}}(A/\phi(B^+)A, -)$ is a group scheme. Thus $A/\phi(B^+)A$ is an R-Hopf algebra. The following is another important example.

Example 4.1.1. Let G be a finite Abelian group. Then the R-algebra RG is an R-Hopf algebra with comultiplication $\Delta : RG \to RG \otimes_R RG$ defined as

$$\Delta\left(\sum_{\tau \in G} a_\tau \tau\right) = \sum_{\tau \in G} a_\tau (\tau \otimes \tau),$$

counit $\epsilon : RG \to R$, given by

$$\epsilon\left(\sum_{\tau \in G} a_\tau \tau\right) = \sum_{\tau \in G} a_\tau,$$

and coinverse $\sigma : RG \to RG$, defined by

$$\sigma\left(\sum_{\tau \in G} a_\tau \tau\right) = \sum_{\tau \in G} a_\tau \tau^{-1}.$$

An R-Hopf algebra H is **cocommutative** if, for all $h \in H$, $\Delta(h) = (t\Delta)(h)$, where t is the twist map. The Hopf algebras $R[X]$, $R[X, X^{-1}]$, and RG given above are cocommutative.

Cocommutativity is an important property in view of the following proposition.

Proposition 4.1.2. *Let H be a cocommutative R-Hopf algebra, and let $F = \text{Hom}_{R\text{-alg}}(H, -)$ denote the corresponding R-group scheme. Then, for all $S \in \text{Ob}(\Im_{R\text{-alg}})$, $F(S)$ is an Abelian group.*

Proof. This is immediate from Proposition 3.1.1. $\square$

Definition 4.1.2. Let H be an R-Hopf algebra with counit map $\epsilon : H \to R$. An **integral** of H is an element $y \in H$ that satisfies

$$xy = \epsilon(x)y$$

for all $x \in H$.

We denote the collection of integrals of H by $\int_H$.

Proposition 4.1.3. *The set of integrals $\int_H$ is an ideal of H.*

Proof. We first show that $\int_H$ is an additive subgroup of H. Let $x, y \in \int_H$. Then, for $z \in H$, we have

$$z(x + y) = zx + zy = \epsilon(z)x + \epsilon(z)y = \epsilon(z)(x + y),$$

and so $\int_H$ is closed under addition. Moreover, $0 \in \int_H$ since $z0 = 0 = \epsilon(z)0$, and $-x \in \int_H$ since $z(-x) = -(zx) = -(\epsilon(z)x) = \epsilon(z)(-x)$. Thus $\int_H \leq H$. Now, for $w \in H$,

$$w(yz) = (wy)z = (\epsilon(w)y)z = \epsilon(w)(yz),$$

and thus $yz \in \int_H$. $\square$

As an example, we consider the group ring RG, where G is a finite Abelian group. We seek to compute $\int_{RG}$.

Proposition 4.1.4. $\int_{RG} = R \sum_{\tau \in G} \tau.$

Proof. Let $S = R \sum_{\tau \in G} \tau$, and let $a \sum_{\tau \in G} \tau \in S$, $a \in R$. Since $\{v\}_{v \in G}$ is an R-basis for RG, an element $x \in RG$ can be written as $x = \sum_{v \in G} x_v v$ for $x_v \in R$. Now

$$\left(\sum_{\upsilon\in G} x_\upsilon\upsilon\right)\left(a\sum_{\tau\in G}\tau\right) = \sum_{\upsilon\in G}\left(x_\upsilon a\upsilon\sum_{\tau\in G}\tau\right)$$

$$= \left(\sum_{\upsilon\in G} x_\upsilon\right)\left(a\sum_{\tau\in G}\tau\right)$$

$$= \epsilon\left(\sum_{\upsilon\in G} x_\upsilon\upsilon\right)\left(a\sum_{\tau\in G}\tau\right),$$

and so $S \subseteq \int_{RG}$.

Now suppose $x = \sum_{\tau\in G} x_\tau\tau \in \int_{RG}$. For $\upsilon \in G$,

$$x = \epsilon(\upsilon)x = \upsilon x = \sum_{\tau\in G} x_\tau\upsilon\tau = \sum_{\tau\in G} x_\tau\rho(\tau),$$

where $\rho : G \to G$ is a permutation of the elements of G. Thus, $x_\tau = a$ for some $a \in R$ and all $\tau \in G$, and so $\int_{RG} \subseteq S$. $\square$

Definition 4.1.3. Let A, B be R-Hopf algebras and let E, F be the corresponding R-group schemes. Then the R-algebra homomorphism $\phi : B \to A$ is a **homomorphism of Hopf algebras** if the corresponding natural transformation $\psi : E \to F$ is a homomorphism of group schemes. Thus $\phi : B \to A$ is a homomorphism of R-Hopf algebras if

$$(\phi \otimes \phi)\Delta_B(b) = \Delta_A(\phi(b))$$

for all $b \in B$.

The canonical surjection of R-algebras $s : A \to A/\phi(B^+)A$ is a Hopf algebra homomorphism. Indeed, the corresponding natural transformation $i : N \to E$ is a homomorphism of group schemes since

$$i_S : N(S) \to E(S)$$

is a group homomorphism for all S. Alternatively, one could show (see §3.2) that

$$\overline{\beta\Delta_A}(s(a)) = (\overline{\alpha})^{-1}(s \otimes s)\Delta_A(a),$$

and so

$$\overline{\alpha}\overline{\beta\Delta_A}(s(a)) = (s \otimes s)\Delta_A(a), \ \forall a \in A.$$

Let A and B be R-Hopf algebras. The tensor product $A \otimes_R B$ is an R-algebra with multiplication defined as $(a \otimes b)(c \otimes d) = ac \otimes bd$. It is also an R-Hopf algebra with comultiplication given as

$$\Delta_{A\otimes B} : A \otimes_R B \to (A \otimes_R B) \otimes_R (A \otimes_R B),$$

where

$$\Delta_{A \otimes B}(a \otimes b) = (I \otimes t \otimes I)(\Delta_A \otimes \Delta_B)(a \otimes b)$$

$$= (I \otimes t \otimes I)\left(\sum_{(a,b)}(a_{(1)} \otimes a_{(2)}) \otimes (b_{(1)} \otimes b_{(2)})\right)$$

$$= \sum_{(a,b)}(a_{(1)} \otimes b_{(1)}) \otimes (a_{(2)} \otimes b_{(2)}).$$

The counit is

$$(\epsilon_A \otimes \epsilon_B) : A \otimes_R B \to R \otimes_R R := R,$$

with $(\epsilon_A \otimes \epsilon_B)(a \otimes b) = \epsilon_A(a)\epsilon_B(b)$, and the coinverse is defined as

$$(\sigma_A \otimes \sigma_B) : A \otimes_R B \to A \otimes_R B,$$

with $(\sigma_A \otimes \sigma_B)(a \otimes b) = \sigma_A(a) \otimes \sigma_B(b)$. We leave it as exercises to show that these maps satisfy the coassociativity, counit, and coinverse properties.

We consider the case $A = B$.

Proposition 4.1.5. *Let A be an R-Hopf algebra. Then multiplication $m : A \otimes_R A \to A$ is a homomorphism of Hopf algebras.*

Proof. We first show that m is an R-algebra map. Let $a, b, c, d \in A$. Then

$$m((a \otimes b)(c \otimes d)) = m(ac \otimes bd)$$

$$= acbd$$

$$= abcd$$

$$= m(a \otimes b)m(c \otimes d),$$

and so m is an R-algebra homomorphism. Moreover,

$$(m \otimes m)\Delta_{A \otimes_R A}(a \otimes b) = (m \otimes m)\sum_{(a,b)}(a_{(1)} \otimes b_{(1)}) \otimes (a_{(2)} \otimes b_{(2)})$$

$$= \sum_{(a,b)} a_{(1)}b_{(1)} \otimes a_{(2)}b_{(2)}$$

$$= \Delta_A(ab)$$

$$= \Delta_A(m(a \otimes b)),$$

and so m is a Hopf algebra map. $\square$

Let H be an R-Hopf algebra. Though the coinverse map $\sigma : H \to H$ is an R-algebra homomorphism, it is not an R-Hopf algebra homomorphism. Indeed, let

$$\alpha : \mathrm{Hom}_{R\text{-alg}}(H \otimes_R H, -) \to \mathrm{Hom}_{R\text{-alg}}(H, -)$$

be the associated map corresponding to comultiplication on H. The group product in $\mathrm{Hom}_{R\text{-alg}}(H, H \otimes_R H)$ is given by the map

$$\alpha_{H \otimes_R H} : \mathrm{Hom}_{R\text{-alg}}(H \otimes_R H, H \otimes_R H) \to \mathrm{Hom}_{R\text{-alg}}(H, H \otimes_R H).$$

Let $I \in \mathrm{Hom}_{R\text{-alg}}(H, H)$ be the identity map, let $I \otimes I \in \mathrm{Hom}_{R\text{-alg}}(H \otimes_R H, H \otimes_R H)$ be defined as $(I \otimes I)(a \otimes b) = a \otimes b$, and let $\mathrm{t}(I \otimes I)$ be given as $a \otimes b \mapsto b \otimes a$. For $x \in A$,

$$\begin{aligned}
(\sigma \otimes \sigma)\Delta(x) &= (I \otimes I)(\sigma \otimes \sigma)\Delta(x) \\
&= \alpha_{H \otimes_R H}(I\sigma \otimes I\sigma)(x) \\
&= \alpha_{H \otimes_R H}((I \otimes I)\mathrm{t})(\sigma(x)) \\
&= (I \otimes I)(\mathrm{t}\Delta(\sigma(x))) \\
&= \mathrm{t}\Delta(\sigma(x)).
\end{aligned}$$

Therefore, σ satisfies

$$(\sigma \otimes \sigma)\Delta(x) = \mathrm{t}\Delta(\sigma(x)), \tag{4.1}$$

and we say that σ is a **Hopf antihomomorphism**. Of course, if H is cocommutative, then σ is a homomorphism of Hopf algebras.

To define the notion of a short exact sequence of Hopf algebras, we translate to group schemes.

Let $\psi : E \to F$ be a homomorphism of R-group schemes with $R[E] = A$ and $R[F] = B$, and suppose that the corresponding homomorphism of Hopf algebras $i : B \to A$ is an injection. The kernel N of ψ is an R-group scheme represented by the R-Hopf algebra $A/i(B^+)A$. Since $N \to E$ is a homomorphism of group schemes, the canonical surjection $s : A \to A/i(B^+)A$ is a homomorphism of R-Hopf algebras. There is a short exact sequence of R-modules

$$0 \to i(B^+)A \to A \xrightarrow{s} A/i(B^+)A \to 0,$$

which is close to what we want but of course cannot be the correct notion of a short exact sequence of Hopf algebras since neither 0 nor $i(B^+)A$ has the structure of an R-Hopf algebra. The following is a correct definition.

Definition 4.1.4. Let A, B, C be R-Hopf algebras. Let $E = \mathrm{Hom}_{R-\text{alg}}(A, -)$, $F = \mathrm{Hom}_{R-\text{alg}}(B, -)$ be the corresponding R-group schemes. Let 1 denote the trivial R-group scheme represented by R. Suppose $s : A \to C$ is a surjection of Hopf algebras and $i : B \to A$ is an injection of Hopf algebras. The sequence

$$R \xrightarrow{\lambda} B \xrightarrow{i} A \xrightarrow{s} C \xrightarrow{\epsilon} R$$

is a **short exact sequence of R-Hopf algebras** if the homomorphism $E \to F$ corresponding to $i : B \to A$ has a kernel represented by C. That is, the sequence above is short exact if there is a Hopf isomorphism $C \cong A/i(B^+)A$.

For example, let G be a finite Abelian group, and let W be a subgroup of G. Then the canonical surjection of groups $G \to \overline{G} = G/W$ induces a surjective homomorphism of R-Hopf algebras $s : RG \to R\overline{G}$. Moreover, there is a Hopf inclusion $i : RW \to RG$. The homomorphism of R-group schemes $E \to F$, $R[E] = RG$, $R[F] = RW$ has a kernel represented by $RG/i(RW)^+RG = R\overline{G}$. Thus there is a short exact sequence of R-Hopf algebras

$$R \to RW \to RG \to R\overline{G} \to R. \tag{4.2}$$

Generally, let M be an R-module, and let $M^* = \operatorname{Hom}_R(M, R)$ denote the collection of linear functionals on M, which we call the **linear dual** of M. The linear dual M^* is an R-module with scalar multiplication defined as $(r \cdot f)(v) = r(f(v))$ for $r \in R$, $f \in M^*$, and $v \in M$.

Let $\phi : M \to N$ be a homomorphism of R-modules. Then there exists a homomorphism of linear duals, $\phi^* : N^* \to M^*$, defined as $\phi^*(f)(v) = f(\phi(v))$, for all $f \in N^*, v \in M$.

If M is free and of finite rank, say m, with basis $\{b_1, b_2, \ldots, b_m\}$, then M^* is a free rank m R-module with **dual basis** $\{f_1, f_2, \ldots, f_m\}$, where $f_i(b_j) = \delta_{ij}$.

If M and N are free R-modules of finite rank, then linear duality behaves well with respect to tensor products.

Proposition 4.1.6. *Let M, N be free R-modules of finite rank m. Then*

$$M^* \otimes_R N^* \cong (M \otimes_R N)^*.$$

Proof. Let $\psi : M^* \otimes_R N^* \to (M \otimes_R N)^*$ be defined as $\psi(f \otimes h)(a \otimes b) = f(a)h(b)$ for $f \in M^*, h \in N^*, a \otimes b \in M \otimes_R N$. Then

$$\psi(rf \otimes h)(a \otimes b) = (rf)(a)h(b) = r(f(a)h(b)) = r\psi(f \otimes h)(a \otimes b)$$

for $r \in R$, so ψ is R-linear. We claim that ψ is an isomorphism of R-modules.

Let $\{a_i\}$ be a basis for M with dual basis $\{f_i\}$, and let $\{b_i\}$ be a basis for N with dual basis $\{h_i\}$. Then the collection of tensors $\{f_i \otimes h_j\}$, $0 \le i, j \le m$, is a basis for $M^* \otimes_R N^*$. For $0 \le i, j \le m$, let

$$\tau_{ij} = \psi(f_i \otimes h_j) \in (M \otimes_R N)^*.$$

Then

$$\tau_{ij}(a_l \otimes b_m) = f_i(a_l)h_j(b_m) = \delta_{il}\delta_{jm},$$

and so the images $\{\tau_{ij}\}$ form a basis for $(M \otimes_R N)^*$. It follows that ψ is an isomorphism of R-modules. $\square$

Proposition 4.1.7. *Let H be a cocommutative R-Hopf algebra that is a free R-module of rank m, and let $H^* = \mathrm{Hom}_R(H, R)$ be the R-module of linear functionals on H. Then H^* is an R-Hopf algebra.*

Proof. We first show that H^* is a commutative ring. Let $\Delta_H : H \to H \otimes_R H$ denote the comultiplication map of H. Then Δ_H yields an R-module map

$$\Delta_H^* : (H \otimes_R H)^* \to H^*.$$

By Proposition 4.1.6, there is an isomorphism

$$\psi : H^* \otimes_R H^* \to (H \otimes_R H)^*,$$

given as $\psi(\alpha \otimes \beta)(a \otimes b) = \alpha(a)\beta(b)$, and we shall henceforth identify $H^* \otimes_R H^*$ with $(H \otimes_R H)^*$. We set $m_{H^*} = \Delta_H^*$, and define multiplication on H^* as

$$(\alpha\beta)(a) = m_{H^*}(\alpha \otimes \beta)(a) = (\alpha \otimes \beta)\Delta_H(a) = \sum_{(a)} \alpha(a_{(1)})\beta(a_{(2)})$$

for $\alpha, \beta \in H^*, a \in H$. The cocommutativity of H implies that $\alpha\beta = \beta\alpha$, and so H^* is a commutative ring. Note that

$$(\alpha\epsilon_H)(a) = \sum_{(a)} \alpha(a_{(1)})\epsilon_H(a_{(2)}) = \alpha\left(\sum_{(a)} a_{(1)}\epsilon_H(a_{(2)})\right) = \alpha(a),$$

and so the unity in H^* is ϵ_H. Moreover, the ring homomorphism $\lambda_{H^*} : R^* := R \to H^*$, with $\lambda_{H^*} = \epsilon_H^*, \epsilon_H^* : R^* \to H^*$, endows H^* with the structure of a commutative R-algebra. (Here we have identified $R^* = \mathrm{Hom}_R(R, R)$ with R.)

Let $m_H : H \otimes_R H \to H$ denote multiplication in H. Then multiplication in $H \otimes_R H$ is given as $m_{H \otimes_R H} = (m_H \otimes m_H)(I \otimes t \otimes I)$. Moreover, multiplication in $H^* \otimes_R H^*$ is given as $m_{H^* \otimes_R H^*} = (m_{H^*} \otimes m_{H^*})(I \otimes t \otimes I)$. Of course, one has $\Delta_{H \otimes_R H}^* = m_{H^* \otimes_R H^*}$, where $\Delta_{H \otimes_R H}$ is comultiplication on $H \otimes_R H$.

Set $\Delta_{H^*} = m_H^*$. We claim that $\Delta_{H^*} : H^* \to H^* \otimes_R H^*$ is an R-algebra map that satisfies the coassociativity property. For $\alpha \in H^*, a, b \in H$,

$$\Delta_{H^*}(\alpha)(a \otimes b) = m_H^*(\alpha)(a \otimes b) = \alpha(m_H(a \otimes b)) = \alpha(ab).$$

We have

$$\Delta_{H^*}(\alpha\beta)(a \otimes b) = (\alpha\beta)(ab)$$
$$= (\alpha \otimes \beta)(\Delta_H(ab))$$
$$= (\alpha \otimes \beta)(\Delta_H(a)\Delta_H(b))$$
$$= (\alpha \otimes \beta)(m_{H\otimes H}(\Delta_H(a) \otimes \Delta_H(b))$$
$$= m^*_{H\otimes H}(\alpha \otimes \beta)(\Delta_H(a) \otimes \Delta_H(b))$$
$$= (\Delta_{H^*} \otimes t \otimes \Delta_{H^*})(\alpha \otimes \beta)(\Delta_H(a) \otimes \Delta_H(b))$$
$$= (\Delta_{H^*}(\alpha) \otimes \Delta_{H^*}(\beta))(\Delta_H \otimes t \otimes \Delta_H)(a \otimes b)$$
$$= (\Delta_{H^*}(\alpha) \otimes \Delta_{H^*}(\beta))(\Delta_{H\otimes H}(a \otimes b))$$
$$= \Delta^*_{H\otimes H}(\Delta_{H^*}(\alpha) \otimes \Delta_{H^*}(\beta))(a \otimes b)$$
$$= m_{H^*\otimes H^*}(\Delta_{H^*}(\alpha) \otimes \Delta_{H^*}(\beta))(a \otimes b)$$
$$= (\Delta_{H^*}(\alpha)\Delta_{H^*}(\beta))(a \otimes b),$$

and so Δ_{H^*} is an algebra map. Now, for all $a, b, c \in H$,

$$(I \otimes \Delta_{H^*})\Delta_{H^*}(\alpha)(a \otimes b \otimes c) = (I \otimes m^*_H)m^*_H(\alpha)(a \otimes b \otimes c)$$
$$= m^*_H(\alpha)((m_H \otimes I)(a \otimes b \otimes c))$$
$$= m^*_H(\alpha)(ab \otimes c)$$
$$= \alpha(m_H(ab \otimes c))$$
$$= \alpha((ab)c)$$
$$= \alpha(a(bc))$$
$$= \alpha(m_H(a \otimes bc))$$
$$= m^*_H(\alpha)((I \otimes m_H)(a \otimes b \otimes c))$$
$$= (m^*_H \otimes I)m^*_H(\alpha)(a \otimes b \otimes c)$$
$$= (\Delta_{H^*} \otimes I)\Delta_{H^*}(a \otimes b \otimes c),$$

and so Δ_{H^*} satisfies the coassociativity property.

For a counit map for H^*, we let $\epsilon_{H^*} = \lambda^*$, where $\lambda^* : H^* \to R^* := R$. Thus $\epsilon_{H^*}(\alpha)(r) = \alpha(\lambda(r)) = \alpha(r \cdot 1) = r\alpha(1)$. We have

$$\epsilon_{H^*}(\alpha\beta)(r) = r\alpha\beta(1)$$
$$= r\alpha(1)\beta(1)$$
$$= \beta(1)(r\alpha(1))$$

$$= \beta(1)(\epsilon_{H^*}(\alpha)(r))$$
$$= (\epsilon_{H^*}(\alpha)(r))\beta(1)$$
$$= \epsilon_{H^*}(\alpha)\epsilon_{H^*}(\beta)(r),$$

and so ϵ_{H^*} is an algebra map. We leave it as an exercise to show that ϵ_{H^*} satisfies the counit property.

Finally, we define the coinverse map as $\sigma_{H^*} = \sigma_H^*$, where $\sigma_H^* : H^* \to H^*$. We have $\sigma_{H^*}(\alpha)(a) = \alpha(\sigma(a))$. We leave it as an exercise to show that σ_{H^*} is an algebra map that satisfies the coinverse property. Thus H^* is an R-Hopf algebra. $\square$

Proposition 4.1.8. *Let H be a cocommutative R-Hopf algebra that is a free R-module of finite rank. Let $F = \mathrm{Hom}_{R\text{-alg}}(H^*, -)$ be the R-group scheme represented by H^*. Then*
$$F(R) = \{a \in H : \Delta_H(a) = a \otimes a, a \neq 0\}.$$

Proof. It is well-known that H can be identified with the double dual H^{**}. For $a \in H$, $a(f) = f(a)$ for $f \in H^*$. One has $F(R) \subseteq H^{**} = H$. Now suppose $a \in H$ with $\Delta_H(a) = a \otimes a$. Then, for $f, g \in H^*$,

$$a(fg) = a(\Delta_H^*(f \otimes g))$$
$$= \Delta_H^*(f \otimes g)(a)$$
$$= (f \otimes g)\Delta_H(a)$$
$$= (f \otimes g)(a \otimes a)$$
$$= f(a)g(a)$$
$$= a(f)a(g),$$

so that a is an element of $F(R)$.

Conversely, suppose that $a \in F(R)$. Then

$$\Delta_H(a)(f \otimes g) = (f \otimes g)\Delta_H(a)$$
$$= a(fg)$$
$$= f(a)g(a)$$
$$= (f \otimes g)(a \otimes a)$$
$$= (a \otimes a)(f \otimes g).$$

Thus the $\Delta_H(a) = a \otimes a$ as linear functionals in $H^{**} \otimes H^{**}$. $\square$

One can dualize a short exact sequence of cocommutative R-Hopf algebras to yield a short exact sequence of duals.

Proposition 4.1.9. *Let A, B, and C be cocommutative R-Hopf algebras that are free over R of finite rank. Let*

$$R \xrightarrow{\lambda} B \xrightarrow{i} A \xrightarrow{s} C \xrightarrow{\epsilon} R$$

be a short exact sequence of R-Hopf algebras. Then there exists a short exact sequence of R-Hopf algebras

$$R \xrightarrow{\epsilon^*} C^* \xrightarrow{s^*} A^* \xrightarrow{i^*} B^* \xrightarrow{\lambda^*} R.$$

Proof. One shows that $i^* : A^* \to B^*$ is a Hopf surjection, $s^* : C^* \to A^*$ is a Hopf injection, and $B^* \cong A^*/s^*((C^*)^+)A^*$ as Hopf algebras. □

4.2 Dedekind Domains

We shall soon be interested in studying Hopf algebras over Dedekind domains. For the convenience of the reader, we review some important facts about Dedekind domains.

A ring R is **Noetherian** if every ideal in R is a finitely generated module over R.

Lemma 4.2.1. *Let R be a Noetherian ring, and let $\prod_{i=1}^{l} R$ denote the product of a finite number of copies of R. Then every ideal I of $\prod_{i=1}^{l} R$ is finitely generated.*

Proof. I is of the form $\prod_{i=1}^{l} J_i$ for ideals J_i of R. Since each J_i is finitely generated, so is I. □

Lemma 4.2.2. *Let R be a Noetherian ring, let M be a finitely generated R-module, and let N be a submodule of M. Then N is finitely generated.*

Proof. Let $m_1, m_2, \ldots, m_l$ be a generating set for M. Let $\prod_{i=1}^{l} R$ denote the product of l copies of R. Then there is a surjective ring homomorphism

$$\phi : \prod_{i=1}^{l} R \to M$$

defined as $\phi(a_1, a_2, \ldots, a_l) = a_1 m_1 + a_2 m_2 + \cdots + a_l m_l$. Now $\phi^{-1}(N)$ is an ideal in $\prod_{i=1}^{l} R$ that is finitely generated by Lemma 4.2.1. Let $\{s_1, s_2, s_3, \ldots, s_k\}$ denote a generating set for $\phi^{-1}(N)$. Then

$$\{\phi(s_1), \phi(s_2), \ldots, \phi(s_k)\}$$

is a generating set for N. □

An integral domain R is **integrally closed** if R consists of the elements of $\mathrm{Frac}(R)$ that are zeros of monic polynomials over R.

Definition 4.2.1. A **Dedekind domain** is an integrally closed, Noetherian integral domain in which every non-zero prime ideal is maximal.

We note that every PID is a Dedekind domain (see [Rot02, Example 11.88]).

Let R be a Dedekind domain, and let P be a non-zero prime ideal of R. Let R_P denote the localization of R at P. Then R_P is a local Dedekind domain with maximal ideal

$$PR_P = \left\{ \sum pa : p \in P, a \in R_P \right\}.$$

Proposition 4.2.1. *With the notation above, the ideal PR_P is principal.*

Proof. Let

$$(PR_P)^{-1} = \{x \in \mathrm{Frac}(R_P) : xPR_P \subseteq R_P\}.$$

Then $(PR_P)^{-1}PR_P$ is an ideal of R_P with

$$PR_P \subseteq (PR_P)^{-1}PR_P \subseteq R_P.$$

Since PR_P is maximal, either $PR_P = (PR_P)^{-1}PR_P$ or $(PR_P)^{-1}PR_P = R_P$.

Since R_P is a Noetherian integral domain that is integrally closed and local, the discussion after the proof of [Se79, Chapter I, Lemma 1] applies to show that $PR_P \neq (PR_P)^{-1}PR_P$. Thus $(PR_P)^{-1}PR_P = R_P$. Now, there exist elements $a_i \in (PR_P)^{-1}$ and $\pi_i \in PR_P$ for which $\sum_{i=1}^{k} a_i \pi_i = 1$. Thus $a_i \pi_i = u$ for some integer i, $1 \leq i \leq k$, and some unit u in R_P. Therefore, $a\pi = 1$ for some $a \in (PR_P)^{-1}$, $\pi \in PR_P$. Let $b \in PR_P$. Then $b = (ba)\pi$ with $ba \in R_P$, and hence PR_P is generated by π. $\square$

Corollary 4.2.1. *Let R be a Dedekind domain, and let P be a non-zero prime ideal of R. Then every non-zero element $x \in R_P$ can be written in the form $x = u\pi^n$, where π is a generator for PR_P, $u \in R_P$ is a unit in R_P, and $n \geq 0$ is an integer.*

Proof. By Proposition 4.2.1, $(\pi) = PR_P$ for some $\pi \in R_P$. We claim that $\cap_{n \geq 1}(\pi)^n = 0$. To this end, let $x \in \cap_{n \geq 1}(\pi)^n$. Then $x = \pi^n a_n$, $a_n \in R_P$, for $n \geq 1$. Now

$$0 = \pi^n a_n - \pi^{n+1} a_{n+1} = \pi^n(a_n - \pi a_{n+1}),$$

and so $a_n = \pi a_{n+1}$ for $n \geq 1$. Consequently, there is an increasing sequence

$$R_P a_1 \subseteq R_P a_2 \subseteq R_P a_3 \subseteq \cdots.$$

Since R is Noetherian, this sequence terminates; that is, $R_P a_{m+1} = R_P a_m$ for some m. Thus, $a_{m+1} \in R_P a_m$, and so $a_{m+1} = r a_m$ for some $r \in R_P$. Now,

$$0 = a_{m+1} - r a_m = a_{m+1} - r\pi a_{m+1} = a_{m+1}(1 - r\pi).$$

Since $1 - r\pi$ is a unit in R_P, $a_{m+1} = 0$. It follows that $x = 0$, and so $\cap_{n \geq 1}(\pi)^n = 0$.

Now suppose that x is a non-zero, non-unit element of R_P. Then $x \in (\pi)$. Necessarily, there exists an integer n for which $x \in (\pi)^n$, $x \notin (\pi)^{n+1}$. Thus, $x = \pi^n a$ for some $a \in R_P$ with $a \notin (\pi)$; that is, a is a unit in R_P. $\square$

The element π for which $(\pi) = PR_P$ is a **uniformizing parameter for** R_P.

Corollary 4.2.2. *Let R be a Dedekind domain, and let P be a non-zero prime ideal of R. Then the local ring R_P is a PID.*

Proof. Clearly, $\{0\}$ and R_P are principal. Let π be a uniformizing parameter for R_p, and let I be a non-trivial proper ideal of R_P. Let S denote the collection of integers $n \geq 1$ for which $\pi^n \in I$. Let m be the smallest integer in S. Then $(\pi^m) = I$, so that I is principal. $\square$

Let R be an integral domain with field of fractions K. Then K is an R-module. A **fractional ideal** of R is a non-zero R-submodule J of K of the form

$$J = cI,$$

where $c \in K^\times$ and I is an ideal of R. For example, $\mathbb{Z}[\frac{1}{2}] = \{n/2 : n \in \mathbb{Z}\}$ is a fractional ideal of $\mathbb{Z}$.

Proposition 4.2.2. *Let R be a Dedekind domain. Then a non-zero submodule J of K is a fractional ideal if and only if it is finitely generated.*

Proof. Let J be a non-zero submodule of K of the form cI, $c \in K$. Then J is finitely generated since I is finitely generated.

Conversely, suppose J is a non-zero submodule of K that is finitely generated as an R-module. Write $J = Rq_1 + Rq_2 + \cdots + Rq_l$ for $q_1, q_2, \ldots, q_l \in K$. There is a generating set for J of the form $\{a_1/q, a_2/q, \ldots, a_l/q\}$ for some $a_i \in R$, $q \in K$. Consequently, $J = cI$, where $c = q^{-1}$ and $I = (a_1, a_2, \ldots, a_l)$. $\square$

The ring of integers in a finite extension of number fields is a Dedekind domain. We spend the remainder of this section proving this fact. We begin with some lemmas.

Lemma 4.2.3. *Let K be a finite extension of $\mathbb{Q}$, and let R be the ring of integers in K. Then every non-zero ideal J of R contains a basis for K over $\mathbb{Q}$.*

Proof. Let $\beta \in K$, $\beta \neq 0$. Then the set $\{1, \beta, \beta^2, \ldots, \beta^l\}$, $l = [K : \mathbb{Q}]$, is linearly dependent over $\mathbb{Q}$, and so there exist integers $a_i \in \mathbb{Z}$, $1 \leq i \leq l$, not all zero, for which

$$a_0 + a_1\beta + a_2\beta^2 + \cdots + a_l\beta^l = 0.$$

Let j be the largest index such that $a_j \neq 0$. Then

$$a_j^{j-1}(a_0 + a_1\beta + a_2\beta^2 + \cdots + a_j\beta^j) = 0,$$

and so

$$a_j^{j-1}a_0 + a_j^{j-2}a_1(a_j\beta) + a_j^{j-3}a_2(a_j\beta)^2 + \cdots + (a_j\beta)^j = 0;$$

thus, $a_j\beta$ is integral over $\mathbb{Z}$. Hence $a_j\beta \in R$.

Next, let $\{b_1, b_2, \ldots, b_l\}$ be a basis for K over $\mathbb{Q}$. Now, by the preceding paragraph, there exist integers c_i for which $\{c_i b_i\}_{i=1}^{l} \subseteq R$. Let a be a non-zero element of J. Then $\{c_i b_i a\}_{i=1}^{l} \subseteq J$ is a basis for $K/\mathbb{Q}$.   □

Next, we fix a non-zero ideal J of R and let S be the collection of all bases $\mathcal{B}$ for $K/\mathbb{Q}$ that are contained in J. By Lemma 4.2.3, S is nonempty.

For each basis $\mathcal{B} = \{b_1, b_2, \ldots, b_l\}$ in S, let

$$N_\mathcal{B} = \mathbb{Z}b_1 \oplus \mathbb{Z}b_2 \oplus \cdots \oplus \mathbb{Z}b_l \subseteq J$$

be the free $\mathbb{Z}$-module with basis $\mathcal{B} = \{b_i\}$. For $a, b \in K$ and $1 \leq i \leq l$, one has

$$abb_i = \sum_{j=1}^{l} q_{i,j} b_j$$

for $q_{i,j} \in \mathbb{Q}$. The map $B : K \times K \to \mathbb{Q}$ defined as

$$B(a,b) = \sum_{i=1}^{l} q_{i,i}$$

is a symmetric non-degenerate bilinear form on K. We compute the discriminant of $N_\mathcal{B}$ with respect to B. We denote this discriminant as $\mathrm{disc}(N_\mathcal{B})$. Since $B : R \times R \to \mathbb{Z}$, $\mathrm{disc}(N_\mathcal{B})$ is generated by a non-zero integer that we identify with $\mathrm{disc}(N_\mathcal{B})$. The collection

$$\{|\mathrm{disc}(N_\mathcal{B})|\}_{\mathcal{B} \in S}$$

is a non-empty set of positive integers and as such has a smallest element. Let $\mathcal{M} = \{m_1, m_2, \ldots, m_l\}$ denote the basis in S that corresponds to the smallest integer $|\mathrm{disc}(N_\mathcal{M})|$.

Lemma 4.2.4. *Let $\mathcal{M} = \{m_1, m_2, \ldots, m_l\}$ denote the basis in S that corresponds to the smallest integer $|\mathrm{disc}(N_\mathcal{M})|$. Then $J = N_\mathcal{M}$.*

Proof. We only need to show that $J \subseteq N_\mathcal{M}$. Let $a \in J$. Since $\{m_i\}$ is a basis for $K/\mathbb{Q}$,

$$a = q_1 m_1 + q_2 m_2 + \cdots + q_l m_l$$

for elements $q_i \in \mathbb{Q}$. We claim that each q_i is an integer. By way of contradiction, let's assume that $q_j \notin \mathbb{Z}$ for some j. Without loss of generality, we can assume that $j = 1$. Note that $q_1 = \eta + \iota$ for some $\eta \in \mathbb{Z}$ and ι with $0 < \iota < 1$. Set $m_1' = a - \eta m_1$ and $m_i' = m_i$ for $2 \leq i \leq l$. Then $\mathcal{M}' = \{m_i'\}$ is a basis for $K/\mathbb{Q}$ that is contained in J.

Let $N_{\mathcal{M}'} = \mathbb{Z}m_1' \oplus \mathbb{Z}m_2' \oplus \cdots \oplus \mathbb{Z}m_l'$. Then $N_{\mathcal{M}'} \subseteq N_{\mathcal{M}}$. The matrix that multiplies the basis $\{m_i\}$ to give the basis $\{m_i'\}$ is

$$
C = \begin{pmatrix}
\iota & q_2 & q_3 & \cdots & q_l \\
0 & 1 & 0 & & 0 \\
0 & 0 & 1 & & 0 \\
\vdots & & & & \vdots \\
0 & 0 & \cdots & 0 & 1
\end{pmatrix}.
$$

Now, by a familiar property of discriminants,

$$
\operatorname{disc}(N_{\mathcal{M}'}) = (\det(C))^2 \operatorname{disc}(N_{\mathcal{M}})
$$

$$
= (\iota^2 \mathbb{Z}) \operatorname{disc}(N_{\mathcal{M}}),
$$

which contradicts the minimality of $|\operatorname{disc}(N_{\mathcal{M}})|$ since $\iota^2 < 1$. Thus each q_i is an integer, and so $J = N_{\mathcal{M}}$. $\qquad\square$

Lemma 4.2.5. *Let J be a non-zero ideal of R. Then R/J is a finite ring.*

Proof. Let $a \in J \cap \mathbb{Z}_{>0}$. (Why is $J \cap \mathbb{Z}_{>0}$ non-empty?) Since the ring homomorphism $R/(a) \to R/J$ is surjective, we only need to show that $R/(a)$ is finite. By Lemma 4.2.4,

$$
J = \mathbb{Z}m_1 \oplus \mathbb{Z}m_2 \oplus \cdots \oplus \mathbb{Z}m_l
$$

for some elements $m_i \in J$. Now

$$
S = \{a_1 m_1 + a_2 m_2 + \cdots + a_l m_l : 0 \le a_i \le a\}
$$

is a set of coset representatives for $R/(a)$. Note that $|S| = a^l < \infty$, so that $|R/(a)| = a^l$. $\qquad\square$

Lemma 4.2.6. *Let $\mathcal{F}$ be a non-empty collection of ideals of R. Then every non-empty set of ideals of R has a maximal element.*

Proof. Let I_0 be an ideal in $\mathcal{F}$, and suppose there is an ideal $I_1 \in \mathcal{F}$ with $I_0 \subset I_1$ (proper inclusion). Next suppose there is an ideal I_2 with $I_1 \subset I_2$. Continuing in this manner, if there is always an ideal I_{i+1} for which $I_i \subset I_{i+1}$, then $I_0 \subset I_1 \subset I_2 \subset \cdots$ is an increasing chain of ideals; thus R/I_0 has an infinite number of ideals, which contradicts Lemma 4.2.5. $\qquad\square$

Lemma 4.2.7. *Let I be an ideal of R. Then I is finitely generated.*

Proof. Let $\mathcal{F}$ denote the collection of all finitely generated ideals that are contained in I. Since $\{0\} \in \mathcal{F}$, $\mathcal{F}$ is non-empty. By Lemma 4.2.6, $\mathcal{F}$ contains a maximal element M. We have $M \subseteq I$. If $M \subset I$, then there exists an element $a \in I \backslash M$.

Let $J = M + Ra$. Then $J \in \mathcal{F}$ and $M \subset J$, which contradicts the maximality of M. Hence $M = I$ and I is finitely generated. $\qquad\square$

Proposition 4.2.3. *Let K be a finite extension of $\mathbb{Q}$ with ring of integers R. Then R is a Dedekind domain.*

Proof. We show that the conditions of Definition 4.2.1 hold. Certainly R is integrally closed by the definition of the ring of integers; moreover, R is Noetherian by Lemma 4.2.7. Let $P \neq 0$ be a prime ideal of R. By Lemma 4.2.5, R/P is a field. Thus P is maximal. $\qquad\square$

4.3 Hopf Modules

The purpose of this section is to establish a fundamental result of R. Larson and M. Sweedler [LS69] that states that for a Dedekind domain R and cocommutative R-Hopf algebra H that is free and of finite rank over R, the linear dual H^* is isomorphic to $H \otimes_R \int_{H^*}$ as H-modules, where $\int_{H^*}$ is the ideal of integrals of H^*.

Definition 4.3.1. Let H be an R-Hopf algebra, and let M be an R-module. Then M is a **left H-comodule** if there exists an R-linear map $\Psi : M \to H \otimes_R M$ for which

(i) $(I \otimes \Psi)\Psi = (\Delta \otimes I)\Psi$ and
(ii) $(\epsilon \otimes I)\Psi(m) = 1 \otimes m, \ \forall m \in M$.

One easily checks that H is a left comodule over itself with the comultiplication map Δ playing the role of Ψ.

Let M be a left H-comodule with structure map $\Psi : M \to H \otimes M$. We adapt the Sweedler notation to write

$$\Psi(\beta) = \sum_{(\beta)} b_{(1)} \otimes \beta_{(2)}$$

for $\beta, \beta_{(2)} \in M, b_{(1)} \in H$. Observe that

$$\Psi(\beta_{(2)}) = \sum_{(\beta_{(2)})} b_{(2)(1)} \otimes \beta_{(2)(2)}.$$

We can extend the Sweedler notation as follows. Let $\beta \in M$. Then

$$(I \otimes \Psi)\Psi(\beta) = (I \otimes \Psi)\left(\sum_{(\beta)} b_{(1)} \otimes \beta_{(2)} \right)$$

$$= \sum_{(\beta, \beta_{(2)})} b_{(1)} \otimes b_{(2)(1)} \otimes \beta_{(2)(2)} \qquad (4.3)$$

and

$$(\Delta \otimes I)\Psi(\beta) = (\Delta \otimes I)\left(\sum_{(\beta)} b_{(1)} \otimes \beta_{(2)}\right)$$

$$= \sum_{(\beta, b_{(1)})} b_{(1)(1)} \otimes b_{(2)(2)} \otimes \beta_{(2)}. \qquad (4.4)$$

By Definition 4.3.1(i), $(I \otimes \Psi)\Psi = (\Delta \otimes I)\Psi$, and so the expressions in (4.3) and (4.4) are equal. The common value in (4.3) and (4.4) will be denoted as

$$\sum_{(\beta)} b_{(1)} \otimes b_{(2)} \otimes \beta_{(3)}.$$

Similarly, the common value of

$$(I \otimes I \otimes \Psi)(I \otimes \Psi)\Psi(\beta) = (I \otimes I \otimes \Psi)(\Delta \otimes I)\Psi(\beta)$$

$$= (I \otimes \Delta \otimes I)(\Delta \otimes I)\Psi(\beta)$$

$$= (I \otimes \Delta \otimes I)(I \otimes \Psi)\Psi(\beta)$$

$$= (\Delta \otimes I \otimes I)(I \otimes \Psi)\Psi(\beta)$$

$$= (\Delta \otimes I \otimes I)(\Delta \otimes I)\Psi(\beta)$$

is denoted as

$$\sum_{(\beta)} b_{(1)} \otimes b_{(2)} \otimes b_{(3)} \otimes \beta_{(4)}.$$

Definition 4.3.2. Let H be an R-Hopf algebra. The R-module M is a **left Hopf module over H** if

(i) M is a left H-module,
(ii) M is a left H-comodule with structure map $\Psi : M \to H \otimes_R M$, and
(iii) the left H-comodule structure map $\Psi : M \to H \otimes_R M$ is a left H-module map, where $H \otimes_R M$ is a left H-module with scalar multiplication

$$h \cdot (k \otimes m) = \sum_{(h)} h_{(1)}k \otimes (h_{(2)} \cdot m).$$

Let R be a Dedekind domain, and let H be a cocommutative R-Hopf algebra that is a free R-module of rank n. By Proposition 4.1.7, H^* is an R-Hopf algebra that is R-free of rank n. The main results of this section are to prove that H^* is a left Hopf module over H and that $H^* \cong H \otimes_R \int_{H^*}$ as H-modules. We start by viewing H^* as a left H-module with scalar product defined as

$$(h \cdot \beta)(k) = \beta(\sigma(h)k) = \sum_{(\beta)} \beta_{(1)}(\sigma(h))\beta_{(2)}(k) \qquad (4.5)$$

for $h, k \in H$, $\beta \in H^*$. In fact, H^* is a left H-module algebra (see the following lemma).

Lemma 4.3.1. *Let $h \in H$, $\beta, \alpha \in H^*$. Then*

$$h \cdot (\beta \alpha) = \sum_{(h)} (h_{(1)} \cdot \beta)(h_{(2)} \cdot \alpha).$$

Proof. Since the comultiplication Δ_{H^*} is an R-algebra homomorphism,

$$
\begin{aligned}
(h \cdot (\beta \alpha))(k) &= (\beta \alpha)(\sigma(h)k) \\
&= \sum_{(\beta, \alpha)} (\beta_{(1)} \alpha_{(1)})(\sigma(h))(\beta_{(2)} \alpha_{(2)})(k)
\end{aligned}
$$

for all $k \in H$. Moreover, $\Delta(\sigma(h)) = \sum_{(h)} \sigma(h_{(1)}) \otimes \sigma(h_{(2)})$; thus,

$$
\begin{aligned}
&\sum_{(\beta, \alpha)} (\beta_{(1)} \alpha_{(1)})(\sigma(h))(\beta_{(2)} \alpha_{(2)})(k) \\
&= \sum_{(\beta, \alpha, h)} \beta_{(1)}(\sigma(h_{(1)}))\alpha_{(1)}(\sigma(h_{(2)}))(\beta_{(2)} \alpha_{(2)})(k) \\
&= \sum_{(\beta, \alpha, h, k)} \beta_{(1)}(\sigma(h_{(1)}))\alpha_{(1)}(\sigma(h_{(2)}))\beta_{(2)}(k_{(1)})\alpha_{(2)}(k_{(2)}) \\
&= \sum_{(\beta, \alpha, h, k)} \beta_{(1)}(\sigma(h_{(1)}))\beta_{(2)}(k_{(1)})\alpha_{(1)}(\sigma(h_{(2)}))\alpha_{(2)}(k_{(2)}) \\
&= \sum_{(h, k)} \beta(\sigma(h_{(1)})k_{(1)})\alpha(\sigma(h_{(2)})k_{(2)}) \\
&= \sum_{(h, k)} (h_{(1)} \cdot \beta)(k_{(1)})(h_{(2)} \cdot \alpha)(k_{(2)}) \\
&= \sum_{(h)} ((h_{(1)} \cdot \beta)(h_{(2)} \cdot \alpha))(k),
\end{aligned}
$$

which proves the lemma. $\square$

The left H-module H^* is also a right H^*-module with scalar product taken to be multiplication on the right in H^*. At the same time, one can define a left H-comodule structure on H^* as follows. Let $\{\alpha_i, b_i\}_{i=1}^n$, $\alpha_i \in H^*$, $b_i \in H$, be a dual basis for H, H^*. For $\beta \in H^*$, we have $\beta = \sum_{i=1}^n b_i(\beta)\alpha_i$. Now H^* is a left H-comodule with structure map

$$\Psi : H^* \to H \otimes H^*$$

defined as

$$\Psi(\beta) = \sum_{i=1}^{n} b_i \otimes \beta \alpha_i. \tag{4.6}$$

This H-comodule structure induces a right H^*-module structure on H^* defined as

$$(\beta)\alpha = m(\alpha \otimes I)\Psi(\beta) = \sum_{i=1}^{m} \alpha(b_i)\beta\alpha_i.$$

The surprising result is that the action $(\beta)\alpha$ is precisely the right multiplication action of H^* on itself since

$$(\beta)\alpha = \sum_{i=1}^{m} \alpha(b_i)\beta\alpha_i = \beta \sum_{i=1}^{m} b_i(\alpha)\alpha_i = \beta\alpha.$$

As a consequence of this, one has the useful formula (in Sweedler notation)

$$\beta\alpha = \sum_{(\beta)} \alpha(b_{(1)})\beta_{(2)}. \tag{4.7}$$

So we consider H^* as both a left H-module through (4.5) and a left H-comodule through (4.6). In addition, there is a left H-module structure on $H \otimes H^*$ defined by the comultiplication on H

$$h(k \otimes \beta) = \sum_{(h)} h_{(1)}k \otimes (h_{(2)} \cdot \beta) \tag{4.8}$$

for $h, k \in H$, $\beta \in H^*$.

Proposition 4.3.1. *(Larson and Sweedler) Let R be a Dedekind domain and let H be a cocommutative R-Hopf algebra that is free and of finite rank n over R. Let H^* be an H-module through (4.5), and let $H \otimes H^*$ be an H-module through (4.8). Then the map $\Psi : H^* \to H \otimes H^*$ (as in (4.6)) is a homomorphism of H-modules; that is, H^* is a Hopf module over H.*

Proof. We show that $\Psi(h \cdot \beta) = h\Psi(\beta)$ for $h \in H$, $\beta \in H^*$. Recall that, for $\beta \in H^*$, $\Psi(\beta) = \sum_{i=1}^{n} b_i \otimes \beta\alpha_i$, where $\{\alpha_i, b_i\}_{i=1}^{n}$ is a dual basis for H^*, H. For $\alpha \in H^*$, one has

$$(h \cdot \beta)\alpha = \left(\left(\sum_{(h)} \epsilon(h_{(1)})h_{(2)} \right) \cdot \beta \right) \alpha \qquad \text{counit prop.}$$

$$= \sum_{(h)}(h_{(2)} \cdot \beta)(\epsilon(h_{(1)})1_H \cdot \alpha)$$

$$= \sum_{(h)}(h_{(3)} \cdot \beta)(\sigma(h_{(1)})h_{(2)} \cdot \alpha)$$

by the coinverse property. Now, for $k \in H$,

$$(\sigma(h_{(1)})h_{(2)} \cdot \alpha)(k) = \alpha(h_{(1)}\sigma(h_{(2)})k) \quad \text{by [Ch00, Proposition 1.11]}$$

$$= \alpha_{(1)}(h_{(1)})\alpha_{(2)}(\sigma(h_{(2)})k)$$

$$= \alpha_{(1)}(h_{(1)})(h_{(2)} \cdot \alpha_{(2)})(k).$$

Thus,

$$(h \cdot \beta)\alpha = \sum_{(h,\alpha)} \alpha_{(1)}(h_{(1)})(h_{(3)} \cdot \beta)(h_{(2)} \cdot \alpha_{(2)})$$

$$= \sum_{(h,\alpha)} \alpha_{(1)}(h_{(1)})(h_{(2)} \cdot (\beta\alpha_{(2)})) \quad \text{by Lemma 4.3.1}$$

$$= \sum_{(h,\alpha)} \alpha_{(1)}(h_{(1)}) \left(h_{(2)} \cdot \left(\sum_{j=1}^{n} \alpha_{(2)}(b_j)\beta\alpha_j \right) \right) \quad \text{by (4.7)}$$

$$= \sum_{(h,\alpha)} \alpha_{(1)}(h_{(1)}) \sum_{j=1}^{n} \alpha_{(2)}(b_j)(h_{(2)} \cdot (\beta\alpha_j))$$

$$= \sum_{(h,\alpha)} \sum_{j=1}^{n} \alpha_{(1)}(h_{(1)})\alpha_{(2)}(b_j)(h_{(2)} \cdot (\beta\alpha_j))$$

$$= \sum_{(h)} \sum_{j=1}^{n} \alpha(h_{(1)}b_j)(h_{(2)} \cdot (\beta\alpha_j)).$$

It follows that

$$b_i \otimes (h \cdot \beta)\alpha_i = b_i \otimes \sum_{(h)} \sum_{j=1}^{n} \alpha_i(h_{(1)}b_j)(h_{(2)} \cdot (\beta\alpha_j))$$

$$= \sum_{(h)} \sum_{j=1}^{n} b_i\alpha_i(h_{(1)}b_j) \otimes h_{(2)} \cdot (\beta\alpha_j).$$

Thus

$$\Psi(h \cdot \beta) = \sum_{i=1}^{n} b_i \otimes (h \cdot \beta)\alpha_i$$

$$= \sum_{i=1}^{n} \sum_{(h)} \sum_{j=1}^{n} b_i \alpha_i (h_{(1)} b_j) \otimes (h_{(2)} \cdot (\beta \alpha_j))$$

$$= \sum_{(h)} \sum_{i=1}^{n} \sum_{j=1}^{n} b_i \alpha_i (h_{(1)} b_j) \otimes (h_{(2)} \cdot (\beta \alpha_j))$$

$$= \sum_{(h)} \sum_{i=1}^{n} h_{(1)} b_i \otimes h_{(2)} \cdot (\beta \alpha_i)$$

$$= h \left(\sum_{i=1}^{n} b_i \otimes \beta \alpha_i \right)$$

$$= h\Psi(\beta). \qquad \qquad \square$$

We next show that $H^* \cong H \otimes_R \int_{H^*}$ as H-modules. We begin with the following characterization of $\int_{H^*}$.

Lemma 4.3.2. $\int_{H^*} = W$, where $W = \{\beta \in H^* : \Psi(\beta) = 1 \otimes \beta\}$.

Proof. Let $\beta \in \int_{H^*}$. Then, for all $\alpha \in H^*$, $\alpha\beta = \epsilon_{H^*}(\alpha)\beta = \alpha(1)\beta$. Thus,

$$\alpha_i \beta = \beta \alpha_i = \alpha_i(1)\beta,$$

where $\{\alpha_i, b_i\}$ is a dual basis. Thus,

$$\Psi(\beta) = \sum_{i=1}^{n} b_i \otimes \beta \alpha_i$$

$$= \sum_{i=1}^{n} b_i \otimes \alpha_i(1)\beta$$

$$= \sum_{i=1}^{n} \alpha_i(1) b_i \otimes \beta$$

$$= 1 \otimes \beta,$$

and thus $\beta \in W$. Conversely, if $\beta \in W$, then, for all $\alpha \in H^*$,

$$\alpha\beta = \alpha(1)\beta = \epsilon_{H^*}(\alpha)\beta$$

by (4.7), and so $\beta \in \int_{H^*}$. $\square$

For $\alpha \in H^*$, let

$$\rho(\alpha) = \sum_{(\alpha)} \sigma(a_{(1)}) \cdot \alpha_{(2)}.$$

Lemma 4.3.3. $\Psi(\rho(\alpha)) = 1 \otimes \rho(\alpha)$.

Proof. We have

$$\Psi(\rho(\alpha)) = \Psi\left(\sum_{(\alpha)} \sigma(a_{(1)}) \cdot \alpha_{(2)}\right)$$

$$= \sum_{(\alpha)} \sigma(a_{(1)}) \Psi(\alpha_{(2)}) \quad \text{by Proposition 4.3.1}$$

$$= \sum_{(\alpha)} \sigma(a_{(1)})(a_{(2)} \otimes \alpha_{(3)})$$

$$= \sum_{(\alpha)} \sigma(a_{(2)}) a_{(3)} \otimes (\sigma(a_{(1)}) \cdot \alpha_{(4)}) \quad \sigma \text{ is a Hopf homomorphism}$$

$$= \sum_{(\alpha)} \epsilon(a_{(2)}) 1 \otimes (\sigma(a_{(1)}) \cdot \alpha_{(3)})$$

$$= 1 \otimes \sum_{(\alpha)} \sigma(\epsilon(a_{(2)}) a_{(1)}) \cdot \alpha_{(3)}$$

$$= 1 \otimes \sum_{(\alpha)} \sigma(a_{(1)}) \cdot \alpha_{(2)}$$

$$= 1 \otimes \rho(\alpha). \qquad \square$$

Now, by Lemma 4.3.2, $\rho(H^*) \subseteq \int_{H^*}$. Moreover, we have the following.

Lemma 4.3.4. *For $h \in H$, $\rho(h \cdot H^*) = \epsilon(h)\rho(H^*)$.*

Proof. Let $\alpha \in H^*$. By Proposition 4.3.1,

$$\Psi(h \cdot \alpha) = \sum_{(h,\alpha)} h_{(1)} a_{(1)} \otimes (h_{(2)} \cdot \alpha_{(2)}).$$

Thus

$$\rho(h \cdot \alpha) = \sum_{(h,\alpha)} \sigma(h_{(1)} a_{(1)}) \cdot (h_{(2)} \cdot \alpha_{(2)})$$

$$= \sum_{(h,\alpha)} (\sigma(a_{(1)}) \sigma(h_{(1)}) h_{(2)} \cdot \alpha_{(2)})$$

$$= \epsilon(h) \sum_{(\alpha)} \sigma(a_{(1)}) \cdot \alpha_{(2)}$$

$$= \epsilon(h)\rho(\alpha) \qquad \square$$

Now, $H \otimes_R \int_{H^*}$ is a left H-module with scalar multiplication defined as $h \cdot (k \otimes \alpha) = hk \otimes \alpha$. This is precisely the restriction of the H-module structure of (4.8) to the subset $H \otimes \int_{H^*}$. Indeed, for $h, k \in H, \alpha \in \int_{H^*}$,

$$h \cdot (k \otimes \alpha) = \sum_{(h)} h_{(1)}k \otimes h_{(2)} \cdot \alpha$$

$$= \sum_{(h)} h_{(1)}k \otimes (h_{(2)} \cdot 1_{H^*})\alpha$$

$$= \sum_{(h)} h_{(1)}k \otimes \epsilon_{H^*}(h_{(2)} \cdot 1_{H^*})\alpha$$

$$= \sum_{(h)} h_{(1)}k \otimes (h_{(2)} \cdot 1_{H^*})(1)\alpha$$

$$= \sum_{(h)} h_{(1)}k \otimes \epsilon_H(\sigma_H(h_{(2)}))\alpha \quad \epsilon_H = 1_{H^*}$$

$$= \sum_{(h)} h_{(1)}k \otimes \epsilon_H(h_{(2)})\alpha \quad \text{[Swe69, Proposition 4.0.1]}$$

$$= \sum_{(h)} h_{(1)}\epsilon_H(h_{(2)})k \otimes \alpha$$

$$= \sum_{(h)} hk \otimes \alpha.$$

Since $\rho(H^*) \subseteq \int_{H^*}$, there exists a map $\varrho : H^* \to H \otimes \int_{H^*}$ defined as

$$\varrho(\alpha) = (I \otimes \rho)\Psi(\alpha) = \sum_{(\alpha)} a_{(1)} \otimes \rho(\alpha_{(2)}).$$

Proposition 4.3.2. *(Larson and Sweedler) Let H^* be an H-module through (4.5), and let $H \otimes \int_{H^*}$ be a left H-module with scalar multiplication defined as $h \cdot (k \otimes \alpha) = hk \otimes \alpha$. Then the map $\varrho : H^* \to H \otimes \int_{H^*}$ is an isomorphism of H-modules.*

Proof. We show that ϱ is an H-module homomorphism. Let $h \in H, \alpha \in \int_H^*$. Then

$$\varrho(h \cdot \alpha) = (I \otimes \rho)\Psi(h \cdot \alpha)$$

$$= (I \otimes \rho) \sum_{(h,\alpha)} h_{(1)}a_{(1)} \otimes (h_{(2)} \cdot \alpha_{(2)}) \quad \text{by Prop. 4.3.1}$$

$$= \sum_{(h,\alpha)} h_{(1)} a_{(1)} \otimes \rho(h_{(2)} \cdot \alpha_{(2)})$$

$$= \sum_{(h,\alpha)} h_{(1)} a_{(1)} \otimes \epsilon(h_{(2)}) \rho(\alpha_{(2)}) \quad \text{by Lemma 4.3.4}$$

$$= \sum_{(h,\alpha)} h_{(1)} \epsilon(h_{(2)}) a_{(1)} \otimes \rho(\alpha_{(2)})$$

$$= \sum_{(\alpha)} h a_{(1)} \otimes \rho(\alpha_{(2)})$$

$$= h \sum_{(\alpha)} a_{(1)} \otimes \rho(\alpha_{(2)})$$

$$= h \varrho(\alpha).$$

Next, we define a map $\varphi : H \otimes \int_{H^*} \to H^*$ by

$$\varphi(h \otimes \alpha) = h \cdot \alpha.$$

The map φ is an H-module homomorphism:

$$\varphi(k(h \otimes \alpha)) = \varphi(kh \otimes \alpha)$$

$$= \sum_{(\alpha)} \alpha_{(1)}(\sigma(kh))\alpha_{(2)}$$

$$= \sum_{(\alpha)} \alpha_{(1)}(\sigma(h)\sigma(k))\alpha_{(2)}$$

$$= \sum_{(\alpha)} \alpha_{(1)}(\sigma(h))\alpha_{(2)}(\sigma(k))\alpha_{(3)}$$

$$= k \cdot \sum_{(\alpha)} \alpha_{(1)}(\sigma(h))\alpha_{(2)}$$

$$= k \cdot (h \cdot \alpha)$$

$$= k \cdot \varphi(h \otimes \alpha).$$

We show that ϱ is an isomorphism of H-modules by showing that $\varrho\varphi = I_{H \otimes \int_{H^*}}$ and that $\varphi\varrho = I_{H^*}$. Let $h \in H$, $\alpha \in \int_{H^*}$. We have

$$\varrho\varphi(h \otimes \alpha) = \varrho(h \cdot \alpha)$$

$$= \sum_{(h)} h_{(1)} \otimes \rho(h_{(2)} \cdot \alpha) \quad \text{since } \Psi(\alpha) = 1 \otimes \alpha$$

$$= \sum_{(h)} h_{(1)} \otimes \epsilon(h_{(2)})\rho(\alpha) \quad \text{by Lemma 4.3.4}$$

$$= \sum_{(h)} h_{(1)}\epsilon(h_{(2)}) \otimes \rho(\alpha)$$

$$= h \otimes \rho(\alpha)$$

$$= h \otimes \alpha \quad \text{since } \alpha \in \smallint_{H^*}.$$

Moreover,

$$\varphi\varrho(\alpha) = \varphi\left(\sum_{(\alpha)} a_{(1)} \otimes \rho(\alpha_{(2)})\right)$$

$$= \sum_{(\alpha)} a_{(1)} \cdot (\sigma(a_{(2)}) \cdot \alpha_{(3)})$$

$$= \sum_{(\alpha)} (a_{(1)}\sigma(a_{(2)})) \cdot \alpha_{(3)}$$

$$= \sum_{(\alpha)} \epsilon(a_{(1)})\alpha_{(2)}$$

$$= \alpha,$$

and so the proof of the proposition is complete. $\qquad\square$

We investigate some important consequences of Proposition 4.3.2. We employ the following lemma.

Lemma 4.3.5. *Let R be a PID, let F be a free R-module of rank m, and let M be a submodule of F. Then M is a free R-module of rank $l \le m$.*

Proof. Our proof is by induction on m. Assume that $m = 1$, and let $\{x_1\}$ be an R-basis for F. Let I be defined as

$$I = \{r \in R : rx_1 \in M\}.$$

Then I is an ideal of R, and since R is a PID, $I = Ra$ for some $a \in R$. If $a = 0$, then $M = 0$, and consequently M has rank 0. If $a \ne 0$, then $M = R(ax_1)$, and so M has rank 1.

For the induction hypothesis, we assume that every submodule of a free R-module of rank $m - 1$ is free and of rank $l \le m - 1$. Let F be a free R-module of rank m on the basis $\{x_1, x_2, \ldots, x_m\}$. Let M be a submodule of F. Put

$$N = M \cap (Rx_1 \oplus Rx_2 \oplus \cdots \oplus Rx_{m-1}).$$

Then N is a submodule of the free module $Rx_1 \oplus \cdots \oplus Rx_{m-1}$, and by the induction hypothesis N is free with rank $\leq m - 1$.

Let

$$I = \{r \in R : \; x = r_1x_1 + r_2x_2 + \cdots + r_{m-1}x_{m-1} + rx_m \in M$$

$$\text{for some } r_1, r_2, \ldots, r_{m-1} \in R\}.$$

Then I is an ideal of R such that $I = Ra$ for some $a \in R$. If $a = 0$, then M is a submodule of the free rank $m - 1$ R-module $Rx_1 \oplus \cdots \oplus Rx_{m-1}$, and so, by the induction hypothesis, M is free and of rank $\leq m - 1$.

If $a \neq 0$, let w be an element of M of the form

$$w = r_1x_1 + r_2x_2 + \cdots + r_{m-1}x_{m-1} + ax_m$$

for some $r_1, r_2, \ldots, r_{m-1} \in R$, and let

$$x = s_1x_1 + s_2x_2 + \cdots + s_{m-1}x_{m-1} + rx_m,$$

for $s_1, s_2, \ldots, s_{m-1}, r \in R$, be an element of M. Then there exists $c \in R$ for which $x - cw \in N$; thus

$$M = N + Rw.$$

Evidently, this sum is isomorphic to the direct sum of R-modules

$$R \oplus R \oplus \cdots \oplus R,$$

where the number of summands l satsifies $l \leq m$. We conclude that a submodule M of a free R-module of rank m is free and of rank $l \leq m$. $\qquad \square$

Proposition 4.3.3. *Let R be a PID, and let H be a cocommutative R-Hopf algebra that is free and of rank n over R. Then there exists an integral $\Lambda \in \int_H$ for which $\int_H = R\Lambda$.*

Proof. Since R is a PID and H is a free R-module of rank n, the submodule $\int_H$ is free and of rank m, $m \leq n$ over R. By Proposition 4.3.2,

$$m \cdot \operatorname{rank}(H^*) = \operatorname{rank}(H) = n,$$

and so $m = 1$ since $\operatorname{rank}(H^*) = \operatorname{rank}(H)$. This says that $\int_H$ is a free rank one R-module, and so there exists an integral $\Lambda \in \int_H$ for which $\int_H = R\Lambda$. $\qquad \square$

An integral Λ for which $R\Lambda = \int_H$ is a **generating integral** for H.

Proposition 4.3.4. *Let R be a PID, and let H be a cocommutative R-Hopf algebra that is free and of rank n over R. Then $H \cong H^*$ as H-modules.*

Proof. By Proposition 4.3.3, there exists a generating integral Λ^* for $\int_{H^*}$. Define a map $\theta : H \to H^*$ by

$$\theta(h) = h \cdot \Lambda^* = \sum_{(\Lambda^*)} \Lambda^*_{(1)}(\sigma(h))\Lambda^*_{(2)}.$$

Then

$$
\begin{aligned}
\theta(hk) &= \sum_{(\Lambda^*)} \Lambda^*_{(1)}(\sigma(hk))\Lambda^*_{(2)} \\
&= \sum_{(\Lambda^*)} \Lambda^*_{(1)}(\sigma(k)\sigma(h))\Lambda^*_{(2)} \\
&= \sum_{(\Lambda^*)} \Lambda^*_{(1)}(\sigma(k))\Lambda^*_{(2)}(\sigma(h))\Lambda^*_{(3)} \\
&= \sum_{(\Lambda^*)} \Lambda^*_{(1)}(\sigma(k))(h \cdot \Lambda^*_{(2)}) \\
&= h \cdot (\Lambda^*_{(1)}(\sigma(k))\Lambda^*_{(2)}) \\
&= h \cdot \theta(k),
\end{aligned}
$$

so that θ is H-linear. We claim that θ is surjective. To this end, let $\varphi : H \otimes \int_{H^*} \to H^*$ be the map defined in the proof of Proposition 4.3.2. For $h \in H$, $\alpha \in \int_{H^*}$,

$$\varphi(h \otimes \alpha) = \varphi(h \otimes r\Lambda^*) = r\varphi(h \otimes \Lambda^*) = r(h \cdot \Lambda^*) = r\theta(h). \qquad (4.9)$$

Let $\beta \in H^*$. Since φ is surjective, there exists an element $h \otimes r\Lambda^* \in H \otimes \int_{H^*}$ for which $\varphi(h \otimes r\Lambda^*) = \beta$. Now,

$$
\begin{aligned}
\theta(rh) &= r\theta(h) \\
&= \varphi(h \otimes r\Lambda^*) \quad \text{by (4.9)} \\
&= \beta,
\end{aligned}
$$

and so θ is surjective. Since H and H^* have the same dimension, θ is bijective. Thus θ is an isomorphism of H-modules. $\qquad \square$

Even if R is not a PID, we still have the following important corollary.

Corollary 4.3.1. H^* *is a locally free rank one H-module.*

Proof. Let P denote a prime ideal of R, and let R_P denote the localization of R at P. Let $H_P = R_P \otimes_R H$. Then R_P is a PID by Corollary 4.2.2, and so $H_P \cong (H_P)^*$ as H_P-modules by Proposition 4.3.4. $\qquad \square$

4.4   Hopf Orders

Let R be an integral domain, let K be its field of fractions, and suppose that G is a finite Abelian group of order l. Then the group ring KG is a K-Hopf algebra with comultiplication $\Delta_{KG} : KG \to KG \otimes_K KG$ defined by $g \mapsto g \otimes g$, counit $\epsilon_{KG} : KG \to K$ defined by $g \mapsto 1$, and coinverse $\sigma_{KG} : KG \to KG$ given by $g \mapsto g^{-1}$, for $g \in G$.

Definition 4.4.1. An **R-order in KG** is an R-submodule A of KG (a submodule of KG as an R-module) that satisfies the conditions

(i) A is finitely generated as an R-module,
(ii) A is closed under the multiplication of KG, and
(iii) $KA = KG$.

Proposition 4.4.1. *Let R be a Dedekind domain, and let A be an R-order in KG. Then A is projective as an R-module.*

Proof. Since $A \subseteq KG$ and KG is a torsion-free R-module, A is torsion-free. Thus, by [Rot02, Corollary 11.107], A is projective as an R-module.   □

Definition 4.4.2. An R-order H in KG for which $\Delta_{KG}(H) \subseteq H \otimes H$ is an **R-Hopf order in** KG.

For example, the integral group ring RG is an R-Hopf order in KG. When $G = 1$, then the Hopf order $R1 = R$ in $K1 = K$ is the **trivial Hopf order**.

Proposition 4.4.2. *Let R be a local Dedekind domain, and let H be an R-Hopf order in KG. Then H is free over R of rank $|G|$.*

Proof. By Proposition 4.4.1, H is projective. Since H is also finitely generated, H is free of rank, say m [Ei95, p. 727, Exercise 4.11]. Moreover, $m = |G|$ since $KH = KG$.   □

Proposition 4.4.3. *Let R be a Dedekind domain and H be an R-Hopf order in KG. Then $RG \subseteq H$.*

Proof. Let P be a non-zero prime ideal of R, and let R_P denote the localization. Put $H_P = R_P \otimes_R H$. We show that $R_{PG} \subseteq H_P$.

By Proposition 4.4.2, H_P is free and of rank $l = |G|$. Let $\{b_1, b_2, b_3, \ldots, b_l\}$ be an R_P-basis for H_P. Let $g \in G$. We show that $R_{PG} \subseteq H_P$. Put

$$I = \{r \in R_P : RG \in H_P\}.$$

Since $KH_P = KG$, $g = \sum_{i=1}^{l} u_i b_i$ for elements $u_i \in K$. If each $u_i \in R_p$, then $R_{PG} \subseteq H_P$.

Otherwise, let $u_1', u_2', \ldots, u_k'$, $1 \le k \le l$, denote the subset of $\{u_i\}_{i=1}^{l}$ for which $u_i' \in K \backslash R_P$, and let $b_1', b_2', \ldots, b_k'$ denote the corresponding basis elements. Then

$$g = h + \sum_{i=1}^{k} u_i' b_i'$$

for some $h \in H_P$. Let $v_i = (u_i')^{-1}$. Note that $v_i \in R_P$ for $1 \leq i \leq k$. Let

$$S = \bigcap_{i=1}^{k} R_P v_i.$$

We claim that $S = I$. We first show that $I \subseteq S$. To this end, let $r \in R_P$ be such that $RG \in H_P$. Then

$$rh + r \sum_{i=1}^{k} u_i' b_i' = rh + \sum_{i=1}^{k} r u_i' b_i' \in H_P,$$

and hence $r u_i' \in R_P$ for all i. Put $s_i = r u_i'$. Then $r = s_i v_i$ for $i = 1, \ldots, k$ with $s_i \in R_P$. It follows that $r \in S$.

Next, let $r \in S$. Then $r = s_i v_i$ for elements $s_i \in R_P$, $i = 1, \ldots, k$. Now

$$rg = rh + r \sum_{i=1}^{k} u_i' b_i'$$

$$= rh + \sum_{i=1}^{k} r u_i' b_i'$$

$$= rh + \sum_{i=1}^{k} (s_i v_i) u_i' b_i'$$

$$= rh + \sum_{i=1}^{k} s_i b_i' \in H_P,$$

and so $r \in I$. It follows that $I = S$.

Next, let

$$J = \{r \in R_P : rg \otimes g \subseteq H_P \otimes_{R_P} H_P\},$$

and let

$$T = \bigcap_{i,j=1}^{k} R_P v_i v_j.$$

Then, by an argument analogous to that above, we have $J = T$.

Now, by Corollary 4.2.1, $v_i = a_i \pi^{n_i}$, where π is a uniformizing parameter for R_P, a_i is a unit of R_P, and $n_i \geq 0$ is an integer. Let $n = \min\{n_i\}$. Then S is a principal ideal of R_P of the form (π^n), and T is a principal ideal of the form $T = (\pi^{2n})$. Thus $S^2 = T$ and so $I^2 = J$. Observe that

$$Ig \otimes g = \Delta_{KG}(Ig) \subseteq \Delta_{KG}(H_P) \subseteq H_P \otimes_{R_P} H_P,$$

and so $I \subseteq J = I^2$. Thus, $I = R_P$, $R_P G \subseteq H_P$, and consequently $R_P G \subseteq H_P$. Thus,

$$\bigcap_{P \text{ prime}} R_P G \subseteq \bigcap_{P \text{ prime}} H_P.$$

Now, by [CF67, Chapter I, §3, Lemma 1], $RG = \bigcap\limits_{P \text{ prime}} R_P G$ and $H = \bigcap\limits_{P \text{ prime}} H_P$;

thus $RG \subseteq H$.   □

Of course, an R-Hopf order H in the Hopf algebra KG is itself an R-Hopf algebra.

Proposition 4.4.4. *Let H be an R-Hopf order in KG, where G is a finite Abelian group of order l. Then H is an R-Hopf algebra.*

Proof. Clearly, H is a ring. Let $\lambda_{KG} : K \to KG$ denote the structure map of the K-algebra KG. Since H is an R-submodule of KG, λ_{KG} restricted to H yields the R-module structure of H; hence H is an R-algebra.

Let $\Delta|_H$, $\epsilon|_H$, and $\sigma|_H$ denote the restrictions of Δ_{KG}, ϵ_{KG}, and σ_{KG} to H, respectively. Since $\Delta_{KG}(H) \subseteq H \otimes_R H$, $\Delta|_H$ serves as a comultiplication map of H, and the coassociativity of Δ_{KG} guarantees that $\Delta|_H$ is coassociative. Moreover, $\epsilon_{KG}(H) \subseteq R$, and so $\epsilon|_H$ is a counit map for H, and the counit property of ϵ_{KG} implies that $\epsilon|_H$ satisfies the counit property.

For an integer $q \geq 2$, define $[q] = m_{KG}^{(q-1)} \Delta_{KG}^{(q-1)}$, where the maps $m_{KG}^{(q-1)}$ and $\Delta_{KG}^{(q-1)}$ are defined as in §3.4. Then, for all $g \in G$,

$$\sigma_{KG}(g) = [l-1](g) = g^{l-1} = g^{-1}.$$

Thus

$$\sigma_{KG}(H) = [l-1](H) = m_{KG}^{(l-1)} \left(\Delta_{KG}^{(l-1)}(H) \right) \subseteq H$$

since $m_{KG}(H \otimes_R H) \subseteq H$ and $\Delta_{KG}(H) \subseteq H \otimes_R H$. It follows that $\sigma|_H$ is a coinverse map for H, and the coinverse property of σ_{KG} implies that σ_H satisfies the coinverse property. Thus H is an R-Hopf algebra.   □

As a Hopf algebra, the R-Hopf order H has an ideal of integrals that can be characterized as follows.

Proposition 4.4.5. *Let $\int_H$ be the ideal of integrals of the Hopf order H in KG. Then $\int_H = \epsilon_H(\int_H)e_0$, where $e_0 = \frac{1}{|G|} \sum_{g \in G} g$.*

Proof. Since H is finitely generated, so is the submodule $\int_H$ by Lemma 4.2.2. Let $\{\Lambda_1, \Lambda_2, \ldots, \Lambda_l\}$ be a generating set for $\int_H$. Now each Λ_i is an integral of KG, and so there exist elements $k_i \in K$ for which $\Lambda_i = k_i e_0$ (by Proposition 4.1.4, $\int_{KG} = K e_0$). It follows that $\int_H = J e_0$, where J is the R-submodule of K generated by $\{k_i\}$. Hence $\epsilon_H(\int_H) = \epsilon_{KG}(\int_H) = \epsilon_{KG}(J e_0) = J$, which proves the proposition.   □

Is there a notion of short exact sequence for Hopf orders in KG? Let G be a finite Abelian group, and let J be a subgroup of G. The sequence

$$1 \to J \to G \overset{s}{\to} \overline{G} \to 1$$

is a short exact sequence of groups where s is the canonical surjection $G \to \overline{G} = G/J$, given as $s(g) = g + J$. There is a short exact sequence of K-modules

$$0 \to (KJ)^+ KG \to KG \to K\overline{G} \to 0$$

that passes to the short exact sequence of K-Hopf algebras

$$K \to KJ \to KG \overset{s}{\to} K\overline{G} \to K;$$

see (4.2). Now, let H be a Hopf order in KG. Then $H'' = s(H)$ is an R-Hopf order in $K\overline{G}$, and $H' = H \cap KJ$ is an R-Hopf order in KJ. (Prove these facts as exercises.) We define the sequence

$$R \to H' \to H \to H'' \to R \tag{4.10}$$

to be a **short exact sequence of R-Hopf orders**.

In general, for a fixed finite Abelian group G, there are many Hopf orders in the group ring KG, and classifying Hopf orders is an active area of research.

For the remainder of this section, we assume the following conditions. For an integer $n \geq 1$ and prime $p \in \mathbb{Z}$, let R be a local Dedekind domain containing a primitive p^nth root of unity, which we denote as ζ_{p^n}. Let K be the field of fractions of R. Let $G = C_{p^n}$ denote the cyclic group of order p^n with $\langle g \rangle = C_{p^n}$, and let $\hat{G} = \hat{C}_{p^n}$ be the character group generated by γ with $\gamma^i(g^j) = \langle \gamma^i, g^j \rangle = \zeta_{p^n}^{ij}$. The minimal idempotents of KG are $\{e_i\}$, $0 \leq i \leq p^n - 1$, where

$$e_i = \sum_{m=0}^{p^n-1} \zeta_{p^n}^{-im} g^m.$$

The set $\{e_i\}$ is a K-basis for KG.

Let $\mathrm{tr} : KG \to K$ denote the trace map given as $\mathrm{tr}(x) = \sum_{i=0}^{p^n-1} \gamma^i(x)$, $x \in KG$. The trace map determines a non-degenerate, symmetric, K-bilinear form on KG defined as

$$B(x, y) = \mathrm{tr}(xy).$$

Let H be an R-Hopf order in KG. By Proposition 4.4.2, H is a free R-module of rank p^n.

The **dual module of** H is the R-module defined as

$$H^D = \{x \in KG : B(x, H) \subseteq R\}.$$

In fact, H^D is free over R of rank p^n.

Proposition 4.4.6. *Suppose* $\{b_1, b_2, \dots, b_{p^n}\}$ *is an R-basis for H. There exists a basis* $\{\beta_1, \beta_2, \dots, \beta_{p^n}\}$ *for H^D, "the dual basis," that satisfies* $B(\beta_i, b_j) = \delta_{ij}$.

Proof. Since $KH = KG$, $\{b_i\}$ is also a basis for KG. Let $\{f_1, f_2 \dots, f_{p^n}\}$ be the dual basis for KG^* satisfying $f_i(b_j) = \delta_{ij}$. The bilinear form B has a unique representation

$$B = \sum_{i,j=1}^{K} A_{i,j} B_{i,j},$$

where $a_{i,j} = B(b_i, b_j)$ and where $B_{i,j}$ are bilinear forms defined as $B_{i,j}(\alpha, \beta) = f_i(\alpha) f_j(\beta)$ for $\alpha, \beta \in KG$.

Since B is non-degenerate, $\{B(-, b_j)\}_{j=1}^{p^n}$ is a basis for KG^*. Consequently, the $p^n \times p^n$ matrix $A = (B(b_i, b_j))$ is invertible. Put $A^{-1} = (\theta_{i,j})$, and set

$$\beta_q = \theta_{q,1} b_1 + \theta_{q,2} b_2 + \cdots + \theta_{q,p^n} b_{p^n}$$

for $1 \le q \le p^n$. Then $\{\beta_1, \beta_2, \dots, \beta_{p^n}\}$ is a basis for M^D with

$$B(\beta_q, b_l) = \sum_{i,j=1}^{p^n} a_{i,j} B_{i,j}(\beta_q, b_l)$$

$$= \sum_{i=1}^{p^n} a_{i,l} f_i(\beta_q)$$

$$= \sum_{i=1}^{p^n} a_{i,l} \theta_{q,i} = \delta_{lq}. \qquad \square$$

Proposition 4.4.7. *Let H be an R-Hopf order in KG. Then* $(H^D)^D = H$.

Proof. Exercise. $\qquad \square$

As an example of a dual module, we let $H = RG$. Then H has R-basis $\{1, g, g^2, \dots, g^{p^n-1}\}$. Since $B(e_i, g^j) = \text{tr}(e_i g^j) = \delta_{ij}$, the dual basis for RG^D is $\{e_0, e_1, \dots, e_{p^n-1}\}$. Consequently,

$$RG^D \cong Re_0 \oplus Re_1 \oplus \cdots \oplus Re_{p^n-1}.$$

Since $KG \cong Ke_0 \oplus Ke_1 \oplus \cdots \oplus Ke_{p^n-1}$, RG^D is the maximal R-order in KG.

Not surprisingly, H^D can be identified with H^*. Let $x = \sum_{i=0}^{p^n-1} a_i g^i, a_i \in K$, and let $y = \sum_{j=0}^{p^n-1} b_j e_j, b_j \in K$, be an element of H. Then

$$
\begin{aligned}
B(x, y) &= B\left(\sum_{i=0}^{p^n-1} a_i g^i, \sum_{j=0}^{p^n-1} b_j e_j\right) \\
&= \sum_{i,j=0}^{p^n-1} a_i b_j B(g^i, e_j) \\
&= \sum_{i,j=0}^{p^n-1} a_i b_j \operatorname{tr}(g^i e_j) \\
&= \sum_{i,j=0}^{p^n-1} a_i b_j \zeta_{p^n}^{ij} \\
&= \sum_{i,j=0}^{p^n-1} a_i b_j \gamma^i(g^j) \\
&= \left\langle \sum_{i=0}^{p^n-1} a_i \gamma^i, \sum_{j=0}^{p^n-1} b_j e_j \right\rangle,
\end{aligned}
$$

and so H^D can be identified with the set

$$
H^* = \{x \in K\hat{C}_{p^n} : \langle x, H \rangle \subseteq R\}.
$$

Indeed, $H^D \cong H^*$ as R-modules; the isomorphism is given by $g^i \mapsto \gamma^i$ for $i = 0, \ldots, p^n - 1$. In what follows, we make the identification $H^D = H^*$.

Proposition 4.4.8. *Suppose $\zeta_{p^n} \in K$, and let H be an R-Hopf order in KG. Then H^D is an R-Hopf order in KG.*

Proof. By definition, H^D is an R-submodule of KG and is free over R of finite rank by Proposition 4.4.6. Since $\Delta_{KG} : H \to H \otimes_R H$, $\Delta_{KG}^* : H^* \otimes_R H^* \to H^*$, and so $H^D = H^*$ is closed under the multiplication of KG. Moreover, $KH^D = KG$, and so H^D is an R-order in KG.

So it remains to show that $\Delta_{KG}(H^D) \subseteq H^D \otimes H^D$. But this follows from the fact that H is closed under the multiplication of KG. Thus H^D is an R-Hopf order in KG. $\qquad\square$

Proposition 4.4.9. *Let H be an R-Hopf order in KG. Then there is an embedding $H \to RG^D$ given by $g \mapsto (1, \zeta_{p^n}, \zeta_{p^n}^2, \ldots, \zeta_{p^n}^{p^n-1})$.*

Proof. By Proposition 4.4.8, H^D is a Hopf order in KG. By Proposition 4.4.3, there is an inclusion $\iota : RG \to H^D$, and so there is a dual map $\iota^* : (H^D)^D = H \to RG^D$, given as $\iota^*(g)(f) = g(\iota(f)) = g(f)$. Noting that

$$g = e_0 + \zeta_{p^n} e_1 + \zeta_{p^n}^2 e_2 + \cdots + \zeta_{p^n}^{p^n-1} e_{p^n-1}$$

completes the proof.   □

One can show that an R-order H in KG is an R-Hopf order without explicitly showing that H is invariant under Δ_{KG}. We employ the discriminant of an R-order H in KG. Let $\{b_1, b_2, \ldots, b_{p^n}\}$ be an R-basis for H. Then the **discriminant of H** with respect to the bilinear form $B(x, y) = \mathrm{tr}(xy)$ is defined as

$$\mathrm{disc}(H) = R\det(A),$$

where A is the $p^n \times p^n$ matrix with ijth entry $B(b_i, b_j)$.

Proposition 4.4.10. *Let $n \geq 1$ be an integer, and suppose K contains a primitive p^nth root of unity. Let G be cyclic of order p^n with character group $\hat{G}$. Suppose H and J are R-orders in KG. If $B(J, H) \subseteq R$ and $\mathrm{disc}(H^D) = \mathrm{disc}(J)$, then $J = H^D$ and both H and J are Hopf orders in KG.*

Proof. Since $\langle J, H \rangle \subseteq R$, $J \subseteq H^D$, and so, by [CF67, Chapter I, §3, Corollary 1], $J = H^D$. Since J is an order over R in KG, J is an R-algebra with operations induced from KG. Consequently, comultiplication on KG restricts to give a comultiplication on $J^D = H$, and so H is a Hopf order. It follows that $H^D = J$ is also a Hopf order.   □

Let H be an R-Hopf order in KG. Then H^* is an R-Hopf order in $K\hat{G}$. The integrals of H and H^* are related by the following proposition due to R. Larson.

Proposition 4.4.11. *Let H be an R-Hopf order in KG with integrals $\int_H$, and let H^* be its linear dual, which is an R-Hopf order in $K\hat{G}$ with integrals $\int_{H^*}$. Then*

$$\epsilon_H(\textstyle\int_H)\epsilon_{H^*}(\textstyle\int_{H^*}) = p^n R.$$

Proof. By Corollary 4.2.2, R is a PID, and so, by Proposition 4.3.3 and Proposition 4.3.4, there exists a generating integral Λ^* for H^* and an H-module isomorphism

$$\theta : H \to H^*$$

with $\theta(h)(a) = \sum_{(\Lambda^*)} \Lambda^*_{(1)}(\sigma(h))\Lambda^*_{(2)}(a)$ for $h, a \in H$. Let $h \in H$ be such that $\theta(h) = 1$. Then, for $a \in H$,

$$\theta(h)(a) = 1(a) = a(1) = \epsilon_H(a). \tag{4.11}$$

We claim that $\sigma_H(h)$ is an integral of H. We have, for $a, b \in H$,

$$\theta(\sigma_H(b)h)(a) = \sum_{(\Lambda^*)} \Lambda^*_{(1)}(\sigma_H(h)b)\Lambda^*_{(2)}(a)$$

$$= \sum_{(\Lambda^*)} \Lambda^*_{(1)}(\sigma_H(h))\Lambda^*_{(2)}(b)\Lambda^*_{(3)}(a)$$

$$= \left(\sum_{(\Lambda^*)} \Lambda^*_{(1)}(\sigma_H(h))\right)\Lambda^*_{(2)}(ba)$$

$$= \theta(h)(ba)$$

$$= \epsilon_H(b)\epsilon_H(a) \quad \text{by (4.11)}$$

$$= \epsilon_H(b)\theta(h)(a)$$

$$= \theta(\epsilon_H(b)h)(a),$$

and so $\theta(\sigma_H(b)h) = \theta(\epsilon_H(b)h)$. Since θ is 1-1, $\sigma_H(b)h = \epsilon_H(b)h$, and so

$$\sigma_H(h)b = \sigma_H(\sigma_H(b)h) = \sigma_H(\epsilon_H(b)h) = \epsilon_H(b)\sigma_H(h).$$

Thus, $\sigma_H(h) \in \int_H$. By the definition of θ,

$$1 = \theta(h)(1) = \sum_{(\Lambda^*)} \Lambda^*_{(1)}(\sigma_H(h))\Lambda^*_{(2)}(1) = \sigma_H(h)(\Lambda^*). \tag{4.12}$$

Set $\Lambda = \sigma_H(h)$. Since Λ is an integral for KG, and since KG has a generating integral $p^n e_0 = \sum_{i=0}^{p^n-1} g^i$, we have $\Lambda = qe_0$ for some $q \in K$. Moreover, $\Lambda^* \in \int_{K\hat{G}}$, and so $\Lambda^* = t\iota_0$, since $\iota_0 = \frac{1}{p^n}\sum_{i=0}^{p^n-1} \gamma^i$ is a generating integral for $K\hat{G}$. Now, by (4.12),

$$1 = \Lambda(\Lambda^*) = qtp^n\langle e_0, \iota_0\rangle = qt,$$

and so

$$\epsilon_H(\Lambda)\epsilon_{H^*}(\Lambda^*) = qp^n\epsilon_H(e_0)t\epsilon_{H^*}(\iota_0)$$

$$= qtp^n$$

$$= p^n,$$

which yields $\epsilon_H(\int_H)\epsilon_{H^*}(\int_{H^*}) = p^n R$. □

We will presently give another useful formula. Let J be a subgroup of G, and consider the short exact sequence of K-Hopf algebras

$$K \to KJ \xrightarrow{i} KG \xrightarrow{s} KM \to K, \quad M = G/J.$$

Let H be an R-Hopf order in KG. Then, recalling (4.10), there is a short exact sequence of Hopf orders

$$R \to H' \to H \to H'' \to R.$$

The sequence

$$K \to KM^* = K\hat{M} \xrightarrow{s^*} KG^* = K\hat{G} \xrightarrow{i^*} KJ^* = K\hat{J} \to K$$

is a short exact sequence of K-Hopf algebras, which yields the short exact sequence of R-Hopf orders

$$R \to (H'')^* \to H^* \to (H')^* \to R, \tag{4.13}$$

which is dual to (4.10).

Proposition 4.4.12. *With the notation above,*

$$\epsilon_H(\textstyle\int_H) = \epsilon_{H'}(\textstyle\int_{H'})\epsilon_{H''}(\textstyle\int_{H''}).$$

Proof. Let

$$R \to H' \xrightarrow{i} H \xrightarrow{s} H'' \to R$$

be a short exact sequence of Hopf orders as in (4.10). By Proposition 4.3.3, H' has generating integral Λ', H has generating integral Λ, and H'' has a generating integral Λ''.

Let Θ be such that $s(\Theta) = \Lambda''$. We claim that $i(\Lambda')\Theta \in \int_H$. Let $a \in H$. Then

$$s(a\Theta) = s(a)\Lambda''$$
$$= \epsilon_{H''}(s(a))\Lambda''$$
$$= s(\epsilon_H(a))\Lambda''$$
$$= \epsilon_H(a)\Lambda'',$$

and so $a\Theta = \epsilon_H(a)\Theta + k$, where $k \in i(H'^+)H$. Thus

$$ai(\Lambda')\Theta = \epsilon_H(a)i(\Lambda')\Theta + ki(\Lambda'),$$

where $ki(\Lambda') \in i(H'^+)i(\Lambda')H = 0$ since $b\Lambda' = \epsilon_{H'}(b)\Lambda' = 0$ for $b \in H'^+$. It follows that $i(\Lambda')\Theta \in \int_H$. Now, $i(\Lambda')\Theta = r\Lambda$ for some $r \in R$, and so

$$r\epsilon_H(\Lambda) = \epsilon_H(i(\Lambda'))\epsilon_H(\Theta)$$
$$= i(\epsilon_{H'}(\Lambda'))s(\epsilon_H(\Theta))$$
$$= \epsilon_{H'}(\Lambda')\epsilon_{H''}(\Lambda''),$$

and so

$$\epsilon_{H'}\left(\smallint_{H'}\right)\epsilon_{H''}\left(\smallint_{H''}\right) \subseteq \epsilon_H\left(\smallint_H\right).$$

A similar argument using the short exact sequence (4.13) shows that

$$\epsilon_{(H'')^*}\left(\smallint_{(H'')^*}\right)\epsilon_{(H')^*}\left(\smallint_{(H')^*}\right) = t\epsilon_{(H)^*}\left(\smallint_{(H)^*}\right)$$

for $t \in R$.

Now, by Proposition 4.4.11,

$$\epsilon_{(H')^*}\left(\smallint_{(H')^*}\right) = (\dim(H'))\epsilon_{H'}\left(\smallint_{H'}\right)^{-1},$$

$$\epsilon_{(H'')^*}\left(\smallint_{(H'')^*}\right) = (\dim(H''))\epsilon_{H''}\left(\smallint_{H''}\right)^{-1},$$

and

$$\epsilon_{(H)^*}\left(\smallint_{(H)^*}\right) = (p^n)\epsilon_H\left(\smallint_H\right)^{-1},$$

and so

$$(tp^n)\epsilon_H\left(\smallint_H\right)^{-1} = (\dim(H'))(\dim(H''))\epsilon_{H'}\left(\smallint_{H'}\right)^{-1}\epsilon_{H''}\left(\smallint_{H''}\right)^{-1}.$$

Thus,

$$\epsilon_H\left(\smallint_H\right) \subseteq \epsilon_{H'}\left(\smallint_{H'}\right)\epsilon_{H''}\left(\smallint_{H''}\right),$$

which completes the proof of the proposition. $\qquad\square$

We also have the following.

Proposition 4.4.13. *Let H be an R-Hopf order in KG. Then*

$$\mathrm{disc}(H) = \epsilon_H\left(\smallint_H\right)^{p^n},$$

where $\mathrm{disc}(H)$ is defined with respect to the bilinear form $B(x, y) = \mathrm{tr}(xy)$.

Proof. Let $\theta : H \rightarrow H^*$ be the map defined as $\theta(h) = h \cdot \Lambda^*$, where Λ^* is a generating integral for H^*. Let $\{h_1, h_2, \dots, h_{p^n}\}$ be a basis for H, and let $\{\alpha_1, \alpha_2, \dots, \alpha_{p^n}\}$ be the dual basis for H^*. We have $\alpha_i(b_j) = \delta_{ij}$. For $i = 1, \dots, p^n$, let $k_i \in H$ be such that $\theta(k_i) = \alpha_i$. Then $\{k_1, k_2, \dots, k_{p^n}\}$ is a basis for H with

$$\theta(k_i)(h_j) = \alpha_i(h_j) = \delta_{ij}.$$

For each l, $1 \leq l \leq p^n$, we have $h_l = \sum_{i=1}^{p^n} r_{i,l}k_i$ for some $r_{i,l} \in R$. Now

$$\theta(h_l)(h_j) = \theta\left(\sum_{i=1}^{p^n} r_{i,l}k_i\right)(h_j)$$

$$= \sum_{i=1}^{p^n} r_{i,l}\theta(k_i)(h_j)$$

$$= \sum_{i=1}^{p^n} r_{i,l} \delta_{ij}$$

$$= r_{j,l}.$$

Observe that the $p^n \times p^n$ matrix $(r_{j,l})$ is invertible.

Since $\sum_{a=0}^{p^n-1} \gamma^a$ is an integral of H^*, $b\Lambda^* = \sum_{a=0}^{p^n} \gamma^a$ for some $b \in R$. Now,

$$r_{j,l} = \theta(h_l)(h_j)$$

$$= (h_l \cdot \Lambda^*)(h_j)$$

$$= \frac{1}{b} \left(h_l \cdot \sum_{a=0}^{p^n-1} \gamma^a \right) (h_j)$$

$$= \frac{1}{b} \left(\sum_{a=0}^{p^n-1} h_l \cdot \gamma^a \right) (h_j)$$

$$= \frac{1}{b} \sum_{a=0}^{p^n-1} \gamma^a (h_l h_j)$$

$$= \frac{1}{b} \mathrm{tr}(h_l h_j),$$

which yields $b r_{j,l} = \mathrm{tr}(h_l h_j)$. Thus

$$b^{p^n} \det((r_{j,l})) R = \mathrm{disc}(H),$$

and so

$$\mathrm{disc}(H) = b^{p^n} R,$$

since $\det((r_{j,l}))$ is a unit of R.

Now, by Proposition 4.4.5, $\int_H = \epsilon_H \left(\int_H \right) e_0$, and since R is a PID, $\epsilon_H \left(\int_H \right) = cR$ for some $c \in R$. Hence $\int_H = cRe_0$.

By Proposition 4.4.11, $\epsilon_H \left(\int_H \right) \epsilon_{H^*} \left(\int_{H^*} \right) = p^n R$, and so

$$\epsilon_H (cRe_0) \epsilon_{H^*} (b^{-1} p^n R \iota_0) = p^n R,$$

which yields $cR = bR$. Consequently,

$$\mathrm{disc}(H) = c^{p^n} R = \epsilon_H \left(\int_H \right)^{p^n}.$$

$\square$

4.5   Chapter Exercises

Exercises for §4.1

1. Let $\epsilon : H \to R$ be the counit map of the R-Hopf algebra H. Let $\lambda : R \to H$ denote the R-algebra structure map. Show that the sequence

$$R \xrightarrow{\lambda} H \xrightarrow{\text{id}} H \xrightarrow{\epsilon} R \to R$$

   is short exact.

2. Let A, B be R-Hopf algebras. Verify that the maps

$$(\epsilon_A \otimes \epsilon_B) : A \otimes_R B \to R \otimes_R R := R$$

   and

$$(\sigma_A \otimes \sigma_B) : A \otimes_R B \to A \otimes_R B,$$

   defined as $(\epsilon_A \otimes \epsilon_B)(a \otimes b) = \epsilon_A(a)\epsilon_B(b)$ and $(\sigma_A \otimes \sigma_B)(a \otimes b) = \sigma_A(a) \otimes \sigma_B(b)$, respectively, satisfy the counit and coinverse properties, respectively.

3. Let A be an R-Hopf algebra. Is the comultiplication $\Delta : A \to A \otimes_R A$ a homomorphism of R-Hopf algebras?

4. Let $\psi : E \to F$ be an epimorphism of R-group schemes with $R[E] = A$, $R[F] = B$. Prove that $\phi : B \to A$ is a Hopf injection.

5. Prove that

$$R \xrightarrow{\lambda} R[Y, Y^{-1}] \xrightarrow{i} R[X, X^{-1}] \xrightarrow{s} RC_p \xrightarrow{\epsilon} R$$

   is a short exact sequence of R-Hopf algebras. Here λ is the R-algebra structure map, i is the inclusion given by $Y \mapsto X^p$, s is the quotient map with kernel $(X^p - 1)$, and ϵ is the counit map of the R-Hopf algebra RC_p.

Exercises for §4.2

6. Prove that a PID is a Dedekind domain.

7. Prove that $\mathbb{Z}[i]$ is a Dedekind domain.

8. Give an example of a Noetherian integral domain that is not a Dedekind domain.

9. Let K be a finite extension of $\mathbb{Q}$ with ring of integers R. Let J be a non-trivial proper ideal of R. Show that J can be factored into a product of prime ideals of R

$$J = P_1^{e_1} P_2^{e_2} \cdots P_k^{e_k},$$

   where e_i are positive integers, by completing the following steps in the proof.

   (a) Show that J is contained in a prime ideal P_1 of R.
   (b) Show that there exist elements $\{q_i\}$ in $\mathbb{Q}$ and elements $\{m_i\}$ in J for which $1 = \sum_{i=1}^{l} q_i m_i$.
   (c) Show that there exists an ideal J_1 of R for which $J = J_1 P_1$.
   (d) Repeat the process with J_1 in the role of J to obtain $J = J_2 P_2 P_1$ for some ideal J_2 and some prime ideal P_2.

(e) Show that this process terminates so that $J = P_l P_{l-1} \cdots P_1$, where each ideal P_i is prime, $1 \le i \le l$.

10. Show that the factorization in Exercise 9 is unique up to a reordering of the factors $P_i^{e_i}$.

Exercise for §4.3

11. Suppose $\zeta_p \in R$, and let C_p denote the cyclic group of order p. Then RC_p^* is an RC_p-module with scalar multiplication defined as $(h \cdot \beta)(k) = \sum_{(\beta)} \beta_{(1)}(\sigma_{RC_p}(h))\beta_{(2)}(k)$ for $h, k \in RC_p$, $\beta \in RC_p^*$. Prove that $\{g^i \cdot \iota_0\}_{i=0}^{p-1}$ is an R-basis for RC_p^*, where $\iota_0 = \frac{1}{p}\sum_{j=0}^{p-1} \gamma^j \gamma^i(g^j) = \zeta_p^{ij}$.

Exercises for §4.4

12. Let R be an integral domain that is a local ring with maximal ideal M. Let K be the field of fractions of R. Let $n \ge 1$ be an integer, and assume that $\zeta_{p^n} \in K$. Suppose that H is an R-Hopf order in KC_{p^n} for which $H \ne RC_{p^n}^*$. Prove that H is a local ring with maximal ideal (M, H^+).

13. Let

$$K \to KJ \to KG \xrightarrow{s} K\overline{G} \to K$$

be the short exact sequence of K-Hopf algebras given in §4.4. Let H be an R-Hopf order in KG. Prove that $H'' = s(H)$ is an R-Hopf order in $K\overline{G}$ and that $H' = H \cap KJ$ is an R-Hopf order in KJ.

Chapter 5
Valuations and Larson Orders

Throughout this chapter, G is a finite Abelian group and K is a finite extension of $\mathbb{Q}$. R. Larson [Lar76] has constructed a collection of Hopf orders in KG that are known as Larson orders. At the core of Larson's construction is the notion of valuations on K.

5.1 Valuations

Definition 5.1.1. An **absolute value on** K is a function $| \ | : K \to \mathbb{R}$ that satisfies, for all $a, b \in K$,

(i) $|a| \geq 0$, with $|a| = 0$ if and only if $a = 0$;
(ii) $|ab| = |a||b|$;
(iii) $|a + b| \leq |a| + |b|$.

For example, if $K = \mathbb{Q}$ and p is a prime of $\mathbb{Z}$, then the p-adic absolute value $| \ |_p$ on $\mathbb{Q}$ is an absolute value. (Prove this as an exercise.)

Definition 5.1.2. Let L/K be a field extension, and suppose $\| \ \|$ is an absolute value on L and $| \ |$ is an absolute value on K. Then $\| \ \|$ is an **extension** of $| \ |$ if $\|x\| = |x|$ for all $x \in K$.

Definition 5.1.3. Let $| \ |$ and $| \ |'$ be absolute values on K. Then $| \ |$ and $| \ |'$ are **equivalent** if, for all $x \in K$, the relation $|x| < 1$ implies $|x|' < 1$.

Using Definition 5.1.3, the absolute values on a field can be partitioned into equivalence classes. For the case $K = \mathbb{Q}$, one can easily show that $| \ |_p$ is not equivalent to $| \ |_q$ if $p \neq q$ and that the ordinary absolute value, which we denote by $| \ |_\infty$, is not equivalent to $| \ |_p$ for all p. (Prove these facts as exercises.)

In fact, we have the Following proposition.

R.G. Underwood, *An Introduction to Hopf Algebras*, DOI 10.1007/978-0-387-72766-0_5,   95
© Springer Science+Business Media, LLC 2011

Proposition 5.1.1. *(A. Ostrowski) An absolute value* | | *on* $\mathbb{Q}$ *is equivalent either to* | $|_p$ *for exactly one prime p or to* | $|_\infty$.

Proof. For a proof, see [FT91, Theorem 9]. □

So, up to equivalence, the collection of absolute values on $\mathbb{Q}$ consists of

$$\{ |\ |_2, |\ |_3, |\ |_5, \ldots, |\ |_\infty \}.$$

Let | | denote an absolute value on the field K. Then the sequence $\{a_n\}$ in K is | |**-Cauchy** if for each $\iota > 0$ there exists an integer $N > 0$ for which

$$|a_m - a_n| < \iota$$

whenever $m, n \geq N$. The **completion of K with respect to** | | is the smallest field extension $K_{|\ |}$ of K in which each Cauchy sequence converges to an element of $K_{|\ |}$. Note that | | extends uniquely to an absolute value (also denoted by | |) on the completion.

The completion of $\mathbb{Q}$ with respect to | $|_p$ is the field of p-adic rationals $\mathbb{Q}_p$, and | $|_p$ extends uniquely to an absolute value | $|_p$ on $\mathbb{Q}_p$. The completion of $\mathbb{Q}$ with respect to | $|_\infty$ is $\mathbb{R}$, and | $|_\infty$ extends (uniquely) to the ordinary absolute value | $|_\infty$ on $\mathbb{R}$.

Let $K = \mathbb{Q}_p(\alpha)$ be a simple algebraic extension of $\mathbb{Q}_p$.

Proposition 5.1.2. *The absolute value* | $|_p$ *on* $\mathbb{Q}_p$ *has a unique extension* | $|_p^*$ *to K. Moreover, K is complete with respect to* | $|_p^*$.

Proof. For a proof, see [CF67, Chapter II, §10, Theorem]. □

But what happens if the base field is not complete? Let $K = \mathbb{Q}(\alpha)$ be a simple algebraic extension of $\mathbb{Q}$, and let $r(x)$ be the irreducible polynomial of α of degree $n = \deg(r(x)) = [K : \mathbb{Q}]$. Let p be a prime of $\mathbb{Z}$, and suppose that $\mathbb{Q}$ is endowed with the absolute value | $|_p$. In what follows, we show how to extend | $|_p$ to the larger field K.

Proposition 5.1.3. *There are at most n extensions of* | $|_p$ *to* $\mathbb{Q}(\alpha)$.

Proof. Let $\lambda : \mathbb{Q}(\alpha) \to \mathbb{Q}_p \otimes_{\mathbb{Q}} \mathbb{Q}(\alpha)$ be the map defined by $a \mapsto 1 \otimes a$ for all $a \in \mathbb{Q}(\alpha)$. Let

$$r(x) = \prod_{j=1}^{l} q_j(x)$$

be the factorization of $r(x)$ over $\mathbb{Q}_p$ into distinct irreducible polynomials $q_j(x) \in \mathbb{Q}_p[x]$. For each j, $1 \leq j \leq l$, let α_j be a zero of $q_j(x)$, and put $K_j = \mathbb{Q}_p(\alpha_j)$. There exists a homomorphism of rings

$$\mu_j : \mathbb{Q}_p \otimes_{\mathbb{Q}} \mathbb{Q}(\alpha) \to \mathbb{Q}_p(\alpha_j),$$

defined by

$$s \otimes \sum_{m=0}^{n-1} r_m \alpha^m \mapsto \sum_{m=0}^{n-1} s(r_m/1)\alpha_j^m.$$

For each j, $1 \leq j \leq l$, let

$$\lambda_j : \mathbb{Q}(\alpha) \rightarrow \mathbb{Q}_p(\alpha_j)$$

be the ring homomorphism defined as $\lambda_j = \mu_j \lambda$. Observe that $\ker(\lambda_j)$ is a proper ideal of the field $\mathbb{Q}(\alpha)$, and so $\ker(\lambda_j) = \{0\}$. Thus λ_j is an injection.

Since $\mathbb{Q}_p$ is complete, $| \ |_p$ extends uniquely to an absolute value $| \ |_{p,j}^*$ on $\mathbb{Q}_p(\alpha_j)$ by Proposition 5.1.2. Now, for each j, $1 \leq j \leq l$, we define an extension $\| \ \|_{p,j}$ of $| \ |_p$ to $\mathbb{Q}(\alpha)$ by the rule

$$\|a\|_{p,j} = |\lambda_j(a)|_{p,j}^*$$

for all $a \in \mathbb{Q}(\alpha)$.

Moreover, by [La84, Chapter XII, §3, Proposition 3.1], an arbitrary absolute value $| \ |$ on $\mathbb{Q}(\alpha)$ extending $| \ |_p$ is equivalent to $\| \ \|_{p,j}$ for some j. And by [La84, Chapter XII, §3, Proposition 3.2], each $\| \ \|_{p,j}$ represents a distinct equivalence class of absolute values on $\mathbb{Q}(\alpha)$. Thus, the extensions of $| \ |_p$ to $\mathbb{Q}(\alpha)$ consist of

$$\{\| \ \|_{p,1}, \| \ \|_{p,2}, \ldots, \| \ \|_{p,l}\},$$

with $l \leq n = [\mathbb{Q}(\alpha) : \mathbb{Q}]$. $\qquad\square$

For example, we take $K = \mathbb{Q}(i)$ and compute the extensions of $| \ |_5$ to K. We form the completion $\mathbb{Q}_5$. By Hensel's Lemma, $\mathbb{Q}_5$ contains the primitive 4th root of unity, i, and so the polynomial $x^2 + 1$, which is irreducible over $\mathbb{Q}$, factors as $x^2 + 1 = (x - i)(x + i)$ over $\mathbb{Q}_5$. Thus there are injections

$$\lambda_1 : \mathbb{Q}(i) \rightarrow \mathbb{Q}_5, \ i \mapsto i;$$

$$\lambda_2 : \mathbb{Q}(i) \rightarrow \mathbb{Q}_5, \ i \mapsto -i.$$

In this case, $| \ |_{5,1}^* = | \ |_{5,2}^* = | \ |_5$. Thus, there are two absolute values on K that extend $| \ |_5$, defined for $a + bi \in \mathbb{Q}(i)$ as

$$\|a + bi\|_{5,1} = |a + bi|_5$$

and

$$\|a + bi\|_{5,2} = |a - bi|_5.$$

We can recover the absolute values given in Proposition 5.1.3 in another way. Let K be a finite extension of $\mathbb{Q}$ with ring of integers R, and let p be a prime of $\mathbb{Z}$. Since R is a Dedekind domain, there is a unique factorization

$$(p) = pR = P_1^{e_1} P_2^{e_2} \cdots P_g^{e_g}$$

of (p) into distinct prime ideals of R.

Proposition 5.1.4. *Each prime factor P_i in the factorization of (p) determines an absolute value on K that extends $|\ |_p$. Moreover, if $P_i \neq P_j$, then the absolute value determined by P_i is not equivalent to the absolute value given by P_j.*

Proof. Let $q = \frac{s}{t}$, for $s \in R$, $t \in R^\times$, be an element of K. Let v be the integer for which $(s) \subseteq P_i^v$, $(s) \nsubseteq P_i^{v+1}$, and let w be such that $(t) \subseteq P_i^w$, $(t) \nsubseteq P_i^{w+1}$. Now define

$$[q]_{p,i} = \frac{p^{w/e_i}}{p^{v/e_i}} = \frac{1}{p^{(v-w)/e_i}} = \left(\frac{1}{p^{1/e_i}}\right)^{v-w}.$$

Then $[\ ]_{p,i}$ is an absolute value on K that extends $|\ |_p$.

Now suppose $P_i \neq P_j$. There exists an element $s \in R$ for which $s \in P_i$, $s \notin P_j$. Thus $\operatorname{ord}_{P_i}(s) = v > 0$, while $\operatorname{ord}_{P_j}(s) = 0$. Thus $[s]_{p,i} < 1$, $[s]_{p,j} = 1$, and so, by Definition 5.1.3, $[\ ]_{p,i} \not\approx [\ ]_{p,j}$. $\square$

The absolute value $[\ ]_{p,i}$ is **discrete** since the set

$$\{\log_N([x]_{p,i}) :\ x \in K^\times\}, \quad N = p^{1/e_i},$$

is the additive subgroup $\mathbb{Z}$ of the real numbers. We shall write the completion $K_{[\ ]_{p,i}}$ as K_{P_i}.

The P_i-**order** of q, denoted by $\operatorname{ord}_{P_i}(q)$, is defined to be the integer $v - w$ in the definition of $[q]_{p,i}$, and the function $\operatorname{ord}_{P_i} :\ K \to \mathbb{Z} \cup \infty$, $q \mapsto \operatorname{ord}_{P_i}(q)$, is the **discrete valuation of K corresponding to the prime ideal** P_i.

We will show that representatives of equivalence classes of absolute values on K extending $|\ |_p$ can be chosen to be discrete absolute values. We need a lemma.

Lemma 5.1.1. *Let $\mu_j :\ \mathbb{Q}_p \otimes_\mathbb{Q} \mathbb{Q}(\alpha) \to \mathbb{Q}_p(\alpha_j)$, $1 \otimes \alpha \mapsto \alpha_j$, be the homomorphism of rings defined in the proof of Proposition 5.1.3. Then there is an isomorphism*

$$\mu : \mathbb{Q}_p \otimes_\mathbb{Q} \mathbb{Q}(\alpha) \to \bigoplus_{j=1}^{l} \mathbb{Q}_p(\alpha_j), \qquad (5.1)$$

defined as $\mu(x) = \bigoplus_{j=1}^{l} \mu_j(x)$.

Proof. Let $y = \sum_{i=0}^{n}(q_i \otimes \beta_i)$ be an element of $\mathbb{Q}_p \otimes_\mathbb{Q} \mathbb{Q}(\alpha)$. Then $y = h(1 \otimes \alpha)$, where $h(x)$ is a polynomial in $\mathbb{Q}_p[x]$. If $r(x) \in \mathbb{Q}[x]$ is the irreducible polynomial for α, then $r(1 \otimes \alpha) = 1 \otimes r(\alpha) = 0$. Now, if $\mu(y) = 0$, then $q_j(x)$ divides $h(x)$ for all j. Thus, $r(x) = \prod_{j=1}^{l} q_j(x)$ divides $h(x)$, and so $y = h(1 \otimes \alpha) = 0$. Thus μ is an injective ring homomorphism.

Now, as vector spaces over $\mathbb{Q}_p$, $\dim(\mathbb{Q}_p \otimes_\mathbb{Q} \mathbb{Q}(\alpha)) = \dim(\bigoplus_{j=1}^{l} \mathbb{Q}_p(\alpha_j))$, and so μ is surjective. Thus μ is a ring isomorphism. $\square$

Proposition 5.1.5. *The collection of prime factors $\{P_i\}_{i=1}^{g}$ in the factorization of (p) is in a 1-1 correspondence with the equivalence classes of absolute values on K that extend $|\ |_p$.*

Proof. Define a map $\Phi : \{P_i\}_{i=1}^{g} \to \{\|\ \|\}_{j=1}^{l}$, where $\Phi(P_i)$ is the representative of the equivalence class containing the discrete absolute value $[\ ]_{p,i}$. If $\Phi(P_i) = \Phi(P_k)$ then $[\ ]_{p,i}$ is equivalent to $[\ ]_{p,j}$, which can only happen if $i = j$. Thus, Φ is an injection.

By Lemma 5.1.1, there is a ring isomorphism $\mu : \mathbb{Q}_p \otimes_{\mathbb{Q}} \mathbb{Q}(\alpha) \to \bigoplus_{j=1}^{l} \mathbb{Q}_p(\alpha_j)$. Put

$$\eta_1 = \mu^{-1}(1,0,0\ldots,0), \ \eta_2 = \mu^{-1}(0,1,0,\ldots,0),\ldots,\eta_l = \mu^{-1}(0,0,0,\ldots,1).$$

Then $\{\eta_i\}$ is the set of primitive idempotents of $\mathbb{Q}_p \otimes_{\mathbb{Q}} \mathbb{Q}(\alpha)$. By [FT91, Chapter III, §1, Theorem 17], the set $\{\eta_i\}_{i=1}^{l}$ is in a 1-1 correspondence with the set of prime ideals $\{P_i\}_{i=1}^{g}$. Consequently, $l = g$, and so Φ is a bijection. $\square$

We have not discussed how the absolute value $|\ |_\infty$ extends to a simple field extension $K = \mathbb{Q}(\alpha)$.

Proposition 5.1.6. *Let $n = [K : \mathbb{Q}]$. There are at most n extensions of $|\ |_\infty$ to K.*

Proof. The proof is similar to the proof of Proposition 5.1.3. Let $\lambda : \mathbb{Q}(\alpha) \to \mathbb{R} \otimes_{\mathbb{Q}} \mathbb{Q}(\alpha)$ be the map defined by $a \mapsto 1 \otimes a$ for $a \in \mathbb{Q}(\alpha)$. Let

$$r(x) = \prod_{j=1}^{l} q_j(x)$$

be the factorization of $r(x)$ over $\mathbb{R}$ into distinct irreducible polynomials $q_j(x)$. For each j, $1 \le j \le l$, let α_j be a zero of $q_j(x)$ and put $K_j = \mathbb{R}(\alpha_j)$. Of course, either $\mathbb{R}(\alpha_j) = \mathbb{R}$ or $\mathbb{R}(\alpha_j) = \mathbb{C}$.

For each j, $1 \le j \le l$, there exists a homomorphism of rings

$$\mu_j : \mathbb{R} \otimes_{\mathbb{Q}} \mathbb{Q}(\alpha) \to \mathbb{R}(\alpha_j),$$

defined by

$$s \otimes \sum_{m=0}^{n-1} r_m \alpha^m \mapsto \sum_{m=0}^{n-1} s r_m \alpha_j^m,$$

and an injection

$$\lambda_j : \mathbb{Q}(\alpha) \to \mathbb{R}(\alpha_j),$$

given as $\lambda_j = \mu_j \lambda$. Since $\mathbb{R}$ is complete, $|\ |_\infty$ extends uniquely to an absolute value on $\mathbb{R}(\alpha_j)$ that is either the ordinary absolute value $|\ |$ on $\mathbb{R}$ or $\mathbb{C}$. Now, for each j, $1 \le j \le l$, we can define an extension $\|\ \|_{\infty,j}$ of $|\ |_\infty$ to $K = \mathbb{Q}(\alpha)$ as

$$\|a\|_{\infty,j} = |\lambda_j(a)|$$

for all $a \in \mathbb{Q}(\alpha)$. By [La84, Chapter XII, §3], the equivalence classes of extensions of $| \; |_\infty$ to K have representatives

$$\{\| \; \|_{\infty,1}, \| \; \|_{\infty,2}, \ldots, \| \; \|_{\infty,l}\},$$

with $l \leq n = [K : \mathbb{Q}]$. $\square$

An extension $\| \; \|_{\infty,j}$ of $| \; |_\infty$ is an **Archimedean absolute value**. Analogous to (5.1), one has a ring isomorphism

$$\mu : \mathbb{R} \otimes_\mathbb{Q} \mathbb{Q}(\alpha) \cong \bigoplus_{i=1}^{s} \mathbb{R} \oplus \bigoplus_{j=1}^{t} \mathbb{C}, \qquad (5.2)$$

defined as $\mu(x) = \bigoplus_{i=1}^{s} \mu_i(x) \oplus \bigoplus_{j=1}^{t} \mu_j(x)$, where μ_i, $1 \leq i \leq s$, are the real embeddings and μ_j, $1 \leq j \leq t$, are the complex embeddings.

5.2 Group Valuations

Definition 5.2.1. Let G be a finite group (not necessarily Abelian!). A **group valuation** is a function $\xi : G \rightarrow \mathbb{Z} \cup \infty$ that satisfies, for all $g, h \in G$,

(i) $\xi(g) \geq 0$,
(ii) $\xi(g) = \infty$ if and only if $g = 1$,
(iii) $\xi(gh) \geq \min\{\xi(g), \xi(h)\}$, and
(iv) $\xi(ghg^{-1}h^{-1}) \geq \xi(g) + \xi(h)$.

Let us review some basic properties of the group valuation ξ on G.

Proposition 5.2.1. *For all $g, h \in G$,*

(i) $\xi(g) = \xi(g^{-1})$ and
(ii) $\xi(ghg^{-1}) = \xi(h)$.

Proof. To prove (i), let $s = |\langle g \rangle|$, and observe that $g^{s-1} = g^{-1}$. Since $g = (g^{s-1})^{s-1}$, Definition 5.2.1(iii) implies that

$$\xi(g) = \xi((g^{s-1})^{s-1}) \geq \xi(g^{s-1}) = \xi(g^{-1}).$$

On the other hand, $\xi(g^{-1}) = \xi(g^{s-1}) \geq \xi(g)$.
For (ii), one has, for all $h, g \in G$,

$$\begin{aligned}
\xi(ghg^{-1}) &= \xi(ghg^{-1}h^{-1}h) \\
&\geq \min\{\xi(ghg^{-1}h^{-1}), \xi(h)\} \quad \text{by Def. 5.2.1(iii)} \\
&\geq \min\{\xi(g) + \xi(h), \xi(h)\} \quad \text{by Def. 5.2.1(iv)} \\
&\geq \xi(h).
\end{aligned}$$

Now, with $k = ghg^{-1}$,

$$\xi(h) = \xi(g^{-1}kg) \geq \xi(k) = \xi(ghg^{-1}) \geq \xi(h),$$

and thus $\xi(h) = \xi(ghg^{-1})$. □

Let $\mathbb{Z}_{\geq 0} = \{r \in \mathbb{Z} : r \geq 0\}$, and let ξ be a group valuation on G. For each $r \in \mathbb{Z}_{\geq 0}$, put

$$G_r = \{g \in G : \xi(g) \geq r\}.$$

Proposition 5.2.2. *For each $r \in \mathbb{Z}_{\geq 0}$, G_r is a normal subgroup of G.*

Proof. We first show that G_r is a subgroup of G. Since $\xi(1) = \infty > r$, $1 \in G_r$. Let $h_r, g_r \in G_r$. Then $\xi(h_r g_r) \geq \min\{\xi(h_r), \xi(g_r)\} \geq r$, and so G_r is closed under the binary operation of G. Also, $\xi(h_r^{-1}) = \xi(h_r) \geq r$, and so $h^{-1} \in G_r$. Thus $G_r \leq G$.

Now, let $gh_r \in gG_r$. Then $\xi(gh_r g^{-1}) = \xi(h_r) \geq r$ by Proposition 5.2.1(ii), and so $gh_r g^{-1} \in G_r$. Thus $gG_r \subseteq G_r g$. Next, let $h_r g \in G_r g$. Then $\xi(g^{-1}h_r g) = \xi(h_r) \geq r$, and so $g^{-1}h_r g \in G_r$. Thus, $G_r g \subseteq gG_r$. □

Let ξ be a group valuation on G with $|G| \geq 2$. Put $m = \max\{\xi(g) : g \in G, g \neq 1\}$.

Proposition 5.2.3. *There is a normal series*

$$1 \subset G_m \subseteq G_{m-1} \subseteq \cdots \subseteq G_1 \subseteq G_0 = G,$$

which can be refined to a composition series

$$1 \subset N_s \subset N_{s-1} \subset \cdots \subset N_1 \subset N_0 = G.$$

Proof. Note that $G_0 = G$, $1 = G_{m+1}$, and $1 \subset G_m$ since there exists at least one $g \in G$ with $\infty > \xi(g) \geq m$. Now, by Proposition 5.2.2, $G_j \triangleleft G$ for $0 \leq j \leq m$, and thus the series

$$1 \subset G_m \subseteq G_{m-1} \subseteq \cdots \subseteq G_1 \subseteq G_0 = G \qquad (5.3)$$

is normal.

We first remove all repeats from the series (5.3) and renumber the indices (if necessary) to obtain the normal series

$$1 \subset G_n \subset G_{n-1} \subset \cdots \subset G_1 \subset G_0 = G. \qquad (5.4)$$

Let i be an index, $0 \leq i \leq n - 1$, for which $G_{i+1} \subset G_i$. Let $Q_i = G_i/G_{i+1}$. If Q_i has no proper non-trivial normal subgroups (that is, if Q_i is simple), then set $N_i = G_i$ and $N_{i+1} = G_{i+1}$, and replace $G_{i+1} \subset G_i$ in the normal series (5.4) with $N_{i+1} \subset N_i$. If Q_i has a proper non-trivial normal subgroup, say M', then necessarily there exists a subgroup M of G_i with $G_{i+1} \subset M \subset G_i$. Set $N_{i-1} = G_i$,

$N_i = M$, and $N_{i+1} = G_{i+1}$, and replace $G_{i+1} \subset G_i$ in the normal series (5.4) with $N_{i+1} \subset N_i \subset N_{i-1}$.

We now repeat this process with the proper containments $N_{i+1} \subset N_i$ and $N_i \subset N_{i-1}$, replacing these if necessary, with refinements as above. Eventually we obtain a normal series

$$1 \subset N_s \subset N_{s-1} \subset \cdots \subset N_1 \subset N_0 = G,$$

which contains the normal series (5.4) and for which each quotient N_j / N_{j+1} is simple, that is, we obtain the composition series for G.   □

The properties given above help to determine the group valuations on a given group. For example, we classify all group valuations on D_3, the dihedral group of order 6, which has group presentation

$$D_3 = \langle a, b : a^3 = 1, b^2 = 1, bab = a^{-1} \rangle.$$

Let $\xi : D_3 \to \mathbb{Z} \cup \infty$ be a group valuation. Then $\xi(1) = \infty$. The non-trivial conjugacy classes of D_3 are $\{a, a^2\}$ and $\{b, ab, a^2b\}$. Thus, by Proposition 5.2.1(ii), we have $\xi(a) = \xi(a^2) = r$ and $\xi(b) = \xi(ab) = \xi(a^2b) = s$ for some $r, s \in \mathbb{Z}_{\geq 0}$. Moreover, since $baba^{-1} = a$, we have

$$r = \xi(a) = \xi(baba^{-1}) \geq \xi(b) + \xi(a) = s + r$$

by Definition 5.2.1(iv). Thus $r \geq s = 0$.

If $r = s = 0$, then we have the trivial group valuation on D_3: $\xi_0(g) = 0$ for all $g \in D_3, g \neq 1$.

If $r > 0$ and $s = 0$, then we have the group valuation ξ_r with $\xi_r(a) = \xi_r(a^2) = r$ and $\xi_r(b) = \xi_r(ab) = \xi_r(a^2b) = 0$. Thus ξ_0 and ξ_r, $r > 0$, are the only group valuations on D_3.

We consider ξ_r with $r = 3$. Then $\max\{\xi_3(g) : g \in G, g \neq 1\} = 3$ and $G_0 = D_3$, $G_1 = G_2 = G_3 = \{1, a, a^2\}$, and $G_4 = 1$, so the normal series is

$$1 \subset \{1, a, a^2\} \subset D_3,$$

which is the composition series for D_3.

In the case where G is a p-group, we have the following proposition.

Proposition 5.2.4. *Let ξ be a group valuation on G, where G is a p-group of order p^n. Then ξ has at most n distinct finite values.*

Proof. Let

$$1 \subset G_m \subseteq G_{m-1} \subseteq \cdots \subseteq G_1 \subseteq G_0 = G$$

be the normal series determined by ξ, and let

$$1 \subset N_s \subset N_{s-1} \subset \cdots \subset N_1 \subset N_0 = G$$

be the composition series. Then $N_i/N_{i+1} \cong C_p$ for $i = 0, 1, \ldots, s$ (here $N_{s+1} = 1$). Note that $s = n - 1$. For each i, $0 \leq i \leq s$, there is a subseries

$$G_{r+1} \subseteq N_{i+1} \subset N_i \subseteq G_r.$$

Let $g \in N_i \backslash N_{i+1}$. Since $g \in G_r$, $\xi(g) \geq r$. But since $g \notin N_{i+1}$, $g \notin G_{r+1}$, and thus $\xi(g) \not\geq r + 1$, and so $\xi(g) = r$. Put $r_i = r$, and let $\{r_i\}$ denote the collection of the r_i as i ranges from 0 to s.

Since each $g \in G$, $g \neq 1$, is in $N_i \backslash N_{i+1}$ for exactly one i, the collection $\{r_i\}_{i=0}^s$ constitutes the values of ξ. At most n of the values in $\{r_i\}_{i=0}^s$ are distinct since $s + 1 = n$.   □

Let K be a simple algebraic extension of $\mathbb{Q}$. How do group valuations relate to discrete valuations on K? Let p be a fixed prime of $\mathbb{Z}$, let P be a prime in the factorization of (p), and let $\operatorname{ord}_P : K \to \mathbb{Z} \cup \infty$ be the corresponding discrete valuation on K. For an ideal I of R, one has $\operatorname{ord}_P(I) = v$, where v is the integer for which $I \subseteq P^v$, $I \nsubseteq P^{v+1}$. We set $e = \operatorname{ord}_P((p))$.

Definition 5.2.2. Let G be a finite group. The group valuation $\xi : G \to \mathbb{Z} \cup \infty$ is *order bounded with respect to* ord_P if

(i) $\xi(G) \subseteq \operatorname{ord}_P(K)$,
(ii) $\xi(g) = 0$ for $|g|$ not a power of p, and
(iii) $\xi(g) \leq e/(p^a - p^{a-1})$ for $|g| = p^a$, $a \geq 1$.

An order-bounded group valuation (obgv) ξ on G is *p*-**adic** if $\xi(g^p) \geq p\xi(g)$ for all $g \in G$.

Counting all of the p-adic order-bounded group valuations on a given group is an interesting problem. For example, let $K = \mathbb{Q}(\zeta_{p^3})$, where ζ_{p^3} is a primitive p^3rd root of unity. The ring of integers of K is $R = \mathbb{Z}[\zeta_{p^3}]$, and one has $(p) = P^{p^2(p-1)}$, with $P = (1 - \zeta_{p^3})$. Let ord_P denote the discrete valuation corresponding to P. Then $e = \operatorname{ord}_P((p)) = p^2(p - 1)$.

Let C_{p^3} denote the cyclic group of order p^3, and let $\xi : C_{p^3} \to \mathbb{Z}$ denote a p-adic obgv on C_{p^3} with respect to ord_P. By Proposition 5.2.4, ξ has at most three distinct finite values $\{r_0, r_1, r_2\}$, satisfying $r_{i+1} \geq r_i \geq 0$ for $i = 0, 1$. Since ξ is order bounded,

$$0 \leq r_i = \xi(g^{p^i}) \leq \frac{e}{p^{2-i}(p-1)} = p^i,$$

and so

$$0 \leq r_0 \leq 1, \quad 0 \leq r_1 \leq p, \quad 0 \leq r_2 \leq p^2.$$

Now, since ξ is a p-adic obgv on C_{p^3}, we have the additional condition

$$r_2 \geq pr_1 \geq p^2 r_0,$$

and from this one finds that there are

$$\frac{p^2(p+1)}{2} + p + 2$$

p-adic obgvs on C_{p^3}.

For more results on counting group valuations, see the paper [KM07] by A. Koch and A. Malagon.

5.3 Larson Orders

Let K be a finite extension of $\mathbb{Q}$ with ring of integers R, let p be a prime of $\mathbb{Z}$, and let P be a prime ideal of R that lies above p; that is, P is in the prime factorization of p. Let ord_P denote the corresponding discrete valuation on K. Then $\mathrm{ord}_P(P) = 1$ and $\mathrm{ord}_P(P^r) = r$ for an integer $r \geq 0$.

Let

$$P^{-r} = (P^r)^{-1} = \{x \in K : xP^r \subseteq R\}.$$

Then, for $y \in P^r$, $y \neq 0$, $yP^{-r} \subseteq R$, and so $P^{-r} \subseteq y^{-1}R$. Since $y^{-1}R$ is a finitely generated module over the Dedekind domain R, P^{-r} is finitely generated by Lemma 4.2.2.

R. Larson [Lar76] has shown that order-bounded group valuations on G give rise to Hopf orders in KG.

Proposition 5.3.1. *Let p be prime, let P be a prime ideal of R that lies above p, let ord_P be the corresponding valuation on K, let G be a finite Abelian group, and let ξ be a group valuation on G that is order bounded with respect to ord_P. Let $A(\xi)$ be the R-subalgebra of KG generated by $P^{-\xi(g)}(g-1)$ for $g \neq 1$. Then $A(\xi)$ is an R-Hopf order in KG.*

Proof. We sketch Larson's proof. Let S denote the set of elements of G that have order p^a for some $a > 0$. The elements of S can be ordered by the rule $g \leq h$ if and only if $\xi(g) \leq \xi(h)$. We have the list of elements of S

$$g_1 \leq g_2 \leq g_3 \leq \cdots \leq g_l.$$

For each i, $1 \leq i \leq l$, let A_{g_i} be the R-subalgebra of $K\langle g_i \rangle$ of the form

$$A_{g_i} = R[P^{-\xi(g_i)}(g_i - 1)].$$

Since $P^{-\xi(g_i)}$ is finitely generated, each A_{g_i} is a finitely generated R-module, and thus the direct sum $\bigoplus_{i=1}^{l} A_{g_i}$ is finitely generated. There exists a surjective map of R-modules

$$s : \bigoplus_{i=1}^{l} A_{g_i} \to A(\xi),$$

which shows that $A(\xi)$ is finitely generated.

One easily verifies the remaining conditions of Definition 4.4.1, and so $A(\xi)$ is an R-order in KG. Also, since $\Delta_{KG}(g) = g \otimes g$ for all $g \in G$,

$$\Delta_{KG}\left(\frac{g-1}{P^{\xi(g)}}\right) = \frac{g-1}{P^{\xi(g)}} \otimes 1 + 1 \otimes \frac{g-1}{P^{\xi(g)}} + (g-1) \otimes \frac{g-1}{P^{\xi(g)}},$$

and so

$$\Delta_{KG}(A(\xi)) \subseteq A(\xi) \otimes_R A(\xi).$$

Thus, $A(\xi)$ is an R-Hopf order in KG. □

A Hopf order H in KG is a **Larson order in** KG if H is of the form $A(\xi)$, where ξ is a p-adic obgv.

We consider Larson orders in KG where $G = C_{p^n}$ is the cyclic group of order p^n generated by g. We consider the case $n = 1$. To characterize a Larson order in KC_p, we need only describe the possible p-adic obgvs on C_p. By Proposition 5.2.4, an obgv ξ on C_p has exactly one finite value: we have $\xi(g) = i$ for $g \neq 1$. Since ξ is order bounded, $0 \leq i \leq e'$, where $e' = e/(p-1)$. Since $\xi(g^p) = \xi(1) = \infty \geq \xi(g)$, ξ is p-adic. The Larson order $H = A(\xi)$ is written

$$H = A(\xi) = R\left[\frac{g-1}{P^i}, \frac{g^2-1}{P^i}, \dots, \frac{g^{p-1}-1}{P^i}\right].$$

Proposition 5.3.2. $H = R\left[\dfrac{g-1}{P^i}\right].$

Proof. Clearly, $R\left[\frac{g-1}{P^i}\right] \subseteq A(\xi)$, so we prove the reverse containment. For $a = 2, 3, \dots, p-1$, one has

$$g^a - 1 = (g-1)^a + \binom{a}{1}(g-1)^{a-1} + \binom{a}{2}(g-1)^{a-2} + \cdots + \binom{a}{a-1}(g-1),$$

and so

$$\frac{g^a - 1}{P^i} \subseteq P^{i(a-1)}\left(\frac{g-1}{P^i}\right)^a + \binom{a}{1}P^{i(a-2)}\left(\frac{g-1}{P^i}\right)^{a-1}$$

$$+ \cdots + \binom{a}{a-1}P^i\left(\frac{g-1}{P^i}\right).$$

Thus $\frac{g^a-1}{P^i} \subseteq R\left[\frac{g-1}{P^i}\right]$, which proves the proposition. □

We shall denote the Larson order of Proposition 5.3.2 by $H(i)$ since it depends on only one valuation parameter i, $0 \le i \le e'$.

Proposition 5.3.3. *Let $H(i)$ be a Larson order in KC_p. Let $e = \mathrm{ord}_P((p))$. Then*

$$\mathrm{ord}_P(\epsilon_{H(i)}(\smallint_{H(i)})) = e - (p-1)i.$$

Proof. We localize at the prime ideal P. Then R_P is an integral domain with $K = \mathrm{Frac}(R_P)$. Over R_P, the Hopf order $H(i) = R\left[\frac{g-1}{P^i}\right]$ is written $H(i)_P = R_P\left[\frac{g-1}{P^i R_P}\right]$.

By Corollary 4.2.2, R_P is a PID, and so there is an element $\pi \in R_P$ for which $(\pi) = PR_P$. Consequently, $H(i)_P = R_P\left[\frac{g-1}{\pi^i}\right]$. Moreover, since $H(i)_P$ is a finitely generated and projective module over the local ring R_P, $H(i)_P$ is free over R_P of rank p. Thus,

$$\left\{ 1, \frac{g-1}{\pi^i}, \left(\frac{g-1}{\pi^i}\right)^2, \ldots, \left(\frac{g-1}{\pi^i}\right)^{p-1} \right\} \tag{5.5}$$

is an R_P-basis for $H(i)_P$.

With respect to this basis, we compute $\mathrm{disc}(H(i)_P/R_P)$. Observe that the matrix

$$M = \begin{pmatrix} 1 & 0 & 0 & \cdots & & 0 \\ 0 & \pi^i & 0 & & & 0 \\ 0 & 0 & \pi^{2i} & & & \vdots \\ & & & & & 0 \\ 0 & 0 & \cdots & & 0 & \pi^{(p-1)i} \end{pmatrix}$$

multiplies the basis (5.5) to yield a basis for $R_P C_p$. One has $(\det(M)) = (\pi)^{ip(p-1)/2}$, and so

$$\mathrm{disc}(R_P C_p) = (\pi)^{ip(p-1)}\mathrm{disc}(H(i)_P/R_P).$$

Now, since $\mathrm{disc}(R_P C_p/R_P) = (pR_P)^p = (\pi)^{pe}$,

$$(\pi)^{pe} = (\pi)^{ip(p-1)}\mathrm{disc}(H(i)_P/R_P)$$

or

$$\mathrm{disc}(H(i)_P/R_P) = (\pi)^{pe-pi(p-1)} = ((\pi)^{e-i(p-1)})^p.$$

Now, by Proposition 4.4.13, $\epsilon_{H(i)_P}\left(\int_{H(i)_P}\right) = (\pi)^{e-i(p-1)}$, which implies that $\mathrm{ord}_P\left(\epsilon_{H(i)}\left(\int_{H(i)}\right)\right) = e - (p-1)i$. $\square$

We next consider Larson orders in $KG = KC_{p^2}$. By Proposition 5.2.4, a group valuation ξ on C_{p^2} has at most two finite values: $\xi(g^a) = i$, for $1 \leq a \leq p^2 - 1$, $a \equiv 0 \pmod{p}$, and $\xi(g^b) = j$, for $1 \leq b \leq p^2 - 1, b \not\equiv 0 \pmod{p}$. Since ξ is order bounded, $0 \leq i \leq e'$ and $0 \leq j \leq e'/p$, and since ξ is p-adic, $pj \leq i$. The Larson order $H = A(\xi)$ can be written

$$H = A(\xi) = R\left[\left\{\frac{g^a - 1}{P^i}, \frac{g^b - 1}{P^j}\right\}\right],$$

where $1 \leq a \leq p^2 - 1, a \equiv 0 \pmod{p}, 1 \leq b \leq p^2 - 1, b \not\equiv 0 \pmod{p}$.

Proposition 5.3.4. $H = A(\xi) = R\left[\frac{g^p-1}{P^i}, \frac{g-1}{P^j}\right]$.

Proof. The proof of Proposition 5.3.2 shows that

$$A(\xi) = R\left[\frac{g^p - 1}{P^i}\right]\left[\left\{\frac{g^b - 1}{P^j}\right\}\right].$$

Certainly $R\left[\frac{g^p-1}{P^i}, \frac{g-1}{P^j}\right] \subseteq A(\xi)$, so it remains to show that $A(\xi) \subseteq R\left[\frac{g^p-1}{P^i}, \frac{g-1}{P^j}\right]$. For $b = 2, 3, \ldots, p^2 - 1, b \not\equiv 0 \pmod{p}$, one has

$$g^b - 1 = (g - 1)^b + \binom{b}{1}(g-1)^{b-1} + \binom{b}{2}(g-1)^{b-2} + \cdot + \binom{b}{b-1}(g-1),$$

and so

$$\frac{g^b - 1}{P^j} \subseteq P^{j(b-1)}\left(\frac{g-1}{P^j}\right)^b + \binom{b}{1}P^{j(b-2)}\left(\frac{g-1}{P^j}\right)^{b-1}$$

$$+ \cdots + \binom{b}{b-1}P^j\left(\frac{g-1}{P^j}\right).$$

Thus $\frac{g^b-1}{P^i} \subseteq R\left[\frac{g^p-1}{P^i}, \frac{g-1}{P^j}\right]$, which proves the proposition. $\square$

We denote the Larson order $H = A(\xi)$ in KC_{p^2} by $H(i, j)$ since it depends on only two valuation parameters, i, j.

Proposition 5.3.5. *Let $H(i, j)$ be a Larson order in KC_{p^2}. Then there exists a short exact sequence of Larson orders*

$$R \to H(i) \to H(i, j) \to H(j) \to R.$$

Proof. There exists a short exact sequence of K-Hopf algebras

$$K \to K\langle g^p \rangle \to KC_{p^2} \overset{s}{\to} K\langle \overline{g} \rangle \to K, \quad s(g^p) = 1,$$

with $s(H(i, j)) = H(j) = R\left[\frac{\overline{g}-1}{p^j} \right]$ and $K\langle g^p \rangle \cap H(i, j) = H(i) = R\left[\frac{g^p-1}{p^i} \right]$. Thus the claimed short exact sequence exists.   □

Proposition 5.3.6. *Let $H(i, j)$ be a Larson order in KC_{p^2}. Then*

$$\text{ord}_P \left(\epsilon_{H(i,j)} \left(\int_{H(i,j)} \right) \right) = 2e - (p-1)(i+j).$$

Proof. In view of the short exact sequence of Proposition 5.3.5, one has

$$\text{ord}_P \left(\epsilon_{H(i,j)} \left(\int_{H(i,j)} \right) \right) = \text{ord}_P \left(\epsilon_{H(i)} \left(\int_{H(i)} \right) \right) + \text{ord}_P \left(\epsilon_{H(j)} \left(\int_{H(j)} \right) \right)$$

by Proposition 4.4.12. Thus $\text{ord}_P \left(\epsilon_{H(i,j)} \left(\int_{H(i,j)} \right) \right) = 2e - (p-1)(i+j)$ by Proposition 5.3.3.   □

For the general case, Larson orders in KC_{p^n} can be described as follows.

Proposition 5.3.7. *Let $n \geq 1$, and let ξ be a p-adic obgv on C_{p^n}. Then the Larson order $A(\xi)$ in KC_{p^n} has the form*

$$A(\xi) = R\left[\frac{g^{p^{n-1}} - 1}{p^{i_1}}, \frac{g^{p^{n-2}} - 1}{p^{i_2}}, \dots, \frac{g^p - 1}{p^{i_{n-1}}}, \frac{g - 1}{p^{i_n}} \right],$$

where i_r, $1 \leq r \leq n$, are integers satisfying $0 \leq p i_r \leq i_{r-1}$ for $2 \leq r \leq n$.

Proof. By Proposition 5.2.4, a p-adic obgv on C_{p^n} has n distinct finite values $\xi(g^{p^{n-r}}) = i_r$, for $r = 1, \cdots, n$, that satisfy $0 \leq p i_r \leq i_{r-1}$ for $2 \leq r \leq n$. We proceed by induction on n. We have already seen that the proposition is true for the cases $n = 1, 2$, and so we assume that the $(n-1)$st case holds. Thus,

$$A(\xi) = R\left[\left\{ \frac{g^a - 1}{P^{\xi(g^a)}} : 1 \leq a \leq p^n - 1, a \equiv 0 \pmod{p} \right\} \right]$$

$$\cdot \left[\left\{ \frac{g^b - 1}{P^{i_n}} : 1 \leq b \leq p^n - 1, b \not\equiv 0 \pmod{p} \right\} \right]$$

$$= R\left[\frac{g^{p^{n-1}} - 1}{P^{i_1}}, \frac{g^{p^{n-2}} - 1}{P^{i_2}}, \dots, \frac{g^p - 1}{P^{i_{n-1}}} \right]$$

$$\cdot \left[\left\{ \frac{g^b - 1}{P^{i_n}} : 1 \leq b \leq p^n - 1, b \not\equiv 0 \pmod{p} \right\} \right].$$

Since

$$\frac{g^b - 1}{P^{i_n}} \subseteq P^{i_n(b-1)}\left(\frac{g-1}{P^{i_n}}\right)^b + \binom{b}{1}P^{i_n(b-2)}\left(\frac{g-1}{P^{i_n}}\right)^{b-1}$$

$$+ \cdots + \binom{b}{b-1}P^{i_n}\left(\frac{g-1}{P^{i_n}}\right),$$

one concludes that

$$A(\xi) = R\left[\frac{g^{p^{n-1}} - 1}{P^{i_1}}, \frac{g^{p^{n-2}} - 1}{P^{i_2}}, \ldots, \frac{g^p - 1}{P^{i_{n-1}}}\right]\left[\frac{g-1}{P^{i_n}}\right].$$

□

Larson orders in KC_{p^n} will be denoted as $H(i_1, i_2, \ldots, i_n)$.

For a finite Abelian group G, not every Hopf order in KG is a Larson order, however. For example, if $K = \mathbb{Q}(\zeta_{p^2})$ and $G = C_{p^2}$ is the cyclic group of order p^2, then RG^* is an R-Hopf order in $K\hat{G}$ that is not Larson.

Proposition 5.3.8. *The R-Hopf order $RC_{p^2}^*$ is not of the form $A(\xi)$ for any p-adic obgv ξ on C_{p^2}.*

Proof. By way of contradiction, we assume that $RC_{p^2}^* = A(\xi)$ for some p-adic obgv ξ on C_{p^2}. By Proposition 5.2.4, an obgv on C_{p^2} has at most two finite values: $\xi(g^{pa}) = i$ for $a = 1, 2, \ldots, p-1$, and $\xi(g^b) = j$ for $(b, p) = 1$. Thus, $0 \le i \le e/(p-1) = p(p-1)/(p-1) = p$, and $0 \le j \le e/p(p-1) = 1$. By Proposition 4.4.11,

$$\mathrm{ord}_P\left(\epsilon_{RC_{p^2}}\left(\int_{RC_{p^2}}\right)\right) + \mathrm{ord}_P\left(\epsilon_{RC_{p^2}^*}\left(\int_{RC_{p^2}^*}\right)\right) = 2e = 2p(p-1),$$

and so, since $\mathrm{ord}_P\left(\epsilon_{RC_{p^2}}\left(\int_{RC_{p^2}}\right)\right) = 2p(p-1)$, $\mathrm{ord}_P\left(\epsilon_{RC_{p^2}^*}\left(\int_{RC_{p^2}^*}\right)\right) = 0$. Now, by Proposition 5.3.6,

$$\mathrm{ord}_P\left(\epsilon_{A(\xi)}\left(\int_{A(\xi)}\right)\right) = 2p(p-1) - (p-1)(i+j),$$

and so $2p(p-1) = (p-1)(i+j)$. Thus, $2p = i + j$. However, by the bounds on i, j, we have $1 + p \ge 2p$, which is impossible. Thus $A(\xi) \ne RC_{p^2}^*$. □

Every Hopf order in KG with G finite determines an obgv on G, however. We prove this fact below for the special case where G is Abelian of order m and $\zeta_m \in K$.

Proposition 5.3.9. *Let G be a finite Abelian group of order m, assume that $\zeta_m \in K$, and let H be an R-Hopf order in KG. Then H determines an order-bounded group valuation on G.*

Proof. Let p be a rational prime, and let P be a prime of R lying above the prime p. Define $\xi(g) = \infty$ if $g = 1$. For each $g \in G$, $g \neq 1$, let $|g| = k$. If k is not a power of p, set $\xi(g) = 0$. If $|g|$ is a power of p, then define

$$I_g = \{x \in K : x(g-1) \in H\},$$
$$I_g^{-1} = \{x \in K : xI_g \in R\}.$$

By Proposition 4.4.3, $RG \subseteq H$, and so $R \subseteq I_g$. Thus I_g^{-1} is an ideal of R. Put $\xi(g) = \operatorname{ord}_P(I_g^{-1})$. Thus $\xi(g) \geq 0$ for all $g \in G$, and so the function $\xi : G \to \mathbb{Z} \cup \infty$ satisfies Definition 5.2.1(i),(ii).

For $g, h \in G$, we have

$$gh - 1 = (g-1)h + h - 1,$$

and so $I_g \cap I_h \subseteq I_{gh}$. Thus,

$$I_{gh}^{-1} \subseteq (I_g \cap I_h)^{-1} = I_g^{-1} + I_h^{-1}.$$

Consequently,

$$\xi(gh) \geq \min\{\xi(g), \xi(h)\},$$

which shows that ξ is a group valuation.

To show that ξ is order bounded with respect to ord_P, let $g \in G$, and let $T_g : KG \to KG$ denote the linear transformation defined as $T_g(a) = ag$ for $a \in KG$. Put $t = m/k$, an integer. Then the characteristic polynomial of T_g is $(x^k - 1)^t$. The eigenvalues of T_g are the kth roots of unity ζ_k^l, $l = 0, \cdots, k-1$, each occurring with multiplicity t.

For each l, there is a set of t linearly independent eigenvectors $\{a_{l,j}\}_{j=1}^t$ for ζ_k^l, so that $T_g(a_{l,j}) = a_{l,j}g = \zeta_k^l a_{l,j}$. The set $\{\zeta_k^l a_{l,j}\}$ for $l = 0, \ldots, k-1$, $j = 1, \ldots, t$, is a K-basis for KG. Now

$$a_{l,j} I_g(g-1) = I_g(\zeta_k^l - 1)a_{l,j}.$$

The Hopf order H is a finitely generated module over the Dedekind domain R, and thus the submodule $I_g(g-1) \subseteq H$ is finitely generated. Consequently, $I_g(\zeta_k^l - 1)$ is an integral ideal of R, and so $R(\zeta_k^l - 1) \subseteq I_g^{-1}$. Thus

$$\xi(g) = \operatorname{ord}_P(I_g^{-1}) \leq \operatorname{ord}_P((\zeta_k^l - 1)).$$

Put $k = p^a$. Observe that

$$e = \operatorname{ord}_P((p))$$
$$= \operatorname{ord}_P((\zeta_k^l - 1)^{p^{a-1}(p-1)})$$
$$= p^{a-1}(p-1)\operatorname{ord}_P(\zeta_k^l - 1),$$

so that

$$\frac{e}{p^{a-1}(p-1)} = \mathrm{ord}_P(\zeta_k^l - 1).$$

Thus $\xi(g) \le \mathrm{ord}_P(\zeta_k^l - 1) = \frac{e}{p^{a-1}(p-1)}$, and so ξ is order bounded. $\qquad\square$

With a little more work, one can show that ξ is a p-adic obgv on G. Thus each R-Hopf order H in KG gives rise to a p-adic obgv on G that we denote as $\Xi(H)$.

Given an R-Hopf order H in KG, $\Xi(H)$, in turn, yields a Larson order $A(\Xi(H))$ that is contained in H. For a given H, $A(\Xi(H))$ is the **largest Larson order contained in** H. The relationship between A and Ξ is given as follows.

Proposition 5.3.10. *Let H be an R-Hopf order in KG. Let ξ be an obgv on G. Then:*

(i) $A(\Xi(H)) \subseteq H$, with equality holding if H is a Larson order.
(ii) $\xi \le \Xi(A(\xi))$, with equality holding if ξ is p-adic.

Proof. Exercise. $\qquad\square$

We can consider Larson orders over complete local fields. Let K be a finite extension of $\mathbb{Q}$, and let R denote the integral closure of $\mathbb{Z}$ in K. Let $H(i_1, i_2, \ldots, i_n)$ denote a Larson order in KC_{p^n}. Let P be a prime ideal of R that lies above p. Let $[\ ]_p$ be the discrete absolute value corresponding to P, and let K_P denote the completion of K with respect to $[\ ]_p$. Note that $[\ ]_p$ restricts to R and the localization R_P. We denote the completion of R_P with respect to $[\ ]_p$ by $\hat{R}_P$. Of course, the completion of R with respect to $[\ ]_p$ is $\hat{R}_P$.

Proposition 5.3.11. *The field K_P is a finite extension of $\mathbb{Q}_p$.*

Proof. Let $(p) = P_1^{e_1} P_2^{e_2} \cdots P_g^{e_g}$ denote the factorization of (p) in R. By [FT91, Chapter III, §1, Theorem 17], there exists a ring isomorphism

$$\mathbb{Q}_p \otimes_{\mathbb{Q}} K \to \bigoplus_{i=1}^{g} K_{P_i}.$$

Since $[\ ]_{p,i}$ is an extension of $|\ |_p$ and K is a field extension of $\mathbb{Q}$, K_{P_i} is a field extension of $\mathbb{Q}_p$. Thus K_{P_i} is a vector space over $\mathbb{Q}_p$, and so $\bigoplus_{i=1}^{g} K_{P_i}$ is a vector space over $\mathbb{Q}_p$. Since $\mathbb{Q}_p \otimes_{\mathbb{Q}} K$ is finite-dimensional over $\mathbb{Q}_p$, the same is true for $\bigoplus_{i=1}^{g} K_{P_i}$. Thus each K_{P_i} is finite-dimensional over $\mathbb{Q}_p$. $\qquad\square$

Proposition 5.3.12. *With the notation above,*

(i) the ring of integers in K_P is the completion of R with respect to $[\ ]_p$;
(ii) the completion of R with respect to $[\ ]_p$ ($= \hat{R}_P$) is a local Dedekind domain with maximal ideal $P\hat{R}_P$; $P\hat{R}_P$ being a principal ideal; and
(iii) each element $x \in K_P$ has a representation $x = u\pi^r$, where π generates $P\hat{R}_P$, $r \in \mathbb{Z}$, and u is a unit in $\hat{R}_P$.

Proof. Let $\mathcal{O}$ denote the integral closure of $\hat{R}_P$ in K_P. Suppose $x = \lim a_n \in \mathcal{O}$. Then x satisfies a monic polynomial with coefficients in $\hat{R}_P$, that is,

$$(\lim a_n)^l = r_0 + r_1(\lim a_n) + r_2(\lim a_n)^2 + \cdots + r_{l-1}(\lim a_n)^{l-1}$$

for $r_i \in \hat{R}_P$. By continuity,

$$\lim a_n^l = \lim(r_0 + r_1 a_n + r_2 a_n^2 + \cdots + r_{l-1} a_n^{l-1}),$$

and thus, for each $\iota > 0$, there exists an integer N for which

$$[a_n^l - (r_0 + r_1 a_n + r_2 a_n^2 + \cdots + r_{l-1} a_n^{l-1})]_p < \iota$$

for $n \geq N$. Consequently, there exists an integer M for which $a_n \in R_P$ for all $n \geq M$. Therefore, $\{a_n\}$ is a $[\ ]_p$-Cauchy sequence in R_P, and so $x \in \hat{R}_P$. This shows that $\mathcal{O} \subseteq \hat{R}_P$. Clearly, $\hat{R}_P \subseteq \mathcal{O}$, which proves (i).

To prove (ii), note that R_P is a local Dedekind domain with maximal ideal $(\pi) = PR_P$, and thus $\hat{R}_P$ is a local Dedekind domain with maximal ideal $(\lim \pi)\hat{R}_P = \pi \hat{R}_P$.

We leave (iii) as an exercise. $\qquad\square$

Observe that the completion of $\mathbb{Q}$ with respect to $|\ |_p$ is the field of p-adic rationals $\mathbb{Q}_p$, and the ring of integers in $\mathbb{Q}_p$ is the completion of $\mathbb{Z}$ with respect to $|\ |_p$, the ring of p-**adic integers**, $\hat{\mathbb{Z}}_{(p)}$. In what follows, we shall use the simpler notation $\mathbb{Z}_p$ to denote the p-adic integers.

By Proposition 5.3.12(iii), the discrete valuation $\mathrm{ord}_P : K \to \mathbb{Z} \cup \infty$ extends to a discrete valuation on K_P, which we denote by ord. For $x \in K_P$, one has $\mathrm{ord}(x) = r$, where $x = u\pi^r$.

Let $H(i_1, \ldots, i_n)$ be a Larson order in KC_{p^n}. We complete R with respect to $[\ ]_p$, and let

$$H(i_1, \ldots, i_n)_P = \hat{R}_P \otimes_P H(i_1, \ldots, i_n).$$

Then

$$H(i_1, \ldots, i_n)_P = \hat{R}_P \left[\frac{g^{p^{n-1}} - 1}{\pi^{i_1}}, \frac{g^{p^{n-2}} - 1}{\pi^{i_2}}, \ldots, \frac{g - 1}{\pi^{i_n}} \right],$$

where $P\hat{R}_P = (\pi)$, and $H(i_1, \ldots, i_n)_P$ is an $\hat{R}_P$-Hopf order in $K_P C_{p^n}$.

Thus Larson orders in $K_P C_p$ appear as $\hat{R}_p \left[\dfrac{g - 1}{\pi^{i_1}} \right]$, and Larson orders in $K_P C_{p^2}$ appear as $\hat{R}_p \left[\dfrac{g^p - 1}{\pi^{i_1}}, \dfrac{g - 1}{\pi^{i_2}} \right]$, for integers i_1, i_2 with $0 \leq i_1, i_2 \leq e'$, $pi_2 \leq i_1$.

5.4   Chapter Exercises

Exercises for §5.1

1. Show that the map $|\ |_p : \mathbb{Q} \to \mathbb{R}$ defined as

$$\left|\frac{a}{b}p^r\right|_p = \frac{1}{p^r},$$

   where $(a, p) = 1$, $(b, p) = 1$, and $r \in \mathbb{Z}$, is an absolute value on $\mathbb{Q}$.
2. Prove that $|\ |_p$ is not equivalent to $|\ |_q$ if $p \neq q$, and that $|\ |_\infty$ is not equivalent to $|\ |_p$ for all p.

Exercises for §5.2

3. Classify all of the group valuations on D_4, the fourth order dihedral group.
4. Classify all of the p-adic order-bounded group valuations (obgvs) on D_3, the third-order dihedral group.
5. Let $n \geq 1$. Compute the number of p-adic obgvs on C_{p^n} when $K = \mathbb{Q}(\zeta_{p^n})$.
6. Let $n \geq 1$. Compute the number of p-adic obgvs on C_p^n when $K = \mathbb{Q}(\zeta_{p^n})$.
7. Let ord_P be a discrete valuation on $K = \mathbb{Q}(\alpha)$. Prove the following.

   (a) $\mathrm{ord}_P : K \to \mathbb{Z} \cup \infty$ is a surjective group homomorphism.
   (b) $\mathrm{ord}_P(xy) = \mathrm{ord}_P(x) + \mathrm{ord}_P(y)$ for all $x, y \in K$.
   (c) $\mathrm{ord}_P(x + y) \geq \min\{\mathrm{ord}_P(x), \mathrm{ord}_P(y)\}$ for all $x, y \in K$.

8. Prove that $x^2 + 1$ factors over $\mathbb{Q}_p$ if and only if $p \equiv 1 \pmod 4$.

Exercises for §5.3

9. Prove that every obgv on C_p is a p-adic obgv.
10. Let $\xi_1, \xi_2 : C_{p^2} \to \mathbb{Z} \cup \infty$ be obgvs on C_{p^2} with $\xi_1(g) \leq \xi_2(g)$ for all $g \in G$. Prove that $A(\xi_1) \subseteq A(\xi_2)$.
11. Let ξ be a p-adic obgv on C_{p^2}. Prove that the sequence $R \to H(i) \to A(\xi) \to H(j) \to R$ is a short exact sequence of Larson orders.
12. Prove that $H(i)$ is a local ring for $0 \leq i < e'$, and compute Spec $H(i)$.
13. Let $F = \mathrm{Hom}_{R\text{-alg}}(H(i), -)$ be the R-group scheme represented by $H(i)$. Compute $F(R)$.
14. Let K be a finite extension of $\mathbb{Q}_p$, and let $H(i), H(j)$ be Larson orders in KC_p with $i < j$.

   (a) Show that there is an inclusion of R-algebras $f : H(i) \to H(j)$.
   (b) Show that $H(j)/f(H(i)^+)H(j)$ is not an R-Hopf order in K.

15. Prove Proposition 5.3.10.

Chapter 6
Formal Group Hopf Orders

In this chapter, we show how to construct Hopf orders using the theory of formal groups. Throughout this chapter, $K = K_P$ is a finite extension of $\mathbb{Q}_p$ with ring of integers $R = \hat{R}_P$ endowed with the discrete valuation ord. We let π denote a uniformizing parameter for R, and put $e = \text{ord}(p)$, $e' = e/(p-1)$.

6.1 Formal Groups

Definition 6.1.1. An n-**dimensional formal group** is an n-tuple of power series

$$F(\overline{x}, \overline{y}) = (F_1(\overline{x}, \overline{y}), F_2(\overline{x}, \overline{y}), \dots, F_n(\overline{x}, \overline{y})) \in R[[\overline{x}, \overline{y}]]^n,$$

in the variables $\overline{x} = x_1, x_2, \dots, x_n$, $\overline{y} = y_1, y_2 \dots, y_n$, that satisfies, for $i = 1, \dots, n$,

  (i) $F_i(\overline{x}, \overline{y}) \equiv x_i + y_i$ modulo degree 2,
 (ii) $F_i(F(\overline{x}, \overline{y}), \overline{z}) = F_i(\overline{x}, F(\overline{y}, \overline{z}))$, and
(iii) $F(\overline{x}, \overline{0}) = \overline{x}$, $F(\overline{0}, \overline{y}) = \overline{y}$.

The formal group F is **commutative** if $F(\overline{x}, \overline{y}) = F(\overline{y}, \overline{x})$. Two formal groups F and G of dimension n are **linearly isomorphic over** R if there is a matrix Θ in $\text{GL}_n(R)$ such that

$$G(\overline{x}, \overline{y}) = \Theta^{-1} F(\Theta(\overline{x}), \Theta(\overline{y})).$$

An n-dimensional formal group $F(\overline{x}, \overline{y})$ is a **polynomial formal group** if each power series $F_i(\overline{x}, \overline{y})$ is a polynomial.

We first consider one-dimensional polynomial formal groups. These are identified with a polynomial $F(x, y)$. We have the following classification.

Proposition 6.1.1. *Let $F(x, y)$ be a one-dimensional polynomial formal group. Then F is of the form*

$$F(x, y) = x + y + axy$$

for some $a \in R$.

R.G. Underwood, *An Introduction to Hopf Algebras*, DOI 10.1007/978-0-387-72766-0_6,   115
© Springer Science+Business Media, LLC 2011

Proof. By Definition 6.1.1(i), $F(x, y) = x + y + g(x, y)$, where $g(x, y)$ is a polynomial in $R[x, y]$ in which all non-zero terms have total degree ≥ 2. Thus, we may write

$$F(x, y) = h(x, y) + k(x, y),$$

where $h(x, y)$ has degree 1 in x and $k(x, y)$ has degree n for some $n \geq 1$ in x. Now, by Definition 6.1.1(ii),

$$h(F(x, y), z) + k(F(x, y), z) = h(x, F(y, z)) + k(x, F(y, z)),$$

and so the left-hand side above has degree n^2 in x, while the right-hand side has degree n in x. These degrees must equal, and so $n = 1$ is the degree of x in $F(x, y)$.

By a similar argument, the degree of y in $F(x, y)$ is also 1. Consequently, F is of the form claimed.   □

Here are some examples of one-dimensional (polynomial) formal groups. Let $F(x, y) = x + y$. Then, as one can easily show, F is a one-dimensional formal group, which is called the **additive formal group**, denoted by G_a.

Another important example is the one-dimensional formal group defined as $F(x, y) = x + y + xy$. Again, it is an easy exercise to show that F satisfies (i), (ii), and (iii) of Definition 6.1.1. This is the one-dimensional **multiplicative formal group**, which we denote by G_m.

The following is an important consequence of Definition 6.1.1.

Proposition 6.1.2. *Let $F(\overline{x}, \overline{y})$ be an n-dimensional formal group. Then there exists an n-tuple of power series in $\overline{x}$,*

$$\sigma(\overline{x}) = (\sigma_1(\overline{x}), \sigma_2(\overline{x}), \dots, \sigma_n(\overline{x})),$$

for which

$$F(\overline{x}, \sigma(\overline{x})) = 0 = F(\sigma(\overline{x}), \overline{x}).$$

Proof. See [Fro68, Chapter 1, §3, Proposition 1].   □

Formal groups are of interest because they allow us to put a group product on various sets of elements. As an example, we consider (π), the maximal ideal in R. We employ the formal group $G_m(x, y) = x + y + xy$ to define a multiplication on (π). For $a\pi, b\pi \in (\pi)$, put

$$a\pi * b\pi = G_m(a\pi, b\pi) = a\pi + b\pi + ab\pi^2.$$

Then (π) is closed under $*$. Moreover, zero serves as an identity element since $G_m(a\pi, 0) = a\pi = G_m(0, a\pi)$ by Definition 6.1.1(iii).

But in order to obtain a group structure on (π), we need to define an inverse for $a\pi$ under $*$. By Proposition 6.1.2, there exists a power series $\sigma(x)$ in $R[[x]]$ for which $G_m(\sigma(x), x) = 0 = G_m(x, \sigma(x))$. Thus $\sigma(x)$ satisfies

$$x + \sigma(x) + x\sigma(x) = 0,$$

and so $\sigma(x) = \dfrac{-x}{1+x}$. As a formal power series,

$$\sigma(x) = -x + x^2 - x^3 + x^4 - \cdots,$$

and so the inverse of $a\pi$ is

$$\sigma(a\pi) = -a\pi + a^2\pi^2 - a^3\pi^3 + \cdots,$$

which is a $[\ ]_p$-Cauchy sequence converging to an element in (π). So (π) endowed with the group product defined by $G_m(x, y)$ is a group, which we denote by P_{G_m}.

Proposition 6.1.3. $P_{G_m} \cong U_1(R) = 1 + R\pi$, the group of principal units in R.

Proof. Define a function $\phi : P_{G_m} \to U_1(R)$ by $a\pi \mapsto 1 + a\pi$. Then ϕ is a bijection. Let $a\pi, b\pi \in (\pi)$. Then

$$\phi(a\pi * b\pi) = \phi(a\pi + b\pi + ab\pi^2)$$
$$= \phi((a + b + ab\pi)\pi)$$
$$= 1 + (a + b + ab\pi)\pi$$
$$= (1 + a\pi)(1 + b\pi)$$
$$= \phi(a\pi)\phi(b\pi),$$

and so ϕ is an isomorphism. $\square$

For an integer $j \geq 1$, there is a subgroup of $U_1(R)$ defined as

$$U_j(R) = \{1 + a\pi^j : a \in R\}.$$

How does one obtain higher-dimensional formal groups? The easiest way is to construct an n-tuple consisting of copies of a given one-dimensional formal group. For example, we have the n-dimensional formal group G_m^n defined as

$$G_m^n(\overline{x}, \overline{y}) = (G_m(x_1, y_1), G_m(x_2, y_2), \dots, G_m(x_n, y_n)),$$

with

$$G_m(x_i, y_i) = x_i + y_i + x_i y_i$$

for $i = 1, \dots, n$.

We can modify G_m^n to create a collection of n-dimensional polynomial formal groups. Let Θ be an $n \times n$ matrix with entries in R with $\det(\Theta) \neq 0$. Consider the n-tuple of polynomials

$$F(\overline{x}, \overline{y}) = \Theta^{-1} G_m^n(\Theta\overline{x}, \Theta\overline{y}) \in K[\overline{x}, \overline{y}]^n.$$

Here $\overline{x}$ is the $n \times 1$ column vector $(x_1, x_2, \ldots, x_n)^T$ (T is the transpose); $\overline{y}$ is defined similarly.

If the polynomials of $F = F(\overline{x}, \overline{y})$ are in $R[\overline{x}, \overline{y}]$, then F is a formal group, which we call the n-**dimensional generically split formal group determined by** Θ, denoted by F_Θ. The formal group F_Θ is linearly isomorphic to G_m^n over K.

In what follows, we specialize to the case where Θ is lower-triangular. To see what kind of structure $\Theta^{-1}G_m^n(\Theta\overline{x}, \Theta\overline{y})$ has, we examine the cases $n = 1, 2$. For $n = 1$, Θ is an element $\theta \neq 0$ of R with $\Theta^{-1} = \theta^{-1}$. Now,

$$\theta^{-1}G_m(\theta x, \theta y) = \theta^{-1}(\theta x + \theta y + \theta^2 xy)$$

$$= x + y + \theta xy,$$

and this is precisely the form of all one-dimensional formal groups given by Proposition 6.1.1.

In dimension 2, we let $\Theta = \begin{pmatrix} \theta_{1,1} & 0 \\ \theta_{2,1} & \theta_{2,2} \end{pmatrix}$ denote a lower-triangular matrix with entries in R with $\det(\Theta) \neq 0$. Let $\Theta^{-1} = \begin{pmatrix} \eta_{1,1} & 0 \\ \eta_{2,1} & \eta_{2,2} \end{pmatrix}$. Now, $G_m^2(\Theta\overline{x}, \Theta\overline{y})$

$$= \begin{pmatrix} \theta_{1,1}x_1 + \theta_{1,1}y_1 + \theta_{1,1}^2 x_1 y_1 \\ \theta_{2,1}x_1 + \theta_{2,2}x_2 + \theta_{2,1}y_1 + \theta_{2,2}y_2 + (\theta_{2,1}x_1 + \theta_{2,2}x_2)(\theta_{2,1}y_1 + \theta_{2,2}y_2), \end{pmatrix},$$

and so
$\Theta^{-1}G_m^2(\Theta\overline{x}, \Theta\overline{y})$

$$= \begin{pmatrix} \eta_{1,1} & 0 \\ \eta_{2,1} & \eta_{2,2} \end{pmatrix}\begin{pmatrix} \theta_{1,1}(x_1 + y_1) + \theta_{1,1}^2 x_1 y_1 \\ \theta_{2,1}(x_1+y_1) + \theta_{2,2}(x_2+y_2) + (\theta_{2,1}x_1 + \theta_{2,2}x_2)(\theta_{2,1}y_1 + \theta_{2,2}y_2) \end{pmatrix}.$$

The first component in this matrix product is

$$x_1 + y_1 + \theta_{1,1}x_1 y_1,$$

while the second component is

$$\eta_{2,1}(\theta_{1,1}(x_1 + y_1) + \theta_{1,2}(x_2 + y_2)) + \eta_{2,1}(\theta_{1,1}x_1 + \theta_{1,2}x_2)(\theta_{1,1}y_1 + \theta_{1,2}y_2)$$

$$+ \eta_{2,2}(\theta_{2,1}(x_1+y_1) + \theta_{2,2}(x_2+y_2)) + \eta_{2,2}(\theta_{2,1}x_1 + \theta_{2,2}x_2)(\theta_{2,1}y_1 + \theta_{2,2}y_2)$$

$$= (\eta_{2,1}\theta_{1,1} + \eta_{2,2}\theta_{2,1})(x_1 + y_1) + \eta_{2,2}\theta_{2,2}(x_2 + y_2) + \eta_{2,1}\theta_{1,1}^2 x_1 y_1$$

$$+ \eta_{2,2}(\theta_{2,1}x_1 + \theta_{2,2}x_2)(\theta_{2,1}y_1 + \theta_{2,2}y_2)$$

$$= x_2 + y_2 + (\eta_{2,1}\theta_{1,1}^2 + \eta_{2,2}\theta_{2,1}^2)x_1 y_1 + \theta_{2,2}x_2 y_2 + \theta_{2,1}x_1 y_2 + \theta_{2,1}x_2 y_1.$$

Now $\eta_{2,2} = \theta_{2,2}^{-1}$ and $\eta_{2,1} = \dfrac{-\theta_{2,1}}{\theta_{1,1}\theta_{2,2}}$, and so the coefficient on $x_1 y_1$ is

$$\frac{-\theta_{2,1}\theta_{1,1}}{\theta_{2,2}} + \frac{\theta_{2,1}^2}{\theta_{2,2}} = \frac{\theta_{2,1}(\theta_{2,1} - \theta_{1,1})}{\theta_{2,2}}.$$

Thus, $\Theta^{-1} G_m^2(\Theta \overline{x}, \Theta \overline{y})$

$$= \begin{pmatrix} x_1 + y_1 + \theta_{1,1} x_1 y_1 \\ x_2 + y_2 + \dfrac{\theta_{2,1}(\theta_{2,1} - \theta_{1,1})}{\theta_{2,2}} x_1 y_1 + \theta_{2,2} x_2 y_2 + \theta_{2,1} x_1 y_2 + \theta_{2,1} x_2 y_1 \end{pmatrix}. \quad (6.1)$$

Of course, in order to define a two-dimensional formal group, we require that every coefficient in the components on the right-hand side of (6.1) be an element of R. The only one that may not be in R is $\dfrac{\theta_{2,1}(\theta_{2,1} - \theta_{1,1})}{\theta_{2,2}}$, but it is not too difficult to find conditions such that this coefficient is in R. For example, as one can verify, the conditions $\mathrm{ord}(\theta_{1,1}) > \mathrm{ord}(\theta_{2,1})$ and $\mathrm{ord}(\theta_{2,1}^2) \geq \mathrm{ord}(\theta_{2,2})$ guarantee that $\dfrac{\theta_{2,1}(\theta_{2,1} - \theta_{1,1})}{\theta_{2,2}} \in R$.

We next consider the general case $n \geq 1$. The goal is to find conditions on the lower-triangular matrix $\Theta \in \mathrm{GL}_n(K)$ such that $\Theta^{-1} G_m^n(\Theta \overline{x}, \Theta \overline{y})$ is defined over R. This is done in the paper [CU03, §1] of L. Childs and R. Underwood, and we review the result here.

Proposition 6.1.4. *Let* $\Theta = (\theta_{i,j}) \in M_n(R)$ *be lower-triangular with* $\det(\Theta) \neq 0$. *Let* q *be a rational number* $0 < q < \frac{p-1}{2p-1}$. *Suppose the entries* $\theta_{i,j}$ *of* Θ *satisfy the conditions*

(i) $\mathrm{ord}(\theta_{i,i}) > \mathrm{ord}(\theta_{i,j}) \geq (1-q)\mathrm{ord}(\theta_{i,i})$ *for all* $i > j$ *with* $\theta_{i,j} \neq 0$ *and*
(ii) $\mathrm{ord}(\theta_{r,r}) \geq p \cdot \mathrm{ord}(\theta_{r+1,r+1})$ *for all* r.
Then the formal group F_Θ *is defined over* R.

We will need the formal group F_Θ constructed in Proposition 6.1.4 soon, but before that we discuss the types of maps that exist between formal groups.

Definition 6.1.2. A **homomorphism** $\phi : F \to G$ of n-dimensional polynomial formal groups is an n-tuple of polynomials

$$\phi(\overline{x}) = (\phi_1(\overline{x}), \phi_2(\overline{x}), \ldots, \phi_n(\overline{x}))$$

for which

$$\phi_i(F(\overline{x}, \overline{y})) = G_i(\phi(\overline{x}), \phi(\overline{y}))$$

for $i = 1, \ldots, n$.

For example, let $F(x, y) = x + y + axy$ be a one-dimensional formal group over R, and let $G = G_m(x, y)$. Then $\phi : F \to G$, $x \mapsto ax$, is a homomorphism of formal groups since

$$a F(x, y) = ax + ay + a^2 xy = G_m(ax, ay).$$

As another example, we describe a homomorphism of F into itself, an **endomorphism** of F. Define $[1](\overline{x}) = \overline{x}$, $[2](\overline{x}) = F(\overline{x}, \overline{x})$, and, for $k > 2$,

$$[k](\overline{x}) = F([k-1](\overline{x}), \overline{x}).$$

Also, define $[-1](\overline{x}) = \sigma(\overline{x})$, $[-2](\overline{x}) = F(\sigma(\overline{x}), \sigma(\overline{x}))$, and, for $k > 2$,

$$[-k](\overline{x}) = F([-k+1](\overline{x}), \sigma(\overline{x})).$$

Here $\sigma(\overline{x})$ is a power series for which $F(\overline{x}, \sigma(\overline{x})) = 0 = F(\sigma(\overline{x}), \overline{x})$.

Proposition 6.1.5. *Let F be an n-dimensional formal group. Then $[k] : F \to F$ is an endomorphism for all non-zero integers k.*

Proof. Exercise.   □

For example, the endomorphism $[3]$ acts on $G_m(x, y) = x + y + xy$ as follows:

$$[1](x) = x = (1 + x) - 1;$$
$$[2](x) = G_m(x, x) = 2x + x^2 = (x + 1)^2 - 1;$$
$$[3](x) = G_m([2](x), x)$$
$$= G_m(2x + x^2, x)$$
$$= 2x + x^2 + x + (2x + x^2)x$$
$$= 3x + 3x^2 + x^3 = (1 + x)^3 - 1.$$

For $k > 0$, one has the following proposition.

Proposition 6.1.6. *Let $k > 0$. Then the endomorphism $[k] : G_m \to G_m$ is defined by $\phi(x) = (1 + x)^k - 1$.*

Proof. Exercise.   □

There is a special type of formal group homomorphism that we will use in the next section to construct Hopf algebras.

Definition 6.1.3. The homomorphism $\phi : F \to G$ is an **isogeny of formal groups** if $R[[\overline{x}]]/(\phi(\overline{x}))$ is a free R-module of finite rank.

For example, $[p] : G_m \to G_m$ is an isogeny since

$$R[[x]]/((1 + x)^p - 1) = R[1 + x]/((1 + x)^p - 1) = R[y]/(y^p - 1) \cong RC_p.$$

Clearly, RC_p is free over R of rank p. In fact, it is an R-Hopf algebra.

Here is an important generalization of the isogeny $[p] : G_m \to G_m$ that we will need in the next section. Recall that $\mathbf{G}_m$ is the R-group scheme represented by the

R-Hopf algebra $R[T, T^{-1}]$, with T indeterminate. We generalize this group scheme to define the group scheme $\mathbf{G}_m^n$ of the form

$$\operatorname{Hom}_{R\text{-alg}}(R[T_1, \ldots, T_n, T_1^{-1}, \ldots, T_n^{-1}],),$$

with T_i, $i = 1, \ldots, n$ indeterminate. For $n \geq 1$, we define a group scheme endomorphism of $\phi_n : \mathbf{G}_m^n \to \mathbf{G}_m^n$ through the R-algebra homomorphism

$$\Psi_n : R[T_1, \ldots, T_n, T_1^{-1}, \ldots, T_n^{-1}] \to R[T_1, \ldots, T_n, T_1^{-1}, \ldots, T_n^{-1}],$$

given as $\Psi_n(T_1) = T_1^p$, $\Psi_n(T_i) = T_i^p T_{i-1}^{-1}$, for $i = 2, \ldots, n$. Translating this to formal groups by setting $x_i = T_i - 1$, we obtain a homomorphism of formal groups

$$\phi_n : G_m^n \to G_m^n,$$

defined as

$$\phi_n(\overline{x}) = \Psi_n \begin{pmatrix} T_1 - 1 \\ \vdots \\ T_n - 1 \end{pmatrix} = \begin{pmatrix} T_1^p - 1 \\ T_1^{-1} T_2^p - 1 \\ \vdots \\ T_{n-1}^{-1} T_n^p - 1 \end{pmatrix} = \begin{pmatrix} (1 + x_1)^p - 1 \\ (1 + x_1)^{-1}(1 + x_2)^p - 1 \\ \vdots \\ (1 + x_{n-1})^{-1}(1 + x_n)^p - 1 \end{pmatrix}.$$

We have $R[[\overline{x}]]/(\phi_n(\overline{x})) \cong RC_{p^n}$; that is, ϕ_n is an isogeny whose kernel is represented by the R-Hopf order RC_{p^n} in KC_{p^n}.

6.2   Formal Group Hopf Orders

In this section, we show how to construct R-Hopf orders in KC_{p^n} from isogenies of formal groups. This collection of Hopf orders will include all of the Larson orders in KC_{p^n}.

Our first step is to use an n-dimensional formal group $F = (F_1, \ldots, F_n)$ to induce a "formal" R-Hopf algebra structure on the ring of power series $R[[\overline{x}]]$. One could imagine comultiplication on $R[[\overline{x}]]$ to be defined as

$$\Delta(x_i) = F_i(\overline{x} \otimes 1, 1 \otimes \overline{x}),$$

for $i = 1, \ldots, n$, but there is a problem here. It may be that $F_i(\overline{x} \otimes 1, 1 \otimes \overline{x}) \notin R[[\overline{x}]] \otimes_R R[[\overline{x}]]$ for some i. For example, if $F_i(\overline{x}, \overline{y}) = x_1 + y_1 + x_1 y_1 + x_1^2 y_1^2 + \cdots$, then

$$F_i(\overline{x} \otimes 1, 1 \otimes \overline{x}) = x_1 \otimes 1 + 1 \otimes x_1 + x_1 \otimes x_1 + x_1^2 \otimes x_1^2 + \cdots,$$

which is not an element of $R[[\overline{x}]] \otimes_R R[[\overline{x}]]$.

To remedy this, we need to replace $R[[\overline{x}]] \otimes_R R[[\overline{x}]]$ with a larger ring, constructed as follows. Let I be the ideal of $R[[\overline{x}]] \otimes R[[\overline{x}]]$ defined as

$$I = (x_1, x_2, \ldots, x_n) \otimes R[[\overline{x}]] + R[[\overline{x}]] \otimes (x_1, x_2, \ldots, x_n).$$

Proposition 6.2.1. *I is a prime ideal of $R[[\overline{x}]] \otimes R[[\overline{x}]]$.*

Proof. We show that $(R[[\overline{x}]] \otimes_R R[[\overline{x}]])/I \cong R$. Since $(x_1, x_2, \ldots, x_n) \otimes 1 \in I$ and $1 \otimes (x_1, x_2, \ldots, x_n) \in I$, the quotient is isomorphic to $R \otimes_R R \cong R$. $\qquad\square$

Put $T = R[[\overline{x}]] \otimes R[[\overline{x}]]$. We endow T with the I-**adic topology**, a basis for which consists of all subsets of the form $a + I^\eta$, where $a \in T$ and $\eta \geq 0$. A sequence $\{a_n\}$ in T is I-**Cauchy** if for each I^η there exists an integer N for which $a_m - a_n \in I^\eta$ whenever $m, n \geq N$.

We complete T with respect to the I-adic topology by adjoining the limits of all Cauchy sequences. The result is the **completed tensor product** and is denoted by $R[[\overline{x}]] \hat{\otimes} R[[\overline{x}]]$. This is the larger ring that we require. For example, $R[[\overline{x}]] \hat{\otimes} R[[\overline{x}]]$ contains the I-convergent sum

$$x_1 \otimes 1 + 1 \otimes x_1 + x_1 \otimes x_1 + x_1^2 \otimes x_1^2 + \cdots,$$

which is not in $T = R[[\overline{x}]] \otimes R[[\overline{x}]]$.

Proposition 6.2.2. *Let $F(\overline{x}, \overline{y}) = (F_1(\overline{x}, \overline{y}), \ldots, F_n(\overline{x}, \overline{y}))$ be an n-dimensional formal group. Then F induces a formal R-Hopf algebra structure on $R[[\overline{x}]]$ (formal in the sense that the comultiplication will be a map $\Delta : R[[\overline{x}]] \to R[[\overline{x}]] \hat{\otimes} R[[\overline{x}]]$).*

Proof. We define comultiplication, counit, and coinverse maps for $R[[\overline{x}]]$ and show that they satisfy the conditions for $R[[\overline{x}]]$ to be a formal R-Hopf algebra. We define comultiplication $\Delta : R[[\overline{x}]] \to R[[\overline{x}]] \hat{\otimes} R[[\overline{x}]]$ by the conditions

$$\Delta(\overline{x}) = (\Delta(x_1), \ldots, \Delta(x_n)),$$

$$\Delta(x_i) = F_i(\overline{x} \hat{\otimes} 1, 1 \hat{\otimes} \overline{x}),$$

$$(I \hat{\otimes} \Delta)(F_i(\overline{x} \hat{\otimes} 1, 1 \hat{\otimes} \overline{x})) = F_i(\overline{x} \hat{\otimes} 1 \hat{\otimes} 1, 1 \hat{\otimes} \Delta(\overline{x})),$$

$$(\Delta \hat{\otimes} I)(F_i(\overline{x} \hat{\otimes} 1, 1 \hat{\otimes} \overline{x})) = F_i(\Delta(\overline{x}) \hat{\otimes} 1, 1 \hat{\otimes} 1 \hat{\otimes} \overline{x}).$$

Now,

$$(I \hat{\otimes} \Delta)(\Delta(x_i)) = (I \hat{\otimes} \Delta)(F_i(\overline{x} \hat{\otimes} 1, 1 \hat{\otimes} \overline{x}))$$

$$= F_i(\overline{x} \hat{\otimes} 1 \hat{\otimes} 1, 1 \hat{\otimes} \Delta(\overline{x}))$$

$$= F_i(\overline{x} \hat{\otimes} 1 \hat{\otimes} 1, 1 \hat{\otimes} F(\overline{x} \hat{\otimes} 1, 1 \hat{\otimes} \overline{x}))$$

$$= F_i(\overline{x} \hat{\otimes} 1 \hat{\otimes} 1, F(1 \hat{\otimes} \overline{x} \hat{\otimes} 1, 1 \hat{\otimes} 1 \hat{\otimes} \overline{x}))$$

$$= F_i(F(\overline{x} \hat{\otimes} 1 \hat{\otimes} 1, 1 \hat{\otimes} \overline{x} \hat{\otimes} 1), 1 \hat{\otimes} 1 \hat{\otimes} \overline{x}) \quad \text{by Def. 6.1.1(ii)}$$

$$= F_i(F(\overline{x} \hat{\otimes} 1, 1 \hat{\otimes} \overline{x}) \hat{\otimes} 1, 1 \hat{\otimes} 1 \hat{\otimes} \overline{x})$$

$$= F_i(\Delta(\overline{x}) \hat{\otimes} 1, 1 \hat{\otimes} 1 \hat{\otimes} \overline{x})$$

$$= (\Delta \hat{\otimes} I)(F_i(\overline{x} \hat{\otimes} 1, 1 \hat{\otimes} \overline{x}))$$

$$= (\Delta \hat{\otimes} I)(\Delta(x_i)),$$

and so Δ is coassociative.

Next, we define the counit map $\epsilon : R[[\overline{x}]] \to R$ by the conditions

$$\epsilon(\overline{x}) = \overline{0},$$

$$(I \hat{\otimes} \epsilon)(F_i(\overline{x} \hat{\otimes} 1, 1 \hat{\otimes} \overline{x})) = F_i(\overline{x} \hat{\otimes} 1, 1 \hat{\otimes} \epsilon(\overline{x})),$$

$$(\epsilon \hat{\otimes} I)(F_i(\overline{x} \hat{\otimes} 1, 1 \hat{\otimes} \overline{x})) = F_i(\epsilon(\overline{x}) \hat{\otimes} 1, 1 \hat{\otimes} \overline{x}).$$

Then

$$\begin{aligned}
m(\epsilon \hat{\otimes} I) \Delta(x_i) &= m(\epsilon \hat{\otimes} I)(F_i(\overline{x} \hat{\otimes} 1, 1 \hat{\otimes} \overline{x})) \\
&= m(F_i(\overline{0} \hat{\otimes} 1, 1 \hat{\otimes} \overline{x})) \\
&= m(1 \hat{\otimes} x_i) \quad \text{by Def. 6.1.1(iii)} \\
&= x_i.
\end{aligned}$$

Likewise, $m(I \hat{\otimes} \epsilon) \Delta(x_i) = x_i$ for all i. Thus ϵ satisfies the counit property.

Finally, let $\sigma(\overline{x}) = (\sigma_1(\overline{x}), \dots, \sigma_n(\overline{x}))$ be the n-tuple of power series with $F(\sigma(\overline{x}), \overline{x}) = 0 = F(\overline{x}, \sigma(\overline{x}))$. Then the coinverse map $\sigma : R[[x]] \to R[[x]]$ is given as $x_i \mapsto \sigma(x_i)$. We leave it as an exercise to formulate the necessary conditions on σ and to show that the coinverse property holds. □

$R[[\overline{x}]]$ together with the formal Hopf algebra structure induced by F is denoted by $R[[\overline{x}]]_F$.

Proposition 6.2.3. Let $\phi : F \to G$, $\phi(\overline{x}) = (\phi_i(\overline{x}), \cdots, \phi_n(\overline{x}))$, be an isogeny of n-dimensional formal groups. Then ϕ induces an R-Hopf algebra structure on the quotient $R[[\overline{x}]]/(\phi(\overline{x}))$.

Proof. Let $\Delta : R[[\overline{x}]] \to R[[\overline{x}]] \hat{\otimes} R[[\overline{x}]]$ denote the comultiplication map for the formal Hopf algebra $R[[\overline{x}]]_F$. One has an isomorphism $\psi : R[[\overline{x}]] \hat{\otimes} R[[\overline{x}]] \to R[[x \hat{\otimes} 1, 1 \hat{\otimes} x]]$, and so Δ induces an R-algebra map

$$\psi \Delta : R[[\overline{x}]] \to R[[\overline{x} \hat{\otimes} 1, 1 \hat{\otimes} \overline{x}]].$$

There exists an isomorphism

$$R[[\overline{x} \hat{\otimes} 1, 1 \hat{\otimes} \overline{x}]] \cong R[[\overline{x} \hat{\otimes} 1]][[1 \hat{\otimes} \overline{x}]],$$

and consequently there exists a surjection

$$R[[\overline{x} \hat{\otimes} 1, 1 \hat{\otimes} \overline{x}]] \to (R[[\overline{x} \hat{\otimes} 1]]/(\phi(\overline{x} \hat{\otimes} 1)))[[1 \hat{\otimes} \overline{x}]].$$

Since ϕ is an isogeny,

$$(R[[\overline{x} \hat{\otimes} 1]]/(\phi(\overline{x} \hat{\otimes} 1)))[[1 \hat{\otimes} \overline{x}]] \cong R[[\overline{x} \hat{\otimes} 1]]/(\phi(\overline{x} \hat{\otimes} 1)) \otimes_R R[[1 \hat{\otimes} \overline{x}]].$$

Thus there is a surjection

$$R[[\overline{x} \hat{\otimes} 1, 1 \hat{\otimes} \overline{x}]] \to R[[\overline{x} \hat{\otimes} 1]]/(\phi(\overline{x} \hat{\otimes} 1)) \otimes_R R[[1 \hat{\otimes} \overline{x}]].$$

Now the map $\psi\Delta$ induces a comultiplication map,

$$R[[\overline{x}]]/(\phi(\overline{x})) \to R[[\overline{x}\,\hat{\otimes}\,1]]/(\phi(\overline{x}\,\hat{\otimes}\,1)) \otimes_R R[[1\,\hat{\otimes}\,\overline{x}]]/(\phi(1\,\hat{\otimes}\,\overline{x}))$$

$$\cong R[[\overline{x}]]/(\phi(\overline{x})) \otimes_R R[[\overline{x}]]/(\phi(\overline{x})),$$

which satisfies coassociativity since Δ does.

The counit and coinverse maps are constructed in a similar fashion. $\qquad\square$

To construct Hopf orders in KC_{p^n}, we modify the formal group G_m^n by the lower-triangular matrices Θ and $\Theta^{(p)}$ in the manner of §6.1. This results in the formal groups F_Θ and $F_{\Theta^{(p)}}$. ($\Theta^{(p)}$ is the matrix whose entries are the pth powers of those of Θ.) We then generalize the homomorphism ϕ_n of §6.1 to give an isogeny $F_\Theta \to F_{\Theta^{(p)}}$ whose kernel is represented by a Hopf order in KC_{p^n}.

The generalization of ϕ_n is defined as

$$f(\overline{x}) = (\Theta^{(p)})^{-1}\Theta\phi_n^\Theta(\overline{x}),$$

where

$$\phi_n^\Theta(\overline{x}) = \Theta^{-1}\phi_n(\Theta\overline{x}).$$

We require certain integrality conditions on $f(\overline{x})$.

Proposition 6.2.4. *Let $\Theta = (\theta_{i,j}) \in M_n(R)$ be lower-triangular with $\det(\Theta) \neq 0$. Suppose $\mathrm{ord}(\theta_{i,i}) > \mathrm{ord}(\theta_{i,j}) \geq (1-q)\mathrm{ord}(\theta_{i,i})$ for all $i > j$ with $\theta_{i,j} \neq 0$, where*

$$0 < q < \frac{p-1}{2p-1}.$$

Suppose also that $\mathrm{ord}(\theta_{r,r}) > 0$ and $\mathrm{ord}(\theta_{r,r}) \geq d \cdot \mathrm{ord}(\theta_{r+1,r+1})$ for all r, with

$$d \geq \frac{p}{1-q} + \frac{q}{1 - \frac{1-q}{p}}$$

and

$$e' > \left(\frac{p}{p-1}\right)\left(1 + \frac{q}{d-1}\right)\mathrm{ord}(\theta_{1,1}).$$

Then $f(\overline{x}) = (\Theta^{(p)})^{-1}\Theta\phi_n^\Theta(\overline{x})$ is in $R[[\overline{x}]]$ and satisfies

$$f(\overline{x}) \equiv \overline{x}^{(p)} \mod \pi R[[\overline{x}]].$$

Proof. See the paper of L. Childs and R. Underwood [CU03, Theorem 2.0]. $\qquad\square$

With these preliminaries established, we can construct our formal group Hopf orders in KC_{p^n}.

Proposition 6.2.5. *Suppose Θ is an $n \times n$ lower-triangular matrix with entries in R for which $\det(\Theta) \neq 0$ and $\mathrm{ord}(\theta_{r,r}) > 0$ for all r. Suppose for all r, and all $s < r$ for which $\theta_{r,s} \neq 0$, there exist numbers q and d such that $\mathrm{ord}(\theta_{r,r}) > \mathrm{ord}(\theta_{r,s}) \geq (1 - q)\mathrm{ord}(\theta_{r,r})$ and $\mathrm{ord}(\theta_{r,r}) \geq d\,\mathrm{ord}(\theta_{r+1,r+1})$, where*

$$0 < q < \frac{p-1}{2p-1},$$

$$\mathrm{ord}(\theta_{1,1}) < \left(\frac{p-1}{p}\right)\left(\frac{d-1}{d-1+q}\right)e',$$

and

$$d \geq \frac{p}{1-q} + \frac{q}{1 - \frac{1-q}{p}}.$$

Then Θ gives rise to an R-Hopf order in KC_{p^n}.

Proof. Put

$$F_\Theta(\overline{x}, \overline{y}) = \Theta^{-1}G_m^n(\Theta\overline{x}, \Theta\overline{y})$$

and

$$F_{\Theta^{(p)}}(\overline{x}, \overline{y}) = (\Theta^{(p)})^{-1}G_m^n(\Theta^{(p)}\overline{x}, \Theta^{(p)}\overline{y}).$$

By Proposition 6.1.5, F_Θ and $F_{\Theta^{(p)}}$ are defined over R, and, by Proposition 6.2.4, $f(\overline{x}) = (\Theta^{(p)})^{-1}\Theta\phi_n^\Theta(\overline{x})$ is defined over R. Moreover, $f : F_\Theta \to F_{\Theta^{(p)}}$ is a homomorphism of generically split n-dimensional polynomial formal groups, and since $f(\overline{x}) \equiv \overline{x}^{(p)} \bmod \pi R[[\overline{x}]]$, f is an isogeny. Applying Proposition 6.2.3, we conclude that the kernel of f is represented by an R-Hopf order in KC_{p^n}. $\quad\square$

The Hopf order given by the matrix Θ of Proposition 6.2.5 is a **formal group Hopf order** and is denoted by H_Θ. The structure of H_Θ is determined as follows. Let $\Theta^{-1} = (\eta_{i,j})$. Then

$$H_\Theta = R[z_1, z_2, \ldots, z_n],$$

where

$$z_1 = \eta_{1,1}(g^{p^{n-1}} - 1)$$

$$z_2 = \eta_{2,1}(g^{p^{n-1}} - 1) + \eta_{2,2}(g^{p^{n-2}} - 1)$$

$$\vdots$$

$$z_n = \eta_{n,1}(g^{p^{n-1}} - 1) + \eta_{n,2}(g^{p^{n-2}} - 1) + \cdots + \eta_{n,n}(g - 1),$$

and $\langle g \rangle = C_{p^n}$.

Note that the matrix Θ^{-1} yields $\dfrac{n(n+1)}{2}$ parameters that describe the Hopf algebra H_Θ.

Remark 6.2.1. Suppose Θ_1 and Θ_2 are lower-triangular matrices that satisfy the hypothesis of Proposition 6.2.5. Suppose $\Theta_2 = \Theta_1 M$, where M is in $\mathrm{GL}_n(R)$.

Then $(\Theta_2)^{-1} = M^{-1}\Theta_1^{-1}$, where M^{-1} is lower-triangular and corresponds to a sequence of row operations on Θ_1^{-1}. Hence, the associated Hopf orders H_{Θ_1} and H_{Θ_2} are equal.

Remark 6.2.2. Suppose Θ is a diagonal matrix that satisfies the hypothesis of Proposition 6.2.5. Then, by Remark 6.2.1, we may assume $\Theta = \mathrm{diag}(\pi^{i_1}, \ldots, \pi^{i_n})$ with $i_{r-1} \geq p i_r$ for $r = 2, \ldots, n$ since in this case we may take $q = 0$ and $d = p$. Now $\Theta^{-1} = \mathrm{diag}(\pi^{-i_1}, \ldots, \pi^{-i_n})$, and H_Θ is a Larson order in KC_{p^n} whose exponents i_r are given by a p-adic obgv on C_{p^n}.

It is fairly easy to satisfy the conditions of Proposition 6.2.5. Let $n \geq 3$.

Proposition 6.2.6. *Let $n \geq 3$, and let K be a finite extension of $\mathbb{Q}_p$ such that*

$$e = \frac{(2p+1)^n + 1}{2}.$$

Let Θ be a lower-triangular $n \times n$ matrix with

$$\theta_{r,s} = \begin{cases} \pi^{(2p+1)^{n-r}} & \text{if } r = s \\ \pi^{(2p+1)^{n-2}-1} & \text{if } r = 2, s = 1 \\ 0 & \text{if } r \neq s. \end{cases}$$

Put

$$d = 2p+1, \quad q = \frac{1}{(2p+1)^{n-2}}.$$

Then Θ gives rise to an R-Hopf order in KC_{p^n}.

Proof. We check that the conditions of Proposition 6.2.5 hold. We have

$$\mathrm{ord}(\theta_{1,1}) = (2p+1)^{n-1}$$

$$= \frac{(2p+1)^n}{2p+1}$$

$$< \frac{(2p+1)^n + 1}{2p+1}$$

$$< \frac{(2p+1)^n + 1}{2p+q}$$

$$= \left(\frac{2}{2p+q}\right)\left(\frac{(2p+1)^n + 1}{2}\right)$$

$$= \frac{2}{2p+q}e$$

$$= \left(\frac{p-1}{p}\right)\left(\frac{2p}{2p+q}\right)e'$$

$$= \left(\frac{p-1}{p}\right)\left(\frac{d-1}{d-1+q}\right)e'.$$

Moreover, $q < 1$, and so $q < 1 - \frac{1-q}{p}$. Thus

$$\frac{p}{1-q} + 1 > \frac{p}{1-q} + \frac{q}{1 - \frac{1-q}{p}}.$$

Also, since $q < 1/2, 2p > \dfrac{p}{1-q}$. Thus

$$d = 2p + 1 > \frac{p}{1-q} + 1 > \frac{p}{1-q} + \frac{q}{1 - \frac{1-q}{p}}.$$

In addition,

$$\operatorname{ord}(\theta_{2,1}) = (2p+1)^{n-2} - 1$$
$$= \frac{(2p+1)^{n-2} - 1}{(2p+1)^{n-2}}(2p+1)^{n-2}$$
$$= \left(1 - \frac{1}{(2p+1)^{n-2}}\right)(2p+1)^{n-2}$$
$$= (1-q)\operatorname{ord}(\theta_{2,2}),$$

and

$$\operatorname{ord}(\theta_{r,r}) = (2p+1)\operatorname{ord}(\theta_{r+1,r+1}).$$

Thus Θ satisfies the conditions of Proposition 6.2.5, and consequently there exists an R-Hopf order in KC_{p^n} of the form H_Θ.   □

To illustrate Proposition 6.2.6, take $p = 5, n = 4$. Let $K = \mathbb{Q}_5(\alpha)$, where $\alpha^{7321} = 5$. We have $5R = \pi^e$ with $e = 7321$. Then $\operatorname{ord}(5) = 7321$. Let Θ be the matrix

$$\Theta = \begin{pmatrix} \pi^{1331} & 0 & 0 & 0 \\ \pi^{120} & \pi^{121} & 0 & 0 \\ 0 & 0 & \pi^{11} & 0 \\ 0 & 0 & 0 & \pi \end{pmatrix}.$$

Then

$$\Theta^{-1} = \begin{pmatrix} \pi^{-1331} & 0 & 0 & 0 \\ -\pi^{-1332} & \pi^{-121} & 0 & 0 \\ 0 & 0 & \pi^{-11} & 0 \\ 0 & 0 & 0 & \pi^{-1} \end{pmatrix},$$

and thus H_Θ is an R-Hopf order in KC_{625} of the form

$$H_\Theta = R\left[\frac{g^{125} - 1}{\pi^{1331}}, \frac{g^{125} - 1}{\pi^{1332}} - \frac{g^{25} - 1}{\pi^{121}}, \frac{g^5 - 1}{\pi^{11}}, \frac{g - 1}{\pi}\right].$$

6.3 Chapter Exercises

Exercises for §6.1

1. Let $F(\overline{x}, \overline{y})$ be a formal group. Show that conditions (i) and (ii) of Definition 6.1.1 together with the condition $F(\overline{x}, \overline{0}) = \overline{x}$ imply that $F(\overline{0}, \overline{y}) = \overline{y}$.
2. Show that $F(x, y) = x + y$ defines a one-dimensional formal group. Show that $F(x, y) = x + y + xy$ defines a one-dimensional formal group.
3. Let $F(x, y) = x + y + x^2 + xy + y^2 + x^3 + x^2y + xy^2 + y^3 + \cdots$. Does $F(x, y)$ determine a one-dimensional formal group?
4. Let $F(x, y)$ be a one-dimensional polynomial formal group. Find a power series $\sigma(x)$ for which $F(x, \sigma(x)) = 0$.
5. Consider the polynomial formal group

$$F(\overline{x}, \overline{y}) = \begin{pmatrix} x_1 + y_1 + \theta_{1,1} x_1 y_1 \\ x_2 + y_2 + \frac{\theta_{2,1}(\theta_{2,1} - \theta_{1,1})}{\theta_{2,2}} x_1 y_1 + \theta_{2,2} x_2 y_2 + \theta_{2,1} x_1 y_2 + \theta_{2,1} x_2 y_1 \end{pmatrix}$$

constructed in §6.1. Find a 2-tuple of power series $\sigma(\overline{x}) = (\sigma_1(\overline{x}), \sigma_2(\overline{x}))$ for which $F(\overline{x}, \sigma(\overline{x})) = 0$.
6. Let Θ be a matrix that satisfies the hypothesis of Proposition 6.1.4. Show that, for any prime p, $p \cdot \mathrm{ord}(\theta_{i,j}) \geq \mathrm{ord}(\theta_{i,i})$ for all $i > j$.
7. Prove Proposition 6.1.5.
8. Prove Proposition 6.1.6.

Exercises for §6.2

9. Finish the proof of Proposition 6.2.2 by formulating the necessary conditions on σ and showing that the coinverse property holds.

Chapter 7
Hopf Orders in KC_p

Let p be a rational prime, let K be a finite extension of $\mathbb{Q}$, and let G be a finite Abelian group. In Chapter 5, we constructed a collection of Hopf orders in KG using p-adic order-bounded group valuations on G. These were called Larson orders. We specialized to the case $G = C_{p^n}$, completed R at the prime P above p, and considered Larson orders over the complete local ring $\hat{R}_P$. In Chapter 6, we constructed a collection of formal group Hopf orders in $K_P C_{p^n}$ and found that the Larson orders in $K_P C_{p^n}$ formed a proper subcollection of the formal group Hopf orders.

In this chapter, we assume that K is a finite extension of $\mathbb{Q}_p$ containing ζ_p, with ring of integers R, uniformizing parameter π, and discrete valuation ord. Note that $(p) = P^{p-1}$, where $P = (1 - \zeta_p)$, and so $e' = \text{ord}(p)/(p-1)$ is an integer. We give a complete classification of Hopf orders in KG, where G is the cyclic group of order p.

7.1 Classification of Hopf Orders in KC_p

Let g be a generator for C_p. Then a Larson order in KC_p can be written

$$H(i) = R\left[\frac{g-1}{\pi^i}\right],$$

where i is an integer $0 \le i \le e'$. In this section, we show that every Hopf order in KC_p is of the form $H(i)$ for some integer i, $0 \le i \le e'$. We begin with the classification of rank p Hopf algebras given by J. Tate and F. Oort [TO70], following closely the notes of L. Childs [Ch00].

Let H be an R-Hopf order in KC_p. For each $n \in \mathbb{F}_p^\times = \{1, 2, 3, \ldots, p-1\}$, there is a Hopf algebra endomorphism $[n] : H \to H$ defined as

$$[n](h) = m^{(n-1)}(\Delta^{(n-1)}(h))$$

R.G. Underwood, *An Introduction to Hopf Algebras*, DOI 10.1007/978-0-387-72766-0_7,   129
© Springer Science+Business Media, LLC 2011

for $h \in H$ (See §3.4). The collection of endomorphisms $E = \{[n] : n \in \mathbb{F}_p^{\times}\}$ is a group under the operation $[n][n'] = [nn']$, with nn' taken modulo p. Clearly $E \cong \mathbb{F}_p^{\times}$, and henceforth we identify $\mathbb{F}_p^{\times}$ with this collection of endomorphisms. We have the group ring $\mathbb{Z}_p \mathbb{F}_p^{\times}$.

By Hensel's Lemma, $\mathbb{Z}_p$ contains the $(p-1)$st roots of unity. For each $[n] \in \mathbb{F}_p^{\times}$, there exists a root of $X^{p-1} - 1$ in $\mathbb{Z}_p$ that is congruent to n modulo p. This defines the **Teichmüller character**, which is the unique multiplicative group homomorphism $\chi : \mathbb{F}_p^{\times} \to \mathbb{Z}_p^{\times}$, where $\chi(n)$ is the root of $X^{p-1} - 1$, which satisfies $\chi(n) \equiv n$ (mod p).

For j, $1 \leq j \leq p-1$, put

$$\eta_j = \frac{1}{p-1} \sum_{n \in \mathbb{F}_p^{\times}} \chi^j(n)^{-1}[n].$$

Then the η_j form a set of pairwise mutually orthogonal minimal idempotents in $\mathbb{Z}_p \mathbb{F}_p^{\times}$ such that

$$\mathbb{Z}_p \mathbb{F}_p^{\times} = \mathbb{Z}_p \eta_1 \oplus \mathbb{Z}_p \eta_2 \oplus \cdots \oplus \mathbb{Z}_p \eta_{p-1}.$$

Let H^+ denote the augmentation ideal of H. From the short exact sequence

$$0 \to H^+ \to H \overset{\epsilon}{\to} R \to 0,$$

one obtains $H = R \oplus H^+$. Moreover, $\mathbb{Z}_p \mathbb{F}_p^{\times}(H^+) = H^+$, and so

$$H^+ = \eta_1(H^+) \oplus \eta_2(H^+) \oplus \cdots \oplus \eta_{p-1}(H^+). \tag{7.1}$$

Lemma 7.1.1. *For i, j, $1 \leq i, j \leq p-1$, we have $\eta_i(H^+)\eta_j(H^+) \subseteq \eta_{i+j}(H^+)$, where the subscript $i + j$ is taken modulo $p - 1$.*

Proof. Let $h \in H^+$. Then $[n]\eta_i(h) = \chi^i(n)(h)$ for all $1 \leq i \leq p-1, n \in \mathbb{F}_p^{\times}$. Thus

$$\eta_i(H^+) = \left\{ h \in H^+ : [n](h) = \chi^i(n)h, \forall n \in \mathbb{F}_p^{\times} \right\}.$$

Let $h_i \in \eta_i(H^+)$, $h_j \in \eta_j(H^+)$. Then, for all $[n] \in \mathbb{F}_p^{\times}$,

$$[n](h_i h_j) = [n](h_i)[n](h_j)$$
$$= \chi^i(n)h\chi^j(n)h_i h_j$$
$$= \chi^{i+j}(n)h_i h_j,$$

so that $h_i h_j \in \eta_{i+j}(H^+)$. $\qquad\square$

With these preliminaries in mind, we give the Tate/Oort classification of Hopf orders in KC_p. We begin with a characterization of the Hopf order $H = RC_p$ in KC_p.

Lemma 7.1.2. *Let*

$$x = - \sum_{n \in \mathbb{F}_p^\times} \chi(n)^{-1}(g^n - 1).$$

Then $RC_p = R[x]$ with $x^p = wx$ for some $w \in R$.

Proof. By (7.1), there is a decomposition

$$RC_p = R \oplus \eta_1(RC_p^+) \oplus \eta_2(RC_p^+) \oplus \cdots \oplus \eta_{p-1}(RC_p^+). \tag{7.2}$$

Let $k = R/\pi R$ be the residue class field of R, and set $kC_p^+ = k \otimes_R RC_p^+$. By direct calculation, $\eta_1(kC_p^+) = k \otimes_R \eta_1(RC_p^+) = kx$, and from this and Lemma 7.1.1 one deduces that $\eta_i(kC_p^+) = kx^i$ for $1 \leq i \leq p - 1$. It follows that $\eta_i(RC_p^+) = Rx^i$ for $1 \leq i \leq p - 1$, so that $RC_p = R[x]$ by (7.2). By Lemma 7.1.1 $x^p \in \eta_1(RC_p^+)$, and thus $x^p = wx$ for some $w \in R$. $\square$

We give a similar characterization for the dual RC_p^D, which by Proposition 4.4.8 is an R-Hopf order in KC_p. Let $\{e_i\}_{i=0}^{p-1}$ denote the set of minimal idempotents in KC_p. Then $RC_p^D = Re_0 \oplus Re_1 \oplus \cdots \oplus Re_{p-1}$.

Lemma 7.1.3. *Let $y = \sum_{n \in \mathbb{F}_p^\times} \chi(n)e_n$. Then $\eta_i((RC_p^D)^+) = Ry^i$, for $i = 1, \ldots, p - 1$, and $RC_p^D = R[y]$ with $y^p = y$.*

Proof. For a proof, see [Ch00, (16.10)]. $\square$

Now let H be an arbitrary Hopf order in KC_p.

Lemma 7.1.4. *$H = R[z]$ with $z^p = bz$ for some $b \in R$, $z \in H$.*

Proof. We have the decomposition

$$H = R \oplus \eta_1(H^+) \oplus \eta_2(H^+) \oplus \cdots \oplus \eta_{p-1}(H^+),$$

with $\eta_1(H^+) = Rz$ for some $z \in H$, by [Ch00, (16.12)].

Let k be the residue class field of R, and let $\overline{k}$ be an algebraic closure of k. By [Wat79, §6.8], $kH = k \otimes_R H$ is either separable as a k-algebra or a local ring. In the separable case, one has

$$\overline{k}H = \overline{k} \otimes_k kH \cong \underbrace{\overline{k} \oplus \overline{k} \oplus \cdots \oplus \overline{k}}_{p}$$

so that $\overline{k}H = \overline{k}C_p^D$. It follows that $H = RC_p^D$, and so $H = R[z]$ with $z^p = bz, b$ in R, by Lemma 7.1.3. If $\overline{k}H$ is local with separable dual, then $\overline{k}H = \overline{k}C_p$, and so $H = R[z]$, with $z^p = bz, b \in R$ by Lemma 7.1.2.

In the case where $\overline{k}H$ and $\overline{k}H^D$ are local, we have $\overline{k}H \cong \overline{k}[t]$ with $t^p = 0$ by [Wat79, §14.4]. As argued in [Ch00, (16.13)], one deduces that

$$\eta_i(\overline{k}H^+) \cong \eta_i(\overline{k}[t]^+) = \overline{k}t^i$$

for $1 \leq i \leq p - 1$, and thus $\eta_i(H^+) = Rz^i$ for $1 \leq i \leq p - 1$, so that $H = R[z]$. Now, by Lemma 7.1.1, $z^p \in \eta_1(H^+) = Rz$, so that $z^p = bz$ for some $b \in R$. $\quad\square$

We can now give the Tate/Oort classification of Hopf orders in KC_p.

Proposition 7.1.1. *Let H be a Hopf order in KC_p. Then $H = R[z]$, where $z^p = bz$ with*

$$z = -\frac{1}{c} \sum_{n \in \mathbb{F}_p^\times} \chi(n)^{-1} (g^n - 1)$$

for some $b, c \in R$.

Proof. By Lemma 7.1.4, $H = R[z]$ for some $z \in H$, with $z^p = bz$, $b \in R$, and, by Lemma 7.1.2, $RC_p = R[x]$ with $x^p = wx$ for $w \in R$ with

$$x = -\sum_{n \in \mathbb{F}_p^\times} \chi(n)^{-1} (g^n - 1).$$

Let L be an extension of K whose ring of integers S contains the $(p-1)$st roots of w and b. Let $c \in S$ be such that $c^{p-1} = w/b$. Then $SC_p = S[x]$ embeds into SH through the relation $x = cz$. It follows that $H = R[z]$ with $z^p = bz$ and

$$z = -\frac{1}{c} \sum_{n \in \mathbb{F}_p^\times} \chi(n)^{-1}(g^n - 1).$$

$\quad\square$

Using the Tate/Oort classification, we can show that every Hopf order in KC_p is a Larson order. Let H be an arbitrary R-Hopf order in KC_p. By Proposition 7.1.1, $H = R[z]$ and $z^p = bz$, with $z = -\frac{1}{c}\sum_{n \in \mathbb{F}_p^\times} \chi(n)^{-1}(g^n - 1)$. We have

$$z = -\frac{1}{c} \sum_{n \in \mathbb{F}_p^\times} \chi(n)^{-1}(g^n - 1)$$

$$= -\frac{(g-1)}{c} \sum_{n \in \mathbb{F}_p^\times} \chi(n)^{-1}(1 + g + g^2 + \cdots + g^{n-1})$$

$$= -v\frac{(g-1)}{c}$$

with

$$v = \sum_{n \in \mathbb{F}_p^\times} \chi(n)^{-1} (1 + g + g^2 + \cdots + g^{n-1}).$$

Observe that $v \in \mathbb{Z}_p C_p \subseteq RC_p$.

Lemma 7.1.5. v *is a unit in* $\mathbb{Z}_p C_p$.

Proof. Let X be indeterminate. By Lemma 7.1.3,

$$\mathbb{Z}_p C_p^D = \mathbb{Z}_p[X]/(X^p - X) \cong \mathbb{Z}_p \oplus \mathbb{Z}_p[\zeta_p].$$

There is an embedding

$$\mathbb{Z}_p C_p \to \mathbb{Z}_p \oplus \mathbb{Z}_p[\zeta_p]$$

defined by $g \mapsto (1, \zeta_p)$. Let $\hat{v}$ denote the image of v under this embedding. We have $\hat{v} = (\alpha, \beta)$, where $\alpha = \sum_{n \in \mathbb{F}_p^\times} \chi(n)^{-1} n$ and $\beta = \sum_{n \in \mathbb{F}_p^\times} \chi(n)^{-1}(1 + \zeta_p + \zeta_p^2 + \cdots + \zeta_p^{n-1})$. If α is not a unit of $\mathbb{Z}_p$, then

$$\sum_{n \in \mathbb{F}_p^\times} \chi(n)^{-1} n \equiv 0 \pmod{p}.$$

Since $\chi(n) \equiv n \pmod{p}$, this says that $p \equiv 1 \pmod{p}$, which is impossible. Thus α is a unit of $\mathbb{Z}_p$.

Moreover, if β is not a unit of $\mathbb{Z}_p[\zeta_p]$, then there exists an integer $k \geq 0$ such that

$$\left(\sum_{n \in \mathbb{F}_p^\times} \chi(n)^{-1}(1 + \zeta_p + \zeta_p^2 + \cdots + \zeta_p^{n-1}) \right)^k \equiv 0 \pmod{(1 - \zeta_p)}.$$

Since $1 - \zeta_p^i \equiv 0 \pmod{(1 - \zeta_p)}$ for all i, $1 \leq i \leq n - 1$,

$$\left(\sum_{n \in \mathbb{F}_p^\times} \chi(n)^{-1} n \right)^k \equiv 0 \pmod{(1 - \zeta_p)},$$

and thus there is an integer $s \geq 0$ with

$$\left(\sum_{n \in \mathbb{F}_p^\times} \chi(n)^{-1} n \right)^{st} \equiv 0 \pmod{p},$$

which again leads to a contradiction. Thus β is a unit in $\mathbb{Z}_p[\zeta_p]$. It follows that $\hat{v}$ is a unit in $\mathbb{Z}_p C_p^D$.

Let $y \in \mathbb{Z}_p C_p^D$ be such that $\hat{v} y = 1$. Since y is integral over $\mathbb{Z}_p$, there exists a monic polynomial $f(X) = X^l + a_1 X^{l-1} + \cdots + a_{l-1} X + a_l$ with $y^l + a_1 y^{l-1} + \cdots + a_{l-1} y + a_l = 0$. Multiplying this equation by $\hat{v}^{l-1}$ yields

$$y + a_1 + a_2 \hat{v} + \cdots + a_l \hat{v}^{l-1} = 0.$$

Hence,

$$y = -a_1 - a_2\hat{v} - \cdots - a_l\hat{v}^{l-1}.$$

Since $v \in \mathbb{Z}_p C_p$, $r = -a_1 - a_2 v - \cdots - a_l v^{l-1} \in \mathbb{Z}_p C_p$ with $vr = 1$. Thus v is a unit of $\mathbb{Z}_p C_p \subseteq RC_p$. $\square$

We can now show that every Hopf order in KC_p is a Larson order.

Proposition 7.1.2. *Let p be a prime number, and let K be a finite extension of $\mathbb{Q}_p$ with ring of integers R and uniformizing parameter π. Suppose $\zeta_p \in K$. Let $e' = \mathrm{ord}(p)/(p-1)$. Let C_p denote the cyclic group of order p, generated by g, and let H be an R-Hopf order in KC_p. Then*

$$H = R\left[\frac{g-1}{\pi^i}\right],$$

where i is an integer $0 \le i \le e'$.

Proof. By Proposition 7.1.1, $H = R\left[\dfrac{-v(g-1)}{c}\right]$, where v is defined as above and c is some element of R. By Lemma 7.1.5, v is a unit of RC_p. Now, by Proposition 4.4.3, $RC_p \subseteq H$, and so v is a unit of H. It follows that $H = R\left[\dfrac{g-1}{c}\right]$.

Let $c = u\pi^i$ for some unit $u \in R$, integer $i \ge 0$, where π is a uniformizing parameter for R. Then $H = R\left[\dfrac{g-1}{\pi^i}\right]$. It remains to show that $0 \le i \le e'$. By Proposition 4.4.9, H embeds into RC_p^D through $g \mapsto (1, \zeta_p, \zeta_p^2, \ldots, \zeta_p^{p-1})$, and consequently $\mathrm{ord}(\zeta_p - 1) = e' \ge i \ge 0$. $\square$

Proposition 7.1.2 can be applied in the following way. Let $\langle g \rangle = C_{p^n}$, so that $\langle g^{p^{n-1}} \rangle = C_p$. Let H be an R-Hopf order in KC_{p^n}, $n \ge 1$. Then $K\langle g^{p^{n-1}} \rangle \cap H$ is an R-Hopf order in KC_p. Consequently, $K\langle g^{p^{n-1}} \rangle \cap H$ is the Larson order

$$H(i) = R\left[\frac{g^{p^{n-1}} - 1}{\pi^i}\right]$$

for some integer $0 \le i \le e'$. This fact is used in the following proposition.

Proposition 7.1.3. *Let $n \ge 1$ be an integer, and suppose that K contains a primitive p^nth root of unity ζ_{p^n}. Let H be an R-Hopf order in KC_{p^n}, let $H(i)$ denote the Larson order as above, and let $\overline{H}$ denote the image of H under the canonical surjection $g^{p^{n-1}} \mapsto 1$. Let $\overline{H}^*$ denote the linear dual of $\overline{H}$. Let $\hat{C}_{p^n} = \langle \gamma \rangle$ denote the character group of C_{p^n}. Set*

$$J = \overline{H}^*\left[\frac{\gamma u - 1}{\pi^{i'}}\right],$$

with $u \in K\langle \gamma^p \rangle$, $i' = e' - i$. If $\langle J, H \rangle \subseteq R$, then $J = H^*$.

Proof. From $\langle J, H \rangle \subseteq R$, one has $J \subseteq H^*$. We show that $J = H^*$ by showing that their discriminants are equal. Let $\alpha = \frac{\gamma u - 1}{\pi^{i'}}$. We have

$$\gamma^p = \frac{(1 + \alpha \pi^{i'})^p}{u^p} \in K\langle \gamma^p \rangle,$$

and so, since $e = (p-1)e' \geq (p-1)i'$,

$$\alpha^p + \sum_{r=1}^{p-1} \binom{p}{r} \alpha^r \pi^{(r-p)i'} = \frac{\gamma^p u^p - 1}{\pi^{pi'}} \in H^* \cap K\langle \gamma^p \rangle.$$

Since $\overline{H}^* = H^* \cap K\langle \gamma^p \rangle$, α satisfies a monic degree p polynomial with coefficients in $\overline{H}^*$. Consequently, if $\{a_v\}_{v=1}^{p^{n-1}}$ is an R-basis for $\overline{H}^*$, then $\{a_v \alpha^k\}$ with $v = 1, 2 \ldots, p^{n-1}$, $k = 0, 1, \ldots, p-1$, is an R-basis for J.

There is a short exact sequence of Hopf orders,

$$R \to H(i) \to H \to \overline{H} \to R,$$

that dualizes to yield the short exact sequence of duals

$$R \to \overline{H}^* \to H^* \to H(i)^* \to R.$$

By Proposition 4.4.11,

$$\epsilon_{H(i)}\left(\int_{H(i)}\right) \epsilon_{H(i)^*}\left(\int_{H(i)^*}\right) = pR,$$

and so, by Proposition 5.3.3, $\epsilon_{H(i)^*}\left(\int_{H(i)^*}\right) = \pi^{(p-1)i} R$. Therefore, by Proposition 4.4.12,

$$\epsilon_{H^*}\left(\int_{H^*}\right) = \epsilon_{\overline{H}^*}\left(\int_{\overline{H}^*}\right) \pi^{(p-1)i} R.$$

Thus,

$$\epsilon_{H^*}\left(\int_{H^*}\right)^{p^n} = \epsilon_{\overline{H}^*}\left(\int_{\overline{H}^*}\right)^{p^n} \pi^{p^n(p-1)i} R.$$

Now, by Proposition 4.4.13,

$$\mathrm{disc}(H^*) = \mathrm{disc}(\overline{H}^*)^p \pi^{p^n(p-1)i} R. \tag{7.3}$$

Let $J_0 = \overline{H}^*[u\gamma - 1] = \overline{H}^*[\gamma]$. Then J_0 is an R-Hopf order in $K\hat{C}_{p^n}$, and we can compute the discriminant of J_0 using the exact sequence

$$R \to \overline{H}^* \to J_0 \to H(0) \to R.$$

We have

$$\mathrm{disc}(J_0) = \mathrm{disc}(\overline{H}^*)^p \pi^{p^n(p-1)e'} R. \tag{7.4}$$

Let

$$M = \mathrm{diag}(\underbrace{1, 1, \ldots, 1}_{p^{n-1}}, \underbrace{\pi^{i'}, \pi^{i'}, \ldots, \pi^{i'}}_{p^{n-1}}, \ldots, \underbrace{\pi^{(p-1)i'}, \pi^{(p-1)i'}, \ldots, \pi^{(p-1)i'}}_{p^{n-1}}).$$

Then M multiplies the basis $\{a_v \alpha^k\}$ of J to give the basis $\{a_v (u\gamma)^k\}$ of J_0. Consequently,

$$\mathrm{disc}(J_0) = \det(M)^2 \mathrm{disc}(J)$$
$$= \pi^{p^n(p-1)i'} \mathrm{disc}(J).$$

Now

$$\mathrm{disc}(J) = \mathrm{disc}(J_0)\pi^{-p^n(p-1)i'} R$$
$$= \mathrm{disc}(\overline{H}^*)^p \pi^{p^n(p-1)e' - p^n(p-1)i'} R \quad \text{by (7.4)}$$
$$= \mathrm{disc}(\overline{H}^*)^p \pi^{p^n(p-1)i} R$$
$$= \mathrm{disc}(H^*) \quad \text{by (7.3).}$$

Since $J \subseteq H^*$ and their discriminants are equal, $J = H^*$. $\square$

By Proposition 4.4.8, the linear dual of the R-Hopf order $H(i)$ in KC_p is an R-Hopf order $H(i)^*$ in $K\hat{C}_p$, $\langle \gamma \rangle = \hat{C}_p$. We compute $H(i)^*$.

Proposition 7.1.4. *Let $H(i)$ be an R-Hopf order in KC_p. Then $H(i)^* = H(i') = R\left[\frac{\gamma-1}{\pi^{i'}}\right]$, where $i' = e' - i$.*

Proof. Let $J = R\left[\frac{\gamma-1}{\pi^{i'}}\right]$ be the Larson order given by the parameter i'. For $a = 1, \ldots, p-1$,

$$\left\langle \frac{\gamma-1}{\pi^{i'}}, \left(\frac{g-1}{\pi^i}\right)^a \right\rangle = \frac{(\zeta_p - 1)^a}{\pi^{i'+ai}},$$

which is in R since $\mathrm{ord}(\zeta_p - 1) = e'$. Thus $J \subseteq H(i)^*$. Now, by Proposition 7.1.3 with $n = 1, u = 1$, one has $H(i') = H(i)^*$. $\square$

Proposition 7.1.3 can also be used to compute the linear dual of a Larson order in KC_{p^2}.

Proposition 7.1.5. *Assume that $\zeta_{p^2} \in K$, let $\hat{C}_{p^2} = \langle \gamma \rangle$ denote the character group of C_{p^2}, and let $H(i, j)$ be a Larson order in KC_{p^2}. Then*

$$H(i, j)^* = A(j', i', \zeta_{p^2}^{-1}) = R\left[\frac{\gamma^p - 1}{\pi^{j'}}, \frac{\gamma u - 1}{\pi^{i'}}\right],$$

where $i' = e' - i$, $j' = e' - j$, and $u = \sum_{m=0}^{p-1} \zeta_{p^2}^{-m} l_m$, where $l_m = \sum_{a=0}^{p-1} \zeta_p^{-ma} \gamma^a$.

Proof. Note that $H(i) = K\langle g^p \rangle \cap H(i, j)$ and $\overline{H(i, j)} = R[\frac{\overline{g}-1}{\pi^j}]$, where $\overline{H(i, j)}$ denotes the image of $H(i, j)$ under the canonical surjection $g^p \mapsto 1$. By Proposition 7.1.4, $\overline{H(i, j)}^* = R[\frac{\gamma^p - 1}{\pi^{j'}}]$. Put

$$J = R\left[\frac{\gamma^p - 1}{\pi^{j'}}\right]\left[\frac{\gamma u - 1}{\pi^{i'}}\right].$$

Then, as one can easily compute, for $0 \leq s, t \leq p - 1$,

$$\langle \gamma^p - 1, (g^p - 1)^s (g - 1)^t \rangle \in \pi^{j' + si + tj} R,$$

and, for $0 \leq s, t \leq p - 1$,

$$\langle \gamma u - 1, (g^p - 1)^s (g - 1)^t \rangle \in \pi^{i' + si + tj} R.$$

Thus $J \subseteq H(i, j)^*$, and so $J = H(i, j)^*$ by Proposition 7.1.3. $\square$

By Proposition 4.4.9, the R-Hopf order $H(i)$ can be embedded into $\bigoplus_{m=0}^{p-1} Re_m$ by the mapping

$$g \mapsto \sum_{m=0}^{p-1} \zeta_p^m e_m, \quad e_m = \frac{1}{p} \sum_{a=0}^{p-1} \zeta_p^{-ma} g^a.$$

Thus every element $h \in H(i)$ can be written as an R-linear combination of the idempotents e_m. We ask: When does a linear combination $\sum_{m=0}^{p-1} a_m e_m$, $a_m \in R$ determine an element $h \in H(i)$? C. Greither [Gr92] has provided the following answer.

Lemma 7.1.6. *Let $a = \sum_{m=0}^{p-1} a_m e_m$, $a_m \in R$. Then $a \in H(i)$ if and only if all of the following conditions hold:*

$$\operatorname{ord}(a_0) \geq 0,$$

$$\operatorname{ord}(a_1 - a_0) \geq i',$$

$$\operatorname{ord}(a_2 - 2a_1 + a_0) \geq 2i',$$

$$\operatorname{ord}(a_3 - 3a_2 + 3a_1 - a_0) \geq 3i',$$

$$\vdots$$

$$\operatorname{ord}\left(\sum_{m=0}^{p-1} \binom{p-1}{m} (-1)^m a_{p-1-m}\right) \geq (p-1)i'.$$

Proof. By Proposition 7.1.4, $H(i)^* = H(i') = R[\frac{\gamma-1}{\pi^{i'}}]$. Thus $a \in H(i)$ if and only if

$$\langle H(i'), a \rangle \subseteq R.$$

But this is equivalent to

$$\left\langle \left(\frac{\gamma-1}{\pi^{i'}}\right)^m, \sum_{m=0}^{p-1} a_m e_m \right\rangle$$

for $m = 0, \ldots, p-1$. Expanding the powers of $\frac{\gamma-1}{\pi^{i'}}$ and using the identity $\langle \gamma^m, e_n \rangle = \delta_{mn}$ yields $a \in H(i)$ if and only if

$$\mathrm{ord}\left(\sum_{m=0}^{k} \binom{k}{m}(-1)^m a_{k-m}\right) \geq ki'$$

for $k = 0, \ldots, p-1$. $\qquad\square$

The next lemma, also due to Greither, shows that we can extend the "partial" sum $\Sigma_{m=0}^{l} a_m e_m$, $l < p-1$ to obtain an element of $H(i)$.

Lemma 7.1.7. *Assume that $l < p-1$. Let $a = \Sigma_{m=0}^{l} a_m e_m$, $a_m \in R$. Suppose that a satisfies*

$$\mathrm{ord}\left(\sum_{m=0}^{k} \binom{k}{m}(-1)^m a_{k-m}\right) \geq ki'$$

for $k = 0, \ldots, l$. Then there exists an element $a_{l+1} \in R$ for which $a' = \Sigma_{m=0}^{l+1} a_m e_m$ satisfies

$$\mathrm{ord}\left(\sum_{m=0}^{k} \binom{k}{m}(-1)^m a_{k-m}\right) \geq ki'$$

for $k = 0, \ldots, l+1$.

Proof. Put $b = \Sigma_{m=1}^{l+1} \binom{l+1}{m}(-1)^m a_{l+1-m}$. Then $b = u\pi^r$ for some unit $u \in R$ and integer $r \geq 0$. Let $a_{l+1} = 1 + \pi^s - u\pi^r$, where $s \geq \min\{r, (l+1)i'\}$. Then

$$\mathrm{ord}\,(a_{l+1} + b) = \mathrm{ord}\left(\sum_{m=0}^{l+1} \binom{l+1}{m}(-1)^m a_{l+1-m}\right) \geq (l+1)i',$$

which proves the proposition. $\qquad\square$

By using Lemma 7.1.7 repeatedly, one can extend the sum $\Sigma_{m=0}^{l} a_m e_m$ to obtain an element of $H(i)$.

7.2 Chapter Exercises

Exercises for §7.1

1. Prove that $\mathbb{F}_p \cong \mathbb{Z}_p/p\mathbb{Z}_p$.
2. Let $\hat{\chi} : \mathbb{Z}_p \to \mathbb{Z}_p$ be the map defined as $\hat{\chi}(a) = \lim_{n\to\infty} a^{p^n}$, $\forall a \in \mathbb{Z}_p$.

   (a) Show that $\hat{\chi}$ is a ring homomorphism.
   (b) Prove that $\hat{\chi}(a) = \chi(a \pmod{p})$, $\forall a \in \mathbb{Z}_p$; that is, show that the image of $\hat{\chi}$ is the same as the image of the Teichmüller character.

3. Let $H(i)$ be a Hopf order in KC_p. Compute the ideal of integrals $\int_{H(i)}$.
4. Let

$$K \to K\langle g^p \rangle \to KC_{p^2} \overset{g^p \mapsto 1}{\to} K\langle \overline{g} \rangle \to K$$

denote the short exact sequence of K-Hopf algebras, and suppose that H is an R-Hopf order in KC_{p^2}. Show that there exists a short exact sequence of R-Hopf orders

$$R \to H(i) \to H \to H(j) \to R,$$

where $H(i)$ and $H(j)$ are Larson orders in KC_p with $i \geq j$.

Chapter 8
Hopf Orders in KC_{p^2}

In this chapter, we assume that K is a finite extension of $\mathbb{Q}_p$, containing ζ_{p^2}, endowed with the discrete valuation ord. Set $e = \text{ord}(p)$, $e' = e/(p-1)$. Let g be a generator for C_{p^2}.

8.1 The Valuation Condition

The K-Hopf algebra KC_{p^2} induces the short exact sequence of K-Hopf algebras

$$K \xrightarrow{\lambda} KC_p \xrightarrow{i} KC_{p^2} \xrightarrow{s} KC_p \xrightarrow{\epsilon} K,$$

where $i : KC_p \to KC_{p^2}$ is the Hopf inclusion and $s : KC_{p^2} \to KC_p$ is the Hopf surjection, given as $g^p \mapsto 1$. Let H denote an R-Hopf order in KC_{p^2}. Since $H' = H \cap KC_p$ is an R-Hopf order in KC_p and $H'' = s(H)$ is an R-Hopf order in KC_p, one has the short exact sequence of Hopf orders

$$R \to H' \to H \to H'' \to R. \tag{8.1}$$

By Proposition 7.1.2, H' and H'' are Larson orders in KC_p of the form

$$H' = H(i) = R\left[\frac{g^p - 1}{\pi^i}\right], \quad H'' = H(j) = R\left[\frac{\overline{g} - 1}{\pi^j}\right], \ \overline{g} = s(g),$$

where i, j are integers satisfying $0 \le i, j \le e'$. Thus (8.1) can be written as

$$R \to H(i) \to H \to H(j) \to R. \tag{8.2}$$

By Proposition 4.3.3, H has a generating integral Λ, which we now compute.

R.G. Underwood, *An Introduction to Hopf Algebras*, DOI 10.1007/978-0-387-72766-0_8, 141
© Springer Science+Business Media, LLC 2011

Proposition 8.1.1. *The ideal of integrals $\int_H$ is of the form $R\Lambda$, where*

$$\Lambda = \frac{p^2}{\pi^{(p-1)(i+j)}} e_0,$$

where $e_0 = \frac{1}{p^2} \sum_{m=0}^{p^2-1} g^m$.

Proof. By Proposition 4.4.5, $\int_H = \epsilon_H \left(\int_H\right) e_0$, and so, by Proposition 4.4.12,

$$\int_H = \epsilon_{H(i)} \left(\int_{H(i)}\right) \epsilon_{H(j)} \left(\int_{H(j)}\right) e_0.$$

Now, by Proposition 5.3.3, $\epsilon_{H(i)} \left(\int_{H(i)}\right) = \frac{p}{\pi^{(p-1)i}} R$, and $\epsilon_{H(j)} \left(\int_{H(j)}\right) = \frac{p}{\pi^{(p-1)j}} R$.
Thus $\int_H = R\Lambda$ with

$$\Lambda = \frac{p^2}{\pi^{(p-1)(i+j)}} e_0. \qquad \qquad \square$$

By Proposition 7.1.4,

$$H(j') = R \left[\frac{\gamma^p - 1}{\pi^{j'}}\right], \qquad H(i') = R \left[\frac{\overline{\gamma} - 1}{\pi^{i'}}\right],$$

and the sequence (8.2) can be dualized (see (4.13)) to form the short exact sequence

$$R \to H(j') \overset{s^*}{\to} H^* \overset{i^*}{\to} H(i') \to R.$$

Definition 8.1.1. Let H be an R-Hopf order in KC_{p^2} that induces the short exact sequence (8.2). Then H satisfies the **valuation condition for** $n = 2$ if either $pj \le i$ or $pi' \le j'$.

In the paper [Lar88], R. Larson proved that every Hopf order in KC_4 satisfies the valuation condition. In [Un94], R. Underwood showed that the valuation condition holds for Hopf orders in KC_{p^2}, $p \ge 2$. Underwood's result also follows from results of N. Byott; see [By93b, §8, Theorem 5]. In this section, we prove that the valuation condition holds for Hopf orders in KC_{p^2}, following closely the proof in [Un94].

We first prove that the valuation condition holds for H that induces short exact sequences (8.2) with $i \ge e/p$. The key is to obtain an R-basis for H^* in a special form.

Lemma 8.1.1. *Let H be an R-Hopf order in KC_{p^2} inducing the short exact sequence (8.2) with $i \ge e/p$. Let $h = \frac{\gamma^p - 1}{\pi^{e'}}$, $k = \frac{\gamma - 1}{\pi^{i'}}$. Put*

$$X = \left(k, k^2, \ldots, k^{p-1}, hk, hk^2, \ldots, hk^{p-1}, \ldots, \right.$$
$$\left. h^{p-1}k, h^{p-1}k^2, \ldots, h^{p-1}k^{p-1}, 1, h, \ldots, h^{p-1}\right).$$

Then there exists an R-basis $\{\alpha_m\}_{m=1}^{p^2}$ for H^* given by the matrix product

$$XM' = (\alpha_1, \alpha_2, \ldots, \alpha_{p^2}),$$

where M' is the $p^2 \times p^2$ matrix

$$M' = \begin{pmatrix} r_{1,1} & r_{1,2} & \cdots & \cdots & r_{1,p^2} \\ 0 & r_{2,2} & \cdots & \cdots & r_{2,p^2} \\ \vdots & \vdots & & & \vdots \\ 0 & 0 & \cdots & \cdots & r_{p^2,p^2} \end{pmatrix},$$

which satisfies the following conditions:

(i) either $r_{l,m} = 0$ or $\mathrm{ord}(r_{l,m}) > \mathrm{ord}(r_{l,m+1})$ for all $1 \le l, m \le p^2 - 1$;

(ii) $\mathrm{ord}(r_{l,m}) \ge aj + bi' - (p-1)i'$ for $1 \le m \le p^2 - p$, where $l = a(p-1) + b$ for $0 \le a \le p-1$, $1 \le b \le p - 1$.

Proof. Since $H(i)$ injects into H, and $RC_{p^2} \subseteq H$, the Larson order $H(i,0)$ is contained in H. Thus

$$H^* \subseteq H(i,0)^* = R\left[\frac{\gamma^p - 1}{\pi^{e'}}, \frac{\gamma u - 1}{\pi^{i'}}\right],$$

where $u = \Sigma_{m=0}^{p-1} \zeta_{p^2}^{-m} l_m$ by Proposition 7.1.5. But, since $i \ge e/p$, Lemma 7.1.6 implies that

$$\frac{u - 1}{\pi^{i'}} = \frac{1}{\pi^{i'}} \sum_{m=0}^{p-1} \left(\zeta_{p^2}^{-m} - 1\right) l_m \in H(e').$$

Thus

$$R\left[\frac{\gamma^p - 1}{\pi^{e'}}, \frac{\gamma u - 1}{\pi^{i'}}\right] = R\left[\frac{\gamma^p - 1}{\pi^{e'}}, \frac{\gamma - 1}{\pi^{i'}}\right],$$

and so $H^* \subseteq H(e',i') = R\left[\frac{\gamma^p-1}{\pi^{e'}}, \frac{\gamma-1}{\pi^{i'}}\right]$. One has that X is an R-basis for $H(e',i')$. Therefore, an R-basis $\{\alpha_m\}_{m=1}^{p^2}$ for H^* is given by the matrix product

$$XM = (\alpha_1, \alpha_2, \ldots, \alpha_{p^2}),$$

where M is a $p^2 \times p^2$ matrix with entries in R. Performing elementary column operations on M, one sees that it is column-equivalent to the matrix

$$M' = \begin{pmatrix} r_{1,1} & r_{1,2} & \cdots & \cdots & r_{1,p^2} \\ 0 & r_{2,2} & \cdots & \cdots & r_{2,p^2} \\ \vdots & \vdots & & & \vdots \\ 0 & 0 & \cdots & \cdots & r_{p^2,p^2} \end{pmatrix},$$

where either $r_{l,m} = 0$ or $\text{ord}(r_{l,m}) > \text{ord}(r_{l,m+1})$ for all $1 \le l, m \le p^2 - 1$. Thus $\{\alpha_m\}_{m=1}^{p^2}$ defined by

$$XM' = (\alpha_1, \alpha_2, \ldots, \alpha_{p^2})$$

is an R-basis for H^*. Note that M' satisfies condition (i) of the lemma.

We now show that M' satisfies condition (ii).

We have

$$\alpha_m = \sum_{l=a(p-1)+b}^{p^2-p} r_{l,m} h^a k^b + \sum_{\iota = p^2-p+1}^{p^2} r_{\iota,m} h^{\iota - p^2 + p - 1}$$

for $m = 1, \ldots, p^2$, $0 \le a \le p-1$, $1 \le b \le p-1$. Since M' is upper-triangular, $r_{l,m} = 0$ for $l = m + 1, \ldots, p^2$. By Proposition 8.1.1, $\Lambda = \frac{p^2}{\pi^{(p-1)(i+j)}} e_0$ is a generating integral for H. Thus, by Proposition 4.3.4, $\alpha_m \cdot \Lambda \in H$ for $m = 1, \ldots, p^2$. Let $d_m = \alpha_m \cdot \Lambda$. Then

$$d_m = \sum_{a(p-1)+b=1}^{p^2-p} \frac{p^2 r_{l,m}}{\pi^{(p-1)(i+j)}} \left(\frac{1}{\pi^{ae'}} \sum_{\eta=0}^{a} \binom{a}{\eta} (-1)^\eta e_{(a-\eta)p} \right) \left(\frac{1}{\pi^{bi'}} \sum_{\iota=0}^{b} \binom{b}{\iota} (-1)^\iota e_{b-\iota} \right)$$

$$+ \sum_{\iota=c}^{p^2} \frac{p^2 r_{\iota,m}}{\pi^{(p-1)(i+j)}} \left(\frac{1}{\pi^{(\iota-c)e'}} \sum_{\eta=0}^{\iota-c} \binom{\iota-c}{\eta} (-1)^\eta e_{(\iota-c-\eta)p} \right),$$

where $c = p^2 - p + 1$.

We have $H(j') \subseteq H^*$, so that $H(j', 0) \subseteq H^*$ and $H \subseteq H(j', 0)^*$. By Proposition 8.1.1, $\varrho = \frac{p^2}{\pi^{(p-1)(e'+j)}} e_0$ is a generating integral for $H(j', 0)^*$, and so, by Proposition 4.3.4, a basis for $H(j', 0)^*$ can be written as

$$\left\{ \left(\frac{\gamma^p - 1}{\pi^{j'}} \right)^a \left(\frac{\gamma - 1}{\pi^0} \right)^b \cdot \varrho \right\} = \left\{ \frac{p^2}{\pi^{(p-1)(e'+j)}} A_a B_b \right\}, \tag{8.3}$$

where

$$A_a = \frac{1}{\pi^{aj'}} \sum_{\eta=0}^{a} \binom{a}{\eta} (-1)^\eta e_{(a-\eta)p}$$

and

$$B_b = \sum_{\iota=0}^{b} \binom{b}{\iota} (-1)^\iota e_{b-\iota}$$

for $0 \le a, b \le p-1$.

Now, since $d_m \in H \subseteq H(j', 0)^*$, d_m can be written as an integral linear combination of the basis (8.3) for $H(j', 0)^*$. Let m be such that $1 \le m \le p^2 - p$. Each term of d_m can be written as

$$\frac{p^2 r_{l,m}}{\pi^{(p-1)(i+j)}} \left(\frac{1}{\pi^{ae'}} \sum_{\eta=0}^{a} \binom{a}{\eta} (-1)^{\eta} e_{(a-\eta)p} \right) \left(\frac{1}{\pi^{bi'}} \sum_{\iota=0}^{b} \binom{b}{\iota} (-1)^{\iota} e_{b-\iota} \right)$$

$$= \frac{r_{l,m} \pi^{aj'} \pi^{e}}{\pi^{(p-1)i} \pi^{bi'} \pi^{ae'}} \left(\frac{p^2}{\pi^{(p-1)(e'+j)}} A_a B_b \right),$$

and so $\dfrac{r_{l,m} \pi^{aj'} \pi^{e}}{\pi^{(p-1)i} \pi^{bi'} \pi^{ae'}} \in R$ or

$$\operatorname{ord}(r_{l,m} \pi^{aj'} \pi^{e}) \geq \operatorname{ord}(\pi^{(p-1)i} \pi^{bi'} \pi^{ae'}),$$

$$\operatorname{ord}(r_{l,m}) + aj' + e \geq (p-1)i + bi' + ae',$$

$$\operatorname{ord}(r_{l,m}) \geq aj + bi' - (p-1)i'. \qquad \square$$

We can now show that the valuation condition holds in the case $i \geq e/p$.

Lemma 8.1.2. *Let H be an R-Hopf order in KC_{p^2} that induces the short exact sequence (8.2). If $i \geq e/p$, then $pi' \leq j'$.*

Proof. Let $\{a_m\}_{m=1}^{p^2}$ denote the R-basis for H^* as constructed in Lemma 8.1.1. Let $\overline{\alpha_m}$ denote the image of α_m under the R-module surjection $H^* \overset{\gamma^p \mapsto 1}{\to} H(i')$. Since $\{\overline{\alpha_m}\}$ is an R-basis for $H(i')$, we can assume that $r_{1,m'} = 1$ for some m', $1 \leq m' \leq p^2$.

Since $H^* = S \oplus H(j')$ for some R-module S, Lemma 8.1.1(i) implies that $m' \leq p^2 - p$, and so, since M' is upper-triangular, we have $r_{l,m'} = 0$ for all $l \geq p^2 - p + 1$. Thus the basis element $\alpha_{m'} \in H^*$ has the form

$$\alpha_{m'} = \sum_{l=a(p-1)+b=1}^{p^2-p} r_{l,m'} h^a k^b = k + \sum_{l=a(p-1)+b=2}^{p^2-p} r_{l,m'} h^a k^b.$$

Thus, H^* contains the pth power

$$\alpha_{m'}^p = k^p + \sum_{l=a(p-1)+b=2}^{p^2-p} r_{l,m'}^p h^{ap} k^{bp} + \sum_{\mathcal{P}} p u_{\mathcal{P}} \prod_{l=a(p-1)+b=1}^{p^2-p} (r_{l,m'} h^a k^b)^{n_l},$$

where the second summation is over all partitions $\mathcal{P}$ of p that are in the form $p = \Sigma_{l=1}^{p^2-p} n_l$, $n_l \geq 0$, and $u_{\mathcal{P}}$ is a unit of R dependent on the partition $\mathcal{P}$.

We claim that

$$\sum_{\mathcal{P}} p u_{\mathcal{P}} \prod_{l=a(p-1)+b=1}^{p^2-p} \left(r_{l,m'} h^a k^b \right)^{n_l} \in H^*.$$

By Lemma 8.1.1(ii), $\operatorname{ord}(r_{l,m'}) \geq aj + bi' - (p-1)i'$, and so there exist elements $s_l \in R$ for which

$$r_{l,m'} = \frac{s_l \pi^{aj} \pi^{bi'}}{\pi^{(p-1)i'}}.$$

Thus

$$\sum_{\mathcal{P}} p u_{\mathcal{P}} \prod_{a(p-1)+b=1}^{p^2-p} (r_{l,m'} h^a k^b)^{n_l} = \sum_{\mathcal{P}} p u_{\mathcal{P}} \prod_{a(p-1)+b=1}^{p^2-p} \left(\frac{s_l \pi^{aj} \pi^{bi'}}{\pi^{(p-1)i'}} h^a k^b \right)^{n_l}$$

$$= \sum_{\mathcal{P}} \frac{p u_{\mathcal{P}}}{\pi^{p(p-1)i'}} \prod_{a(p-1)+b=1}^{p^2-p} s_l^{n_l} (\pi^{aj} h^a \pi^{bi'} k^b)^{n_l}$$

$$= \sum_{\mathcal{P}} \frac{p u_{\mathcal{P}}}{\pi^{p(p-1)i'}} \prod_{a(p-1)+b=1}^{p^2-p} s_l^{n_l} \left(\frac{\gamma^p - 1}{\pi^{j'}} \right)^{an_l} \left(\frac{\gamma - 1}{\pi^0} \right)^{bn_l},$$

which is in H^* since $i \geq e/p$ implies $e'/p \geq i'$, which in turn yields $\operatorname{ord}(p) = e \geq p(p-1)i'$. We conclude that

$$k^p + \sum_{l=a(p-1)+b=2}^{p^2-p} r_{l,m'}^p h^{ap} k^{bp} \in H^*,$$

and thus

$$\left(\frac{\gamma-1}{\pi^{i'}} \right)^p + \sum_{a(p-1)+b=2}^{p^2-p} r_{l,m'}^p \left(\frac{\gamma^p - 1}{\pi^{e'}} \right)^{ap} \left(\frac{\gamma-1}{\pi^{i'}} \right)^{bp}$$

$$= \frac{\gamma^p - 1}{\pi^{pi'}} + \frac{1}{\pi^{pi'}} \sum_{\iota=1}^{p} \binom{p}{\iota} (-1)^\iota \gamma^{p-\iota}$$

$$+ \sum_{a(p-1)+b=2}^{p^2-p} \frac{s_l^p}{\pi^{p(p-1)i'}} \left(\left(\frac{\gamma^p - 1}{\pi^{j'}} \right)^p \right)^a \left(\left(\frac{\gamma - 1}{\pi^0} \right)^p \right)^b$$

is in H^*.

Now $\pi^{-pi'} \sum_{\iota=1}^{p} \binom{p}{\iota} (-1)^\iota \gamma^{p-\iota} \in H^*$, and all of the terms in the second summation above with $l = a(p-1) + b \geq p$ are in H^*. Thus

$$\frac{\gamma^p - 1}{\pi^{pi'}} + \sum_{b=2}^{p-1} \frac{s_l^p}{\pi^{p(p-1)i'}} ((\gamma - 1)^p)^b \in H^*.$$

Since this quantity is in $K\hat{C}_{p^2} \cap H^* = H(j')$, and since $\left\{ \left(\frac{\gamma^p - 1}{\pi^{j'}} \right)^b \right\}$, $b = 0, \ldots, p - 1$, is an R-basis for $H(j')$, we conclude that $pi' \leq j'$. $\qquad\square$

We next consider the case $i < e/p$.

Lemma 8.1.3. *Let H be an R-Hopf order in KC_{p^2} that induces the short exact sequence (8.2). If $i < e/p$, then $j' \geq e/p$.*

Proof. Since $H(i)$ injects into H, the Larson order $H(i, 0) \subseteq H$. Thus $H^* \subseteq H(i, 0)^* = A\left(e', i', \zeta_{p^2}^{-1}\right) = R\left[\frac{\gamma^p - 1}{\pi^{e'}}, \frac{\gamma u - 1}{\pi^{i'}}\right]$. Note that $A\left(e', i', \zeta_{p^2}^{-1}\right)$ is not Larson since $i < e/p$ implies $i' > e'/p$. By Lemma 7.1.6, u is a unit in $R\left[\frac{\gamma^p - 1}{\pi^{e'}}\right]$, and thus $A\left(e', i', \zeta_{p^2}^{-1}\right) = R\left[\frac{\gamma^p - 1}{\pi^{e'}}, \frac{\gamma - u^{-1}}{\pi^{i'}}\right]$.

Let $h = \frac{\gamma^p - 1}{\pi^{e'}}, k = \frac{\gamma - u^{-1}}{\pi^{i'}}$. Then

$$Y = \left(k, k^2, \ldots, k^{p-1}, hk, hk^2, \ldots, hk^{p-1}, \ldots, \right.$$
$$\left. h^{p-1}k, h^{p-1}k^2, \ldots, h^{p-1}k^{p-1}, 1, h, \ldots, h^{p-1}\right)$$

is an R-basis for $A(e', i', \zeta_{p^2}^{-1})$. Therefore an R-basis $\{\alpha_m\}$ for H^* is given by the matrix product

$$YM = \left(\alpha_1, \alpha_2, \ldots, \alpha_{p^2}\right),$$

where M is an upper-triangular $p^2 \times p^2$ matrix with entries $r_{l,m} \in R$.

The condition $i' > e'/p$ implies $pi' > e' \geq j'$, and so H^* is not a Larson order, yet it does contain a largest Larson order $A(\Xi(H^*))$. Thus H^* contains $\frac{\gamma - 1}{\pi^t}$ for some $t \geq 0$. There exist elements $s_m \in R$ with

$$\frac{\gamma - 1}{\pi^t} = \sum_{m=1}^{p^2} s_m \alpha_m,$$

and so there exist elements $q_l \in R$ for which

$$\frac{\gamma - 1}{\pi^t} = \sum_{l=a(p-1)+b=1}^{p^2-p} q_l h^a k^b + \sum_{l=c}^{p^2} q_l h^{l-c}, \quad c = p^2 - p + 1.$$

One has $q_l = 0$ for $2 \leq l \leq p^2 - p$, and so

$$\frac{\gamma - 1}{\pi^t} = q_1 k + \sum_{l=c}^{p^2} q_l h^{l-c}, \quad c = p^2 - p + 1$$

$$= q_1 \frac{\gamma - 1}{\pi^{i'}} + q_1 \frac{1 - u^{-1}}{\pi^{i'}} + \sum_{l=c}^{p^2} q_l h^{l-c}, \quad c = p^2 - p + 1.$$

If $\mathrm{ord}(q_1) > i'$, then $\frac{\gamma-1}{\pi^{i'}} \notin H^*$, and thus $\mathrm{ord}(q_1) \le i'$. Moreover,

$$\sum_{\iota=c}^{p^2} q_\iota h^{\iota-c} \in H^* \cap KC_{p^2} = H(j'),$$

and so $q_1 \frac{1-u^{-1}}{\pi^{i'}} \in H(j')$. Consequently, $1 - u^{-1} \in H(j')$. Now, by Lemma 7.1.6, $\mathrm{ord}(1 - \zeta_{p^2}) = e'/p \ge j$; that is, $j' \ge e/p$. $\square$

We now show that H in KC_{p^2} satisfies the valuation condition for $n = 2$.

Proposition 8.1.2. *Let H be an R-Hopf order in KC_{p^2} that induces the short exact sequence of (8.2). Then either $pj \le i$ or $pi' \le j'$.*

Proof. Suppose $i \ge e/p$. Then, by Lemma 8.1.2, $pj' \le i'$. On the other hand, if $i < e/p$, then $j' \ge e/p$ by Lemma 8.1.3, and thus Lemma 8.1.2 can be applied to the dual short exact sequence to obtain $pj \le i$. $\square$

8.2 Some Cohomology

In this section, we give a review of cohomology of groups, which will be needed in the subsequent section.

Let C_m denote the cyclic group of order m, generated by τ, let G be an Abelian group, and let G^m denote the subgroup of G generated by $\{g^m : g \in G\}$. An **extension of G by C_m** is a short exact sequence of groups

$$E : 1 \to G \xrightarrow{i} G' \xrightarrow{s} C_m \to 1.$$

Two extensions E_1, E_2 are **equivalent** if there exists an isomorphism $G_1' \to G_2'$ such that the diagram

$$
\begin{array}{ccccccccc}
E_1 : & 1 \to & G & \to & G_1' & \to & C_m & \to & 1 \\
 & & \| & & \downarrow & & \| & & \\
E_2 : & 1 \to & G & \to & G_2' & \to & C_m & \to & 1
\end{array}
\tag{8.4}
$$

commutes. Let $E(G, C_m)$ denote the collection of equivalence classes of extensions of G by C_m. In this section, we compute $E(G, C_m)$ following [Rot02, §10.3].

A **cocycle** is a function $f : C_m \times C_m \to G$ that satisfies the conditions

$$f(1, \tau^j) = f(\tau^i, 1) = 1 \tag{8.5}$$

and

$$f\left(\tau^i, \tau^j\right) f\left(\tau^{i+j}, \tau^k\right) = f\left(\tau^j, \tau^k\right) f\left(\tau^i, \tau^{j+k}\right) \tag{8.6}$$

for $i, j, k = 0, \dots, m-1$.

Proposition 8.2.1. *The collection of cocycles, denoted by $C(G, C_m)$, is an Abelian group under the product*

$$(f*g)(\tau^i, \tau^j) = f(\tau^i, \tau^j)g(\tau^i, \tau^j),$$

with identity element $f(\tau^i, \tau^j) = 1$, for all i, j.

Proof. Exercise.   □

We can identify a subgroup of $C(G, C_m)$ as follows. Let $h : C_m \to G$ be a function with $h(1) = 1$. Let $f_h : C_m \times C_m \to G$ be defined as

$$f_h(\tau^i, \tau^j) = h(\tau^i)(h(\tau^{i+j}))^{-1}h(\tau^j)$$

for $i, j = 0, \ldots, m - 1$. Then it is routine to check that f_h is a cocycle. We write the cocycle f_h as ∂h and call it the **coboundary of** h. Let $B(G, C_m)$ denote the collection of coboundaries

$$B(G, C_m) = \{\partial h : h : C_m \to G, h(1) = 1\}.$$

Then $B(G, C_m)$ is a subgroup of $C(G, C_m)$.

Proposition 8.2.2. *The quotient group $C(G, C_m)/B(G, C_m)$ is in a 1-1 correspondence with the equivalence classes of extensions $E(G, C_m)$.*

Proof. Let f_1, f_2 be cocycles that satisfy the condition

$$f_1\left(\tau^i, \tau^j\right)\left(f_2(\tau^i, \tau^j)\right)^{-1} = \partial h\left(\tau^i, \tau^j\right) = h(\tau^i)\left(h(\tau^{i+j})\right)^{-1}h(\tau^j)$$

for some $\partial h \in B(G, C_m)$. Then

$$h(\tau^{i+j})f_1(\tau^i, \tau^j) = h(\tau^i)h(\tau^j)f_2(\tau^i, \tau^j). \tag{8.7}$$

On the Cartesian product $G_1' = G \times C_m$, define a multiplication

$$(a, \tau^i) \cdot_{f_1} (b, \tau^j) = (abf_1(\tau^i, \tau^j), \tau^{i+j}),$$

and on $G_2' = G \times C_m$ define a multiplication

$$(a, \tau^i) \cdot_{f_2} (b, \tau^j) = (abf_2(\tau^i, \tau^j), \tau^{i+j}).$$

Then there exist extensions of groups

$$E_1 : 1 \to G \to G_1' \to C_m \to 1$$

and

$$E_2 : 1 \to G \to G_2' \to C_m \to 1.$$

We claim that E_1 and E_2 are equivalent. To this end, define a map $h : G'_1 \to G'_2$ by the rule $h((a, \tau^i)) = (ah(\tau^i), \tau^i)$. Then

$$
\begin{aligned}
h((a, \tau^i) \cdot_{f_1} (b, \tau^j)) &= h((abf_1(\tau^i, \tau^j), \tau^{i+j})) \\
&= (abh(\tau^{i+j}) f_1(\tau^i, \tau^j), \tau^{i+j}) \\
&= (abh(\tau^i) h(\tau^j) f_2(\tau^i, \tau^j), \tau^{i+j}) \quad \text{by (8.7)} \\
&= (ah(\tau^i), \tau^i) \cdot_{f_2} (bh(\tau^j), \tau^j) \\
&= h((a, \tau^i)) \cdot_{f_2} h((b, \tau^j)),
\end{aligned}
$$

and so h is an isomorphism that makes the diagram (8.4) commute.

Consequently, there is a well-defined map

$$
\Psi : C(G, C_m)/B(G, C_m) \to E(G, C_m),
$$

where $\Psi(fB(G, C_m))$ is the equivalence class represented by

$$
E : 1 \to G \to G' \to C_m \to 1,
$$

where $G' = G \times C_m$ is endowed with the binary operation

$$
(a, \tau^i) \cdot_f (b, \tau^j) = (abf(\tau^i, \tau^j), \tau^{i+j}). \tag{8.8}
$$

Evidently, Ψ is a bijection.   □

The values of a given cocycle $f : C_m \times C_m \to G$ can be arranged in an $m \times m$ matrix M_f whose i, jth entry is $f(\tau^i, \tau^j) = a_{i,j}$ for $i, j = 0, \ldots, m - 1$. Note that the entries in the first row and first column are all 1 by cocycle property (8.5). Moreover, since the binary operation on $G \times C_m$ is Abelian,

$$
(a, \tau^i) \cdot (b, \tau^j) = (b, \tau^j) \cdot (a, \tau^i),
$$

and so

$$
(abf(\tau^i, \tau^j), \tau^{i+j}) = (baf(\tau^j, \tau^i), \tau^{j+i}),
$$

which implies that

$$
f(\tau^i, \tau^j) = f(\tau^j, \tau^i)
$$

for all $i, j = 0, \ldots, m - 1$. Thus M_f is symmetric. We prove the following.

Proposition 8.2.3. *Let C_m denote the cyclic group of order m generated by τ, let G be an Abelian group, and let G^m denote the subgroup of G generated by $\{g^m : g \in G\}$. Then there is a group isomorphism*

$$
\Phi : G/G^m \to C(G, C_m)/B(G, C_m)
$$

defined as

$$\Phi(wG^m) = f_w B(G, C_m),$$

where $f_w : C_m \times C_m \to G$ *is the cocycle in* $C(G, C_m)$ *whose values are given by the matrix*

$$M_{f_w} = \begin{pmatrix} 1 & 1 & 1 & \cdots & 1 \\ 1 & 1 & \cdots & 1 & w \\ 1 & \cdots & 1 & w & w \\ \vdots & & & & \vdots \\ 1 & w & w & \cdots & w \end{pmatrix}.$$

Proof. We first show that Φ is well-defined on cosets of G/G^m. Suppose $w \in vG^m$. Then $wv^{-1} = g^m$ for some $g \in G$, and the cocycle $f_{wv^{-1}} : C_m \times C_m \to G$ satisfies $f_{wv^{-1}} = \partial h$, where $h : C_m \to G$ is defined by $h(\tau^i) = g^i$. Thus $f_{wv^{-1}} \in B(G, C_m)$, and so Φ is well-defined.

Now,

$$\Phi(vG^m wG^m) = \Phi(vwG^m)$$
$$= f_{vw} B(G, C_m)$$
$$= f_v B(G, C_m) f_w B(G, C_m),$$

and so Φ is a group homomorphism. Next, suppose that $\Phi(wG^m) = \Phi(vG^m)$. Then $f_{wv^{-1}} \in B(G, C_m)$. Thus there exists a function $h : C_m \to G$, $h(1) = 1$ for which $\partial h = f_{wv^{-1}}$. Thus $wv^{-1} = g^m$ for some $g \in G$, and so $wG^m = vG^m$, which shows that Φ is an injection.

To show that Φ is surjective, let $f B(G, C_m)$ be a coset in $C(G, C_m)/B(G, C_m)$. Assuming that this coset can be written in the form $f_w B(G, C_m)$ for some $w \in G$, we have $\Psi(wG^m) = f B(G, C_m)$, and so Φ is surjective. So the remainder of this proof will be concerned with proving that the coset $f B(G, C_m)$ can be written in this form.

We know that f has matrix

$$M_f = \begin{pmatrix} 1 & 1 & 1 & \cdots & 1 \\ 1 & a_{1,1} & a_{1,2} & \cdots & a_{1,m-1} \\ 1 & a_{2,1} & a_{2,2} & \cdots & a_{2,m-1} \\ \vdots & \vdots & \vdots & & \vdots \\ \vdots & \vdots & \vdots & & \vdots \\ 1 & a_{m-1,1} & a_{m-1,2} & \cdots & a_{m-1,m-1} \end{pmatrix}$$

for elements $a_{i,j} \in G$ with M_f symmetric. In the second row of M_f, let l be the smallest integer for which $a_{1,l} \neq 1$. Define a function $h : C_m \to G$ by

$$h(\tau^i) = \begin{cases} 1 & \text{for } i = 0, \ldots, l \\ a_{1,l} & \text{for } i = l+1, \ldots, m-1. \end{cases}$$

Then ∂h is a coboundary with matrix $M_{\partial h}$ whose first and second rows and columns, up to and including the codiagonal, have the form

$$M_{\partial h} = \begin{pmatrix}
1 & 1 & \cdots & 1 & 1 & 1 & \cdots & \cdots & 1 \\
1 & 1 & \cdots & 1 & a_{1,l}^{-1} & 1 & \cdots & & 1 & * \\
\vdots & \vdots & & * & * & * & * & * & * & * \\
1 & 1 & & * & * & * & * & * & * & * \\
1 & a_{1,l}^{-1} & & * & * & * & * & * & * & * \\
1 & 1 & & * & * & * & * & * & * & * \\
\vdots & \vdots & & * & * & * & * & * & * & * \\
\vdots & 1 & & * & * & * & * & * & * & * \\
1 & * & & * & * & * & * & * & * & *
\end{pmatrix}.$$

Here the entry $a_{1,l}^{-1}$ is in the $(1, l)$th and $(l, 1)$th places. It follows that f is congruent modulo $B(G, C_m)$ to a cocycle whose matrix is of the form

$$M_f = \begin{pmatrix}
1 & 1 & 1 & 1 & 1 & \cdots & 1 \\
1 & 1 & 1 & 1 & \cdots & 1 & * \\
1 & 1 & * & * & * & * & * \\
1 & 1 & * & * & * & * & * \\
1 & \vdots & * & * & * & * & * \\
\vdots & 1 & * & * & * & * & * \\
1 & * & * & * & * & * & *
\end{pmatrix}.$$

Consequently, we can assume, without loss of generality, that the matrix M_f of the cocycle f is in the form above.

By the cocycle condition (8.6),

$$f(\tau, \tau) f(\tau^2, \tau^k) = f(\tau, \tau^k) f(\tau, \tau^{1+k})$$

for $k, 0 \le k \le m-1$. But $f(\tau, \tau) = f(\tau, \tau^k) = f(\tau, \tau^{1+k}) = 1$ for $1 \le k \le m-3$, and so

$$f(\tau^2, \tau^k) = 1$$

for $k = 1, \ldots, m-3$. Now, again by (8.6),

$$f(\tau^2, \tau) f(\tau^3, \tau^k) = f(\tau, \tau^k) f(\tau^2, \tau^{1+k})$$

for all k. But, as we have seen above, $f(\tau^2, \tau) = f(\tau^2, \tau^k) = f(\tau^2, \tau^{1+k}) = 1$ for $1 \le k \le m - 4$. Thus

$$f(\tau^3, \tau^k) = 1$$

for $k = 1, \ldots, m - 4$. Continuing in this manner, we see that

$$f(\tau^{m-2}, \tau^k) = 1$$

for $k = 1, \ldots, m - (m - 2 + 1) = 1$. It follows that the matrix for f is of the form

$$M_f = \begin{pmatrix} 1 & 1 & 1 & 1 & \cdots & 1 \\ 1 & 1 & 1 & \cdots & 1 & * \\ 1 & 1 & \cdots & 1 & * & * \\ 1 & \vdots & & * & * & * \\ \vdots & 1 & * & * & * & * \\ 1 & * & * & * & * & * \end{pmatrix}.$$

But what about the matrix entries below the main codiagonal? We show that they are all equal. By (8.6),

$$f(\tau, \tau) f(\tau^2, \tau^{m-1}) = f(\tau, 1) f(\tau, \tau^{m-1}).$$

But, as we have seen, this implies that

$$f(\tau^2, \tau^{m-1}) = f(\tau, \tau^{m-1}).$$

Moreover,

$$f(\tau, \tau^2) f(\tau^3, \tau^{m-1}) = f(\tau, \tau) f(\tau^2, \tau^{m-1}),$$

and so

$$f(\tau^3, \tau^{m-1}) = f(\tau^2, \tau^{m-1}).$$

Hence

$$f(\tau^{k+1}, \tau^{m-1}) = f(\tau^k, \tau^{m-1})$$

for $k = 1, \ldots, m - 2$. By (8.6),

$$f(\tau, \tau^k) f(\tau^{k+1}, \tau^{m-2}) = f(\tau, \tau^{m-2+k}) f(\tau^k, \tau^{m-2})$$

for all k. Since $f(\tau, \tau^k) = f(\tau, \tau^{m-2+k}) = 1$ for $k = 2, \ldots, m - 2$,

$$f(\tau^{k+1}, \tau^{m-2}) = f(\tau^k, \tau^{m-2})$$

for $k = 2, \ldots, m - 2$. Continuing in this manner, we see that the entries below the main codiagonal in a column are equal.

We next show that all of the entries below the main codiagonal are equal to a common value. By (8.6),

$$f(\tau,\tau)f(\tau^2,\tau^{m-2}) = f(\tau,\tau^{m-1})f(\tau,\tau^{m-2})$$

for all k, and thus

$$f(\tau^2,\tau^{m-2}) = f(\tau,\tau^{m-1}).$$

By (8.6),

$$f(\tau,\tau^2)f(\tau^3,\tau^{m-3}) = f(\tau,\tau^{m-1})f(\tau^2,\tau^{m-3}),$$

and thus

$$f(\tau^3,\tau^{m-3}) = f(\tau,\tau^{m-1}).$$

Continuing in this manner, we conclude that

$$f(\tau^k,\tau^{m-k}) = f(\tau,\tau^{m-1})$$

for $k = 2,\dots,m-1$. Now

$$f(\tau^i,\tau^j) = f(\tau^l,\tau^k)$$

for all $i + j > m - 1$, $l + k > m - 1$, and so all of the entries below the main codiagonal are equal to a common value, say w. Thus the matrix of f is in the form

$$\begin{pmatrix} 1 & 1 & 1 & \cdots & 1 \\ 1 & 1 & \cdots & 1 & w \\ 1 & \cdots & 1 & w & w \\ \vdots & & \vdots & & \vdots \\ 1 & w & w & \cdots & w \end{pmatrix}.$$

This completes the proof of the proposition.   □

We summarize with the following proposition.

Proposition 8.2.4. *There is a 1-1 correspondence between the group G/G^m and the equivalence classes in $E(G,C_m)$. Specifically, a coset $wG^m \in G/G^m$ corresponds to an equivalence class of extensions represented by*

$$E_w : 1 \to G \to G'_w \to C_m \to 1,$$

where the group operation on the set $G'_w = G \times C_m$ is given as

$$(a,\tau^i) \cdot (b,\tau^j) = \begin{cases} (ab,\tau^{i+j}) & \text{if } i + j < m \\ (abw,\tau^{i+j-m}) & \text{if } i + j \geq m. \end{cases}$$

Proof. By Proposition 8.2.3 and Proposition 8.2.2, there exists a 1-1 correspondence $G/G^m \to E(G, C_m)$, where the coset wG^m corresponds to the matrix M_{f_w}. The definition of the binary operation follows from (8.8). □

8.3 Greither Orders

In this section, we compute the algebraic structure of the Hopf order H in the short exact sequence (8.2) assuming that $i \geq pj$, following closely the work of C. Greither [Gr92].

We shall employ the cohomology of the previous section in the following context. Let i, j be integers with $0 \leq i, j \leq e'$, and $pj \leq i$. Let $\mathbb{D}_i$ be the R-group scheme represented by the Larson order

$$H(i) = R\left[\frac{\tau - 1}{\pi^i}\right], \quad \langle \tau \rangle = C_p,$$

and let $\mathbb{E}_j$ denote the functor represented by the polynomial algebra

$$R\left[\frac{X - 1}{\pi^j}, X^{-1}\right],$$

where X is indeterminate.

Proposition 8.3.1. *The R-algebra $R\left[\frac{X-1}{\pi^j}, X^{-1}\right]$ is an R-Hopf algebra with co-multiplication, counit, and coinverse maps induced from the corresponding maps of the K-Hopf algebra $K[X, X^{-1}]$. Consequently, $\mathbb{E}_j$ is an R-group scheme that over K appears as $\mathbf{G}_{m,K} = \mathrm{Hom}_{K\text{-alg}}(K[X, X^{-1}], -)$.*

Proof. One readily computes

$$\Delta_{K[X,X^{-1}]}\left(\frac{X-1}{\pi^j}\right) = \frac{X-1}{\pi^j} \otimes 1 + 1 \otimes \frac{X-1}{\pi^j} + (X-1) \otimes \frac{X-1}{\pi^j}$$

and $\Delta_{K[X,X^{-1}]}(X^{-1}) = X^{-1} \times X^{-1}$. Moreover,

$$\epsilon_{K(X,X^{-1}]}\left(R\left[\frac{X-1}{\pi^j}, X^{-1}\right]\right) \subseteq R$$

and

$$\sigma_{K[X,X^{-1}]}\left(R\left[\frac{X-1}{\pi^j}, X^{-1}\right]\right) \subseteq R\left[\frac{X-1}{\pi^j}, X^{-1}\right]$$

since $X, \frac{X^{-1}-1}{\pi^j} \in R\left[\frac{X-1}{\pi^j}, X^{-1}\right]$. □

Observe that $\mathbb{E}_j(R) = U_j(R)$. An **extension of** $\mathbb{E}_j$ **by** $\mathbb{D}_i$ is a short exact sequence of group schemes

$$E : 1 \to \mathbb{E}_j \xrightarrow{i} G \xrightarrow{s} \mathbb{D}_i \to 1;$$

that is, an extension of $\mathbb{E}_j$ by $\mathbb{D}_i$ is a sequence as above in which $\mathbb{D}_i$ is the quotient sheaf of G by $\mathbb{E}_j$ (see Definition 3.3.3).

Observe that if E is an extension of $\mathbb{E}_j$ by $\mathbb{D}_i$, then it is not obvious that

$$E(S) : 1 \to \mathbb{E}_j(S) \xrightarrow{i_S} G(S) \xrightarrow{s_S} \mathbb{D}_i(S) \to 1$$

is a short exact sequence of groups for an R-algebra S. M. Demazure and P. Gabriel [DG70, III, §6, 2.5] have provided the following proposition.

Proposition 8.3.2. *Let i, j be integers that satisfy $0 \le i \le e'$, $0 \le j \le pe'$. Then*

$$E : 1 \to \mathbb{E}_j \xrightarrow{i} G \xrightarrow{s} \mathbb{D}_i \to 1$$

is an extension of $\mathbb{E}_j$ by $\mathbb{D}_i$ if and only if

$$E(S) : 1 \to \mathbb{E}_j(S) \xrightarrow{i_S} G(S) \xrightarrow{s_S} \mathbb{D}_i(S) \to 1$$

is an extension of $\mathbb{E}_j(S)$ by $\mathbb{D}_i(S)$ in the sense of §8.2 for each R-algebra S.

In view of Proposition 8.3.2, two extensions, E_1, and E_2 such are **equivalent** if there exists an isomorphism of group schemes $G_1 \to G_2$ such that the diagram

$$E_1(S) : 1 \to \mathbb{E}_j(S) \to G_1(S) \to \mathbb{D}_i(S) \to 1$$
$$\quad\quad\quad \| \quad\quad\quad \downarrow \quad\quad\quad \|$$
$$E_2(S) : 1 \to \mathbb{E}_j(S) \to G_2(S) \to \mathbb{D}_i(S) \to 1$$

commutes for each R-algebra S.

Let $E(\mathbb{E}_j, \mathbb{D}_i)$ denote the collection of equivalence classes of extensions of $\mathbb{E}_j$ by $\mathbb{D}_i$. We seek to compute $E(\mathbb{E}_j, \mathbb{D}_i)$.

In view of the cohomology already discussed, we should consider "cocycles modulo coboundaries." But what plays the role of the cocycles $C(G, C_m)$? Let $\mathbb{D}_i \times \mathbb{D}_i$ denote the product of group schemes. Then, by Proposition 2.4.2, the representing algebra of $\mathbb{D}_i \times \mathbb{D}_i$ is $H(i) \otimes_R H(i)$.

Definition 8.3.1. The natural transformation $f : \mathbb{D}_i \times \mathbb{D}_i \to \mathbb{E}_j$ is a **cocycle** if, for each R-algebra S and $x, y, z \in \mathbb{D}_i(S)$

$$f_S(x, y)(X) f_S(xy, z)(X) = f_S(y, z)(X) f_S(x, yz)(X),$$
$$f_S(x, 1)(X) = f_S(1, y)(X) = 1.$$

The collection of cocycles is denoted by $C(\mathbb{E}_j, \mathbb{D}_i)$.

Analogous to Proposition 8.2.1, $C(\mathbb{E}_j, \mathbb{D}_i)$ is a group under the binary operation

$$(f_S * g_S)(x, y)(X) = f_S(x, y)(X) g_S(x, y)(X)$$

for each R-algebra S, $x, y \in \mathbb{D}_i(S)$.

We seek to characterize these cocycles in terms of the cocycles we've already developed. By Yoneda's Lemma, $f \in C(\mathbb{E}_j, \mathbb{D}_i)$ corresponds to an R-algebra homomorphism,

$$\phi_f : R\left[\frac{X-1}{\pi^j}, X^{-1}\right] \to H(i) \otimes_R H(i),$$

that is determined by the two conditions

$$X \mapsto \sum_{m=0}^{p-1} \sum_{n=0}^{p-1} a_{m,n}(e_m \otimes e_n) \in U(H(i) \otimes H(i))$$

and

$$\frac{\left(\sum_{m=0}^{p-1} \sum_{n=0}^{p-1} a_{m,n}(e_m \otimes e_n)\right) - 1}{\pi^j} \in H(i) \otimes H(i).$$

There is an injection

$$\iota : H(i) \otimes_R H(i) \to \bigoplus_{m=0}^{p-1} Re_m \otimes_R \bigoplus_{n=0}^{p-1} Re_n,$$

and the map $\iota\phi_f$ is an R-algebra homomorphism

$$\iota\phi_f : R\left[\frac{X-1}{\pi^j}, X^{-1}\right] \to \bigoplus_{m=0}^{p-1} Re_m \otimes_R \bigoplus_{m=0}^{p-1} Re_m = RC_p^* \otimes_R RC_p^*.$$

Consequently, the algebra homomorphism $\iota\phi_f$ corresponds to a natural transformation

$$\tilde{f} : \mathrm{Hom}_{R\text{-alg}}(RC_p^* \otimes_R RC_p^*, -) \to \mathbb{E}_j.$$

By Proposition 4.1.8, $\mathrm{Hom}_{R\text{-alg}}(RC_p^* \otimes_R RC_p^*, R) = C_p \times C_p$, and so

$$\tilde{f}_R : C_p \times C_p \to \mathbb{E}_j(R) = U_j(R).$$

Note that

$$\tilde{f}_R(\tau^m, \tau^n)(X) = (\tau^m, \tau^n)(\iota \phi_f)(X)$$

$$= (\tau^m, \tau^n) \left(\sum_{m'=0}^{p-1} \sum_{n'=0}^{p-1} a_{m',n'}(e_{m'} \otimes e_{n'}) \right)$$

$$= a_{m,n}.$$

In this manner, f gives rise to a function

$$\hat{f} : C_p \times C_p \to U_j(R),$$

defined as $(\tau^m, \tau^n) \mapsto a_{m,n}$.

We can now prove the following proposition.

Proposition 8.3.3. *The natural transformation $f : \mathbb{D}_i \times \mathbb{D}_i \to \mathbb{E}_j$ is a cocycle in $C(\mathbb{E}_j, \mathbb{D}_i)$ if and only if $\hat{f}$ is a cocycle in $C(U_j(R), C_p)$.*

Proof. Suppose that the natural transformation $f : \mathbb{D}_i \times \mathbb{D}_i \to \mathbb{E}_j$ is a cocycle with algebra homomorphism ϕ_f. Then, for all $x, y, z \in \mathbb{D}_i(R)$,

$$f_R(x, y)(X) f_R(xy, z)(X) = f_R(y, z)(X) f_R(x, yz)(X).$$

Consequently, for all $l, m, 0 \le l, m \le p - 1$,

$$\tilde{f}_R(\tau^l, \tau^m)(X) \tilde{f}_R(\tau^{l+m}, \tau^n)(X) = \tilde{f}_R(\tau^m, \tau^n) \tilde{f}_R(\tau^l, \tau^{m+n})(X),$$

so that

$$a_{l,m} a_{l+m,n} = a_{m,n} a_{l,m+n},$$

where $m + n$ and $l + m$ are taken modulo p. Thus

$$\hat{f}(\tau^l, \tau^m) \hat{f}(\tau^{l+m}, \tau^n) = \hat{f}(\tau^m, \tau^n) \hat{f}(\tau^l, \tau^{m+n}),$$

and hence $\hat{f}$ satisfies cocycle condition (8.6).

Moreover, since $f : \mathbb{D}_i \times \mathbb{D}_i \to \mathbb{E}_j$ is a cocycle,

$$f_R(x, 1)(X) = 1 = f_R(1, y)(X)$$

for all $x, y \in \mathbb{D}_i(R)$. Thus, for all $0 \le l, m \le p - 1$,

$$\tilde{f}_R(\tau^l, 1)(X) = 1 = \tilde{f}_R(1, \tau^m)(X),$$

so that

$$a_{l,0} = 1 = a_{0,m}$$

for all $0 \le l, m \le p-1$. Thus $\hat{f}$ satisfies the cocycle condition (8.5). It follows that $\hat{f}$ is in $C(G, C_m)$ with $G = U_j(R)$, $C_m = C_p$.

Now suppose that $\hat{f} : C_p \times C_p \to U_j(R)$ is a cocycle obtained from the natural transformation $f : \mathbb{D}_i \times \mathbb{D}_i \to \mathbb{E}_j$. Then, for all l, m, n, $0 \le l, m, n \le p-1$, one has

$$a_{l,m} a_{l+m,n} = a_{l,m+n} a_{m,n},$$

where $m+n$ and $l+m$ are taken modulo p. Thus,

$$\sum_{l=0}^{p-1} \sum_{m=0}^{p-1} \sum_{n=0}^{p-1} a_{l,m} a_{l+m,n} (e_l \otimes e_m \otimes e_n) = \sum_{l=0}^{p-1} \sum_{m=0}^{p-1} \sum_{n=0}^{p-1} a_{l,m+n} a_{m,n} (e_l \otimes e_m \otimes e_n),$$

which yields

$$\left(\sum_{l=0}^{p-1} \sum_{m=0}^{p-1} a_{l,m} (e_l \otimes e_m \otimes 1) \right) \left(\sum_{k=0}^{p-1} \sum_{n=0}^{p-1} a_{k,n} (\Delta(e_k) \otimes e_n) \right)$$

$$= \left(\sum_{m=0}^{p-1} \sum_{n=0}^{p-1} a_{m,n} (1 \otimes e_m \otimes e_n) \right) \left(\sum_{l=0}^{p-1} \sum_{k=0}^{p-1} a_{l,k} (e_l \otimes \Delta(e_k)) \right).$$

Thus, for $x, y, z \in \mathbb{D}_i(S)$,

$$(x \otimes y \otimes z) \left(\sum_{l=0}^{p-1} \sum_{m=0}^{p-1} a_{l,m} (e_l \otimes e_m \otimes 1) \right) (x \otimes y \otimes z) \left(\sum_{k=0}^{p-1} \sum_{n=0}^{p-1} a_{k,n} (\Delta(e_k) \otimes e_n) \right)$$

$$= (x \otimes y \otimes z) \left(\sum_{m=0}^{p-1} \sum_{n=0}^{p-1} a_{m,n} (1 \otimes e_m \otimes e_n) \right) (x \otimes y \otimes z) \left(\sum_{l=0}^{p-1} \sum_{k=0}^{p-1} a_{l,k} (e_l \otimes \Delta(e_k)) \right),$$

which implies

$$(x,y)\phi_f(X)(xy,z)\phi_f(X) = (y,z)\phi_f(X)(x,yz)\phi_f(X).$$

Thus

$$f_S(x,y)(X) f_S(xy,z)(X) = f_S(y,z)(X) f_S(x,yz)(X).$$

Now, assume the condition $1 = a_{m,0}$ for $m = 0, \ldots, p - 1$, and let $x \in \mathbb{D}_i(S)$. Then

$$1 = x \left(\sum_{m=0}^{p-1} a_{m,0} e_m \right)$$

$$= (x \otimes \lambda_{S \epsilon H(i)}) \left(\sum_{m=0}^{p-1} \sum_{n=0}^{p-1} a_{m,n} (e_m \otimes e_n) \right)$$

$$= (x, 1) \phi_f(X)$$

$$= f_S(x, 1)(X).$$

In a similar manner, the condition $1 = a_{0,n}$, for $n = 0, \ldots, p - 1$, yields $f_S(1, y)(X) = 1$ for $y \in \mathbb{D}_i(S)$. Consequently, f is a cocycle. $\qquad\square$

We identify the group $C(\mathbb{E}_j, \mathbb{D}_i)$ with the subgroup C of $C(U_j(R), C_p)$ defined as

$$C = \{ g \in C(U_j(R), C_p) : g = \hat{f} \text{ for some } f \in C(\mathbb{E}_j, \mathbb{D}_i) \}.$$

We next consider coboundaries. Let $h : \mathbb{D}_i \to \mathbb{E}_j$ be a natural transformation that satisfies $h_S(1) = 1$, $\forall S$. Let $f_h : \mathbb{D}_i \times \mathbb{D}_i \to \mathbb{E}_j$ be the natural transformation defined as

$$(f_h)_S(x, y)(X) = h_S(x)(X)(h_S(xy))^{-1}(X) h_S(y)(X).$$

Then f_h satisfies the cocycle conditions of Definition 8.3.1. We denote this cocycle by ∂h. The collection of all cocycles of the form

$$\{ \partial h : h : \mathbb{D}_i \to \mathbb{E}_j \text{ is a natural transformation}, h_S(1) = 1, \forall S \}$$

is a subgroup of $C(\mathbb{E}_j, \mathbb{D}_i)$ denoted by $B(\mathbb{E}_j, \mathbb{D}_i)$. $B(\mathbb{E}_j, \mathbb{D}_i)$ is the collection of **coboundaries**.

Observe that the hat operation on cocycles can be applied to the natural transformation $h : \mathbb{D}_i \to \mathbb{E}_j$. By Yoneda's Lemma, h corresponds to an R-algebra homomorphism,

$$\phi_h : R \left[\frac{X - 1}{\pi^j}, X^{-1} \right] \to H(i),$$

which is determined by the conditions

$$X \mapsto \sum_{m=0}^{p-1} a_m e_m \in U(H(i))$$

and

$$\frac{(\sum_{m=0}^{p-1} a_m e_m) - 1}{\pi^j} \in H(i).$$

There is an injection

$$\iota : H(i) \rightarrow \bigoplus_{m=0}^{p-1} Re_m,$$

and the map $\iota\phi_h$ is an R-algebra homomorphism

$$\iota\phi_h : R\left[\frac{X-1}{\pi^j}, X^{-1}\right] \rightarrow \bigoplus_{m=0}^{p-1} Re_m = RC_p^*$$

that corresponds to a natural transformation

$$\tilde{h} : \mathrm{Hom}_{R\text{-alg}}(RC_p^*, -) \rightarrow \mathbb{E}_j$$

with

$$\tilde{h}_R : \mathrm{Hom}_{R\text{-alg}}(RC_p^*, R) = C_p \rightarrow \mathbb{E}_j(R) = U_j(R).$$

Note that

$$\tilde{h}_R(\tau^m)(X) = (\tau^m)(\iota\phi_h)(X)$$

$$= (\tau^m)\left(\sum_{m'=0}^{p-1} a_{m'} e_{m'}\right)$$

$$= a_m.$$

In this manner, h gives rise to a function

$$\hat{h} : C_p \rightarrow U_j(R),$$

defined as $\tau^m \mapsto a_m$.

One can compute the coboundary of $\hat{h}$ (as defined in §8.2) to obtain $\partial\hat{h}$, which is in $B(U_j(R), C_p)$.

Proposition 8.3.4. *The cocycle $f : \mathbb{D}_i \times \mathbb{D}_i \rightarrow \mathbb{E}_j$ is a coboundary ($f = \partial h$, for some h) if and only if the cocycle $\hat{f}$ is a coboundary ($\hat{f} = \partial\hat{h}$).*

Proof. Let $h : \mathbb{D}_i \rightarrow \mathbb{E}_j$ be a natural transformation with algebra map $\phi_h :$ $R\left[\frac{X-1}{\pi^j}, X^{-1}\right] \rightarrow H(i)$ given by $X \mapsto \sum_{m=0}^{p-1} a_m e_m \in U(H(i))$, $\sum_{m=0}^{p-1} a_m e_m \in 1 + \pi^j H(i)$.

Let $f \in C(\mathbb{E}_j, \mathbb{D}_i)$ be a cocycle that satisfies

$$f_R(x, y)(X) = h_R(x)(X)(h_R(xy)(X))^{-1}h_R(y)(X)$$

for all $x, y \in \mathbb{D}_i(R)$. Then, for $0 \leq m, n \leq p - 1$,

$$\tilde{f}_R(\tau^m, \tau^n)(X) = \tilde{h}_R(\tau^m)(X)(\tilde{h}_R(\tau^{m+n})(X))^{-1}\tilde{h}_R(\tau^n)(X).$$

Thus

$$a_{m,n} = a_m(a_{m+n})^{-1}a_m$$

for all $0 \leq m, n \leq p - 1$ ($m + n$ taken modulo p), and so $\hat{f} = \partial\hat{h}$.

Now suppose $\partial\hat{h} = \hat{f}$; that is, suppose that

$$a_{m,n} = a_m(a_{m+n})^{-1}a_n$$

for $0 \leq m, n \leq p - 1$ ($m + n$ taken modulo p). Then

$$\sum_{m,n=0}^{p-1} a_{m,n}(e_m \otimes e_n) = \sum_{m,n=0}^{p-1} a_m(a_{m+n})^{-1}a_n(e_m \otimes e_n)$$

$$= \sum_{m=0}^{p-1} a_m \sum_{n=0}^{p-1} a_n \sum_{k=0}^{p-1} a_k^{-1} \frac{1}{p} \sum_{l=0}^{p-1} \zeta_p^{-lk}(\zeta_p^{ml+nl})(e_m \otimes e_n)$$

$$= \sum_{m=0}^{p-1} a_m \sum_{n=0}^{p-1} a_n \sum_{k=0}^{p-1} a_k^{-1} \frac{1}{p} \sum_{l=0}^{p-1} \zeta_p^{-lk}(\tau^l \otimes \tau^l)(e_m \otimes e_n)$$

$$= \sum_{m=0}^{p-1} a_m(e_m \otimes 1) \sum_{k=0}^{p-1} a_k^{-1} \Delta_{H(i)}(e_k) \sum_{n=0}^{p-1} a_n(1 \otimes e_n).$$

$$(8.9)$$

Let ϕ_f denote the algebra map of f. Let $x, y \in \mathbb{D}_i(S)$. Then the relation (8.9) implies that

$$(x, y)\phi_f(X) = (x)\phi_h(X)((xy)\phi_h(X))^{-1}(y)\phi_h(X),$$

which yields

$$f_S(x, y)(X) = h_S(x)(X)(h_S(xy)(X))^{-1}h_S(y)(X),$$

and hence $\partial h = f$.   □

By Proposition 8.3.4, we can identify the group $B(\mathbb{E}_j, \mathbb{D}_i)$, with the subgroup of $B(U_j(R), C_p)$ defined as

$$B = \{\partial k \ : \ k = \hat{h} \text{ for some } h : \mathbb{D}_i \to \mathbb{E}_j\}.$$

Analogous to Proposition 8.2.2, there is a bijection

$$\Psi : C/B \to E(\mathbb{E}_j, \mathbb{D}_i),$$

and through Ψ one can define a group operation on $E(\mathbb{E}_j, \mathbb{D}_i)$: for E_1, E_2, let

$$E_1 * E_2 = \Psi(xy),$$

where $\Psi(x) = E_1$ and $\Psi(y) = E_2$. Clearly, $E(\mathbb{E}_j, \mathbb{D}_i) \cong C/B$.
We want to characterize the quotient C/B.

Proposition 8.3.5. $C/B \cong U_{pi'+j}(R)/U^p_{i'+j}(R)$.

Proof. Let $\hat{f} \in C$. By §8.2, $\hat{f}$ has matrix

$$M_{\hat{f}} = \begin{pmatrix} 1 & 1 & 1 & \cdots & 1 \\ 1 & a_{1,1} & a_{1,2} & \cdots & a_{1,m-1} \\ 1 & a_{2,1} & a_{2,2} & \cdots & a_{2,m-1} \\ \vdots & \vdots & \vdots & & \vdots \\ \vdots & \vdots & \vdots & & \vdots \\ 1 & a_{m-1,1} & a_{m-1,2} & \cdots & a_{m-1,m-1} \end{pmatrix}$$

for elements $a_{i,j} \in U_j(R)$ with $M_{\hat{f}}$ symmetric.
Since

$$\sum_{m=0}^{p-1} \sum_{n=0}^{p-1} a_{m,n}(e_m \otimes e_n) \in 1 + \pi^j (H(i) \otimes H(i)),$$

$$\left\langle (\gamma - 1)^a \otimes (\gamma - 1)^b, \sum_{m=0}^{p-1} \sum_{n=0}^{p-1} a_{m,n}(e_m \otimes e_n) \right\rangle \in \pi^{(a+b)i'} R$$

for $0 \le a, b \le p - 1, a + b > 0$. In the second row of $M_{\hat{f}}$, let l be the smallest integer for which $a_{1,l} \ne 1$. Now,

$$\left\langle \gamma - 1 \otimes (\gamma - 1)^l, \sum_{m=0}^{p-1} \sum_{n=0}^{p-1} a_{m,n}(e_m \otimes e_n) \right\rangle = 1 - a_{1,l} \in \pi^{(1+l)i'} R. \quad (8.10)$$

Consider the partial sum $\sum_{n=0}^{l} a_{1,n} e_n$. Then, by (8.10), this sum satisfies the first $l+2$ conditions for membership in $H(i)$ (see Lemma 7.1.6). By using Lemma 7.1.7, we can extend this sum to an element $y \in H(i)$.

The element y determines an R-algebra map

$$\phi_y : R\left[\frac{X-1}{\pi^j}, X^{-1}\right] \to H(i), \quad X \mapsto y,$$

which corresponds to a natural transformation $s = h_{\phi_y} : \mathbb{D}_i \to \mathbb{E}_j$ and a function

$$\hat{s} : C_p \to U_j(R).$$

Now $\partial \hat{s}$ is a cocycle in C, and $\hat{f} \cdot \partial \hat{s}$ has a matrix whose second row satisfies

$$a_{1,0} = a_{2,0} = \cdots = a_{1,l} = 1.$$

As in §8.2, we see that $\hat{f} \cdot \partial \hat{h} = f_w$ for some $\partial \hat{h} \in B$, where f_w has $p \times p$ matrix

$$M_w = \begin{pmatrix} 1 & 1 & 1 & \cdots & 1 \\ 1 & 1 & \cdots & 1 & w \\ 1 & \cdots & 1 & w & w \\ \vdots & & \vdots & & \vdots \\ 1 & w & w & \cdots & w \end{pmatrix}$$

for $w \in U_j(R)$ and so each element $\hat{f} \in C$ is congruent modulo B to a cocycle in C of the form f_w.

Next, one shows that a cocycle f_w in $C(U_j(R), C_p)$ with matrix in the form M_w, $w \in U_j(R)$, corresponds to a cocycle in C if and only if $\mathrm{ord}(1-w) \geq pi' + j$; that is, if and only if $w \in U_{pi'+j}(R)$. Moreover, f_w is trivial in C if and only if $w \in U_{i'+j}^p(R)$. (See Greither's proof in [Gr92] for details.) It follows that $C/B \cong U_{pi'+j}(R)/U_{i'+j}^p(R)$. □

As a consequence of Proposition 8.3.5,

$$E(\mathbb{E}_j, \mathbb{D}_i) \cong U_{pi'+j}(R)/U_{i'+j}^p(R).$$

Let $E_{\text{gen-triv}}(\mathbb{E}_j, \mathbb{D}_i)$ denote the collection of **generically trivial** extensions; that is, those elements in $E(\mathbb{E}_j, \mathbb{D}_i)$ that over K appear as

$$1 \to \mathbf{G}_{m,K} \to \mathbf{G}_{m,K} \times \mu_{p,K} \to \mu_{p,K} \to 1.$$

Proposition 8.3.6. $E_{\text{gen-triv}}(\mathbb{E}_j, \mathbb{D}_i) \cong U_{i'+(j/p)}(R)/U_{i'+j}(R)$.

Proof. Over K, one has an isomorphism

$$E(\mathbf{G}_{m,K}, \boldsymbol{\mu}_{p,K}) \cong K^{\times}/(K^{\times})^p,$$

and so the generically trivial extensions correspond to units w that are pth roots in K. □

Let $\mathbb{D}_j = \mathrm{Hom}_{R\text{-alg}}(H(j), -)$ with $H(j) = R\left[\frac{\eta-1}{\pi^j}\right]$, $\eta^p = 1$, and let $E(\mathbb{D}_j, \mathbb{D}_i)$ denote the extensions of $\mathbb{D}_j$ by $\mathbb{D}_i$. Let $E_{\text{gen-triv}}(\mathbb{D}_j, \mathbb{D}_i)$ denote the collection of extensions that over K appear as

$$1 \to \boldsymbol{\mu}_{p,K} \to \boldsymbol{\mu}_{p,K} \times \boldsymbol{\mu}_{p,K} \to \boldsymbol{\mu}_{p,K} \to 1.$$

We want to replace $\mathbb{E}_j$ with $\mathbb{D}_j$ in Proposition 8.3.6 and compute $E_{\text{gen-triv}}(\mathbb{D}_j, \mathbb{D}_i)$. The group schemes $\mathbb{E}_j$ and $\mathbb{D}_i$ are related, and so this computation is possible. Let

$$\mathbf{G}'_{m,K} = \mathrm{Hom}_{K\text{-alg}}(K[Y, Y^{-1}], -),$$

with Y indeterminate, and let $\mathbb{E}_{pj} = \mathrm{Hom}_{R\text{-alg}}\left(R\left[\frac{Y-1}{\pi^{pj}}, Y^{-1}\right], -\right)$. Then there exists a short exact sequence of R-group schemes

$$1 \to \mathbb{D}_i \to \mathbb{E}_j \overset{p}{\to} \mathbb{E}_{pj} \to 1 \tag{8.11}$$

that is induced from the short exact sequence of K-group schemes

$$1 \to \boldsymbol{\mu}_{p,K} \to \mathbf{G}_{m,K} \overset{p}{\to} \mathbf{G}'_{m,K} \to 1$$

of §3.4.

We now state C. Greither's main result [Gr92, II, Corollary 3.6(b)].

Proposition 8.3.7. *(Greither) Let i, j be integers with $0 \le i, j \le e'$, $pj \le i$.*

(i) There is an isomorphism

$$E_{\text{gen-triv}}(\mathbb{D}_j, \mathbb{D}_i) \cong (U_{i'+(j/p)}(R) \cap U_{(i'/p)+j}(R))/U_{i'+j}(R),$$

where the coset $vU_{i'+j}(R)$ corresponds to an equivalence class of extensions represented by the extension

$$E_v : 1 \to \mathbb{D}_j \to G \to \mathbb{D}_i \to 1.$$

(ii) The extension E_v corresponds to a short exact sequence of R-Hopf orders

$$E'_v : R \to H(i) \to R\left[\frac{\tau-1}{\pi^i}, \frac{\eta a_{v-1} - 1}{\pi^j}\right] \to H(j) \to R,$$

where $a_{v-1} = \sum\limits_{m=0}^{p-1} v^{-m} e_m$, where e_m are the minimal idempotents in $KC_p = K\langle\tau\rangle$. The middle term is an R-Hopf order in $K(C_p \times C_p)$.

Proof. Proof of (i). The short exact sequence (8.11) induces the exact sequence

$$1 \to E_{\text{gen-triv}}(\mathbb{D}_j, \mathbb{D}_i) \to E_{\text{gen-triv}}(\mathbb{E}_j, \mathbb{D}_i) \xrightarrow{p} E_{\text{gen-triv}}(\mathbb{E}_{pj}, \mathbb{D}_i).$$

Consequently,

$$E_{\text{gen-triv}}(\mathbb{D}_j, \mathbb{D}_i) = \ker\left(E_{\text{gen-triv}}(\mathbb{E}_j, \mathbb{D}_i) \xrightarrow{p} E_{\text{gen-triv}}(\mathbb{E}_{pj}, \mathbb{D}_i) \right).$$

Now, by Proposition 8.3.6,

$$E_{\text{gen-triv}}(\mathbb{E}_j, \mathbb{D}_i) \cong U_{i'+(j/p)}(R)/U_{i'+j}(R)$$

and

$$E_{\text{gen-triv}}(\mathbb{E}_{pj}, \mathbb{D}_i) \cong U_{i'+j}(R)/U_{i'+pj}(R),$$

from which (i) follows.

We next prove (ii). Let E_v be the generically trivial extension of $\mathbb{D}_j$ by $\mathbb{D}_j$ determined by the element $v \in U_{i'+(j/p)}(R) \cap U_{(i'/p)+j}(R)$. Then, over K, E_v appears as

$$\begin{aligned}
E_{v,K} : \; 1 &\to \text{Hom}_{K\text{-alg}}(H(j) \otimes_R K, -) \\
&\to \text{Hom}_{K\text{-alg}}(H(j) \otimes_R K, -) \times \text{Hom}_{K\text{-alg}}(H(i) \otimes_R K, -) \\
&\to \text{Hom}_{K\text{-alg}}(H(i) \otimes_R K, -) \to 1.
\end{aligned}$$

The group structure of

$$\text{Hom}_{K\text{-alg}}(H(j) \otimes_R K, -) \times \text{Hom}_{K\text{-alg}}(H(i) \otimes_R K, -)$$

is given by a cocycle $\hat{f} \in C$ whose matrix is M_{v^p}.

The extension $E_{v,K}$ is equivalent to the trivial extension

$$1 \to \mu_{p,K} \to \mu_{p,K} \times \mu_{p,K} \to \mu_{p,K} \to 1,$$

and so the natural transformation $h : \mathbb{D}_i \to \mathbb{E}_j$ given as $X \mapsto \sum_{m=0}^{p-1} v^m e_m$ determines a group scheme isomorphism

$$h : \text{Hom}_{K\text{-alg}}((H(j) \otimes_R H(i)) \otimes_R K, -) \to \text{Hom}_{K\text{-alg}}(K\langle \eta \rangle \otimes K\langle \tau \rangle, -),$$

defined as $h_S(x, y) = (x h_S(y), y)$, for a K-algebra S and $x \in \text{Hom}_{K\text{-alg}}(H(j) \otimes_R K, S)$, $y \in \text{Hom}_{K\text{-alg}}(H(i) \otimes_R K, S)$.

Note that $K\langle\eta\rangle \otimes K\langle\tau\rangle \cong K[\eta, \tau]$ and $(H(j) \otimes_R H(i)) \cong R\left[\frac{\eta - 1}{\pi^j}, \frac{\tau - 1}{\pi^i}\right]$. Thus the isomorphism h yields the K-algebra homomorphism

$$\varrho : K[\eta, \tau] \to K\left[\frac{\eta - 1}{\pi^j}, \frac{\tau - 1}{\pi^i}\right]$$

with $\varrho(\eta) = \eta a_{v-1}$, $\varrho(\tau) = \tau$. Thus

$$\varrho\left(R\left[\frac{\eta - 1}{\pi^j}, \frac{\tau - 1}{\pi^i}\right]\right) = R\left[\frac{\tau - 1}{\pi^i}, \frac{\eta a_{v-1} - 1}{\pi^j}\right],$$

which is the representing algebra of the group scheme G. Consequently, $R\left[\frac{\tau-1}{\pi^i}, \frac{\eta a_{v-1}-1}{\pi^j}\right]$ is an R-Hopf algebra.

We leave it to the reader to show that $R\left[\frac{\tau-1}{\pi^i}, \frac{\eta a_{v-1}-1}{\pi^j}\right]$ is an R-Hopf order in $K(C_p \times C_p)$ that induces the short exact sequence

$$R \to H(i) \to R\left[\frac{\tau - 1}{\pi^i}, \frac{\eta a_{v-1} - 1}{\pi^j}\right] \to H(j) \to R.$$

$\square$

But how do we obtain Hopf orders in KC_{p^2} from the extensions of Proposition 8.3.7? The key is to endow the collection of short exact sequences of Hopf orders with a group product, which we describe as follows. Let

$$E^{(1)} : R \to H(i) \to H_1 \xrightarrow{s_1} H(j) \to R,$$

$$E^{(2)} : R \to H(i) \to H_2 \xrightarrow{s_2} H(j) \to R,$$

be short exact sequences of Hopf algebras. Recalling that the tensor product of two Hopf algebras is again a Hopf algebra, we obtain a short exact sequence of R-Hopf algebras,

$$R \to H(i) \otimes_R H(i) \to H_1 \otimes_R H_2 \xrightarrow{s_1 \otimes s_2} H(j) \otimes_R H(j) \to R.$$

There is a unique map ϕ that makes the following diagram commute:

$$
\begin{array}{ccc}
\operatorname{coker}(\Delta_{H(j)}) & = & \operatorname{coker}(\Delta_{H(j)}) \\
\uparrow \phi & & \uparrow
\end{array}
$$

$$
\begin{array}{ccccc}
R \to H(i) \otimes_R H(i) \to & H_1 \otimes_R H_2 & \xrightarrow{s_1 \otimes s_2} & H(j) \otimes_R H(j) \to R \\
\| \quad\quad \| & \uparrow & & \uparrow \Delta_{H(j)} \quad \| \\
R \to H(i) \otimes_R H(i) \to & \ker(\phi) & \to & H(j) \quad\quad \to R
\end{array}
$$

And there is a map ψ that makes the following diagram commute:

$$
\begin{array}{ccccccccc}
R & \to & H(i) \otimes_R H(i) & \to & \ker(\phi) & \to & H(j) & \to & R \\
\| & & \downarrow m & & \downarrow & & \| & & \| \\
R & \to & H(i) & \overset{\psi}{\to} & \ker(\phi)/\ker(m) & \to & H(j) & \to & R
\end{array}
$$

The short exact sequence in the bottom row is the **Baer product**

$$E = E^{(1)} * E^{(2)}.$$

The Baer product endows the collection of short exact sequences with the structure of a group.

Now, in our case, since $pj \le i$, there exists an extension of Larson orders

$$E_0: \ R \to H(i) \to R\left[\frac{\tau - 1}{\pi^i}, \frac{\eta - 1}{\pi^j}\right] \to H(j) \to R, \ \eta^p = \tau.$$

The Baer product $E_0 * E_v'$ is an extension

$$R \to H(i) \to R\left[\frac{g^p - 1}{\pi^i}, \frac{ga_{v^{-1}} - 1}{\pi^j}\right] \to H(j) \to R,$$

where $g = \eta, g^p = \eta^p = \tau$, whose middle term is an R-Hopf order in KC_{p^2} that we call a **Greither order**. A Greither order is determined by two valuation parameters i and j and one unit parameter $u = v^{-1}$, and will be denoted by $A(i, j, u)$.

Now suppose that H is an R-Hopf order in KC_{p^2}, that induces the short exact sequence

$$E: R \to H(i) \to H \to H(j) \to R,$$

where $pj \le i$. Then the Baer product $E_0^{-1} * E$ is a short exact sequence of the form

$$R \to H(i) \to H' \to H(j) \to R,$$

where H' is an R-Hopf order in $K(C_p \times C_p)$. There is a corresponding short exact sequence of R-group schemes

$$1 \to \mathbb{D}_j \to \mathrm{Hom}_{R\text{-alg}}(H', -) \to \mathbb{D}_i \to 1,$$

which is genericaly trivial. Now, by Proposition 8.3.7, $E_0^{-1} * E = E_v'$ for some $v \in U_{i'+(j/p)}(R) \cap U_{(i'/p)+j}(R)$. Thus $E = E_0 * E_v'$, and so, H is a Greither order.

Thus we have solved the problem that was stated at the beginning of this section: for $pj \le i$, the middle term in the short exact sequence

$$R \to H(i) \to H \to H(j) \to R$$

is a Greither order $A(i, j, u)$, where u represents a class in the quotient $(U_{i'+(j/p)}$ $(R) \cap U_{(i'/p)+j}(R))/U_{i'+j}(R)$.

If $j' \geq pi'$, then Greither's result may be used to give the structure of H^*.

How does the collection of Greither orders relate to the Larson orders in KC_{p^2}? It is not too hard to prove the following.

Proposition 8.3.8. *Let $A(i, j, u)$ be a Greither order. Then $A(i, j, u)$ is the Larson order $H(i, j)$ in KC_{p^2} if and only if $\mathrm{ord}(1 - u) \geq i' + j$.*

Proof. Exercise. $\qquad\square$

Not all Hopf orders in KC_{p^2} are Greither orders, however. For example, in the short exact sequence

$$R \to H(e') \to RC_{p^2}^* \to H(e') \to R,$$

one has $e' \not\geq pe'$.

But in view of the valuation condition (Proposition 8.1.2), a given R-Hopf order in KC_{p^2} is either of the form $A(i, j, u)$ or $A(i, j, u)^*$. So, to give a full account of the structure of Hopf orders in KC_{p^2}, one needs to obtain the algebraic structure of $A(i, j, u)^*$.

We need a lemma. Let $\langle g \rangle = C_{p^2}$, $\langle \gamma \rangle = \hat{C}_{p^2}$, and let $\langle\,,\,\rangle : K\hat{C}_{p^2} \times KC_{p^2} \mapsto K$ denote the duality map.

Lemma 8.3.1. *Let e_i denote the minimal idempotents of KC_p, and let $\hat{e}_j$ denote the minimal idempotents of $K\hat{C}_p$. Then*

$$\langle \hat{e}_k \gamma^{pc+d}, e_j g^{pa+b} \rangle = \zeta_{p^2}^{(pa+b)(pc+d)}$$

if $j = d$ and $k = b$, and is 0 otherwise.

Proof. Exercise. $\qquad\square$

Proposition 8.3.9. *Assume that K contains ζ_{p^2}, and let*

$$A(i, j, u) = R\left[\frac{g^p - 1}{\pi^i}, \frac{ga_u - 1}{\pi^j}\right]$$

be a Greither order in KC_{p^2}. Let $B = R\left[\frac{\gamma^p - 1}{\pi^{j'}}\right]$, and let

$$J = B\left[\frac{\gamma a_{\tilde{u}} - 1}{\pi^{i'}}\right],$$

where $\tilde{u} = \zeta_{p^2}^{-1} u^{-1}$, $a_{\tilde{u}} = \sum_{m=0}^{p-1} \tilde{u}^m \rho_m$, and $\rho_m = \frac{1}{p} \sum_{l=0}^{p-1} \zeta_p^{-m} \gamma^{pl}$. Then $J = A(i, j, u)^*$.

Proof. By Proposition 7.1.3, it suffices to show that

$$\langle J, A(i, j, u)\rangle \subseteq R,$$

and this is equivalent to the conditions

$$\langle \gamma^p - 1, (g^p - 1)^q (ga_u - 1)^r \rangle \in \pi^{qi+rj+j'} R \tag{8.12}$$

for $q, r = 0, \ldots, p - 1, q + r > 0$, and

$$\langle \gamma a_{\tilde{u}} - 1, (g^p - 1)^q (ga_u - 1)^r \rangle \in \pi^{qi+rj+i'} R \tag{8.13}$$

for $q, r = 0, \ldots, p - 1, q + r > 0$. One quickly sees that (8.12) is equivalent to
$(\zeta_p - 1)^r \in \pi^{rj+j'} R$ for $r = 1, \ldots, p - 1$, which holds.

To show that (8.13) holds, let $S = \langle \gamma a_{\tilde{u}} - 1, (g^p - 1)^q (ga_u - 1)^r \rangle$, let $\displaystyle\sum_{c,d=0}^{q,r}$

denote $\displaystyle\sum_{c=0}^{q} \sum_{d=0}^{r}$, and let $C(c, d) = \binom{q}{c}\binom{r}{d}(-1)^{q-c}(-1)^{r-d}$. Then

$$S = \sum_{c,d=0}^{q,r} C(c,d)\langle \gamma a_{\tilde{u}} - 1, g^{pc}(ga_u)^d \rangle$$

$$= \sum_{c,d=0}^{q,r} C(c,d)\langle \gamma a_{\tilde{u}}, g^{pc}(ga_u)^d \rangle - \sum_{c,d=0}^{q,r} C(c,d)\langle 1, g^{pc}(ga_u)^d \rangle$$

$$= \sum_{c,d=0}^{q,r} C(c,d)\langle \gamma a_{\tilde{u}}, g^{pc}(ga_u)^d \rangle - \sum_{c,d=0}^{q,r} C(c,d)$$

$$= \sum_{c,d=0}^{q,r} C(c,d)\langle \gamma a_{\tilde{u}}, a_{u^d} g^{pc+d} \rangle - \sum_{c,d=0}^{q,r} C(c,d)$$

$$= \sum_{c,d=0}^{q,r} C(c,d) \sum_{i,j} u^{di} \tilde{u}^{j} \langle \hat{e}_j \gamma, e_i g^{pc+d} \rangle - \sum_{c,d=0}^{q,r} C(c,d)$$

$$= \sum_{c,d=0}^{q,r} C(c,d) u^d \tilde{u}^d \zeta_{p^2}^{pc+d} - \sum_{c,d=0}^{q,r} C(c,d) \quad \text{(by Lemma 8.3.1)}$$

$$= \sum_{c,d=0}^{q,r} C(c,d)(u\tilde{u}\zeta_{p^2})^d \zeta_p^c - \sum_{c,d=0}^{q,r} C(c,d)$$

$$= \sum_{c,d=0}^{q,r} C(c,d)\zeta_p^c - \sum_{c,d=0}^{q,r} C(c,d) \quad \text{since } u\tilde{u}\zeta_{p^2} = 1$$

$$= \sum_{c,d=0}^{q,r} C(c,d)(\zeta_p^c - 1).$$

Now $S = 0$ unless $q \geq 1$, $r = 0$. In this case, $S = (\zeta_p - 1)^q$, which is in $\pi^{qi'+i} R$. Thus (8.13) holds.   □

And so we have shown that an arbitrary R-Hopf order in KC_{p^2} can be written in the form $A(i, j, u)$ for some integers $0 \leq i, j \leq e'$ and unit $u \in R$.

The largest Larson order in $A(i, j, u)$ can be computed as follows.

Proposition 8.3.10. *Let $A(i, j, u)$ be an R-Hopf order in KC_{p^2}. Then*

$$H(i, l) = A(\Xi(A(i, j, u)))$$

is the Larson order $H(i, l)$, where $l = j$ if $\mathrm{ord}(1 - u) \geq i' + j$ and $l = i - e' + \mathrm{ord}(1 - u)$ otherwise.

Proof. If $\mathrm{ord}(1 - u) \geq i' + j$, then $v = u^{-1}$ corresponds to the trivial class in $E_{\mathrm{gen\text{-}triv}}(\mathbb{D}_j, \mathbb{D}_i)$. Thus, the Baer product $E_0 * E_v' = E_0$, which says that $A(i, j, u) = H(i, j)$. On the other hand, suppose that $\mathrm{ord}(1 - u) < i' + j$. We have

$$\frac{ga_u - 1}{\pi^j} = g\left(\frac{a_u - 1}{\pi^j}\right) + \frac{g - 1}{\pi^j}.$$

Now, by Lemma 7.1.6, $l = \mathrm{ord}(1 - u) - i'$ is the largest integer for which $\frac{a_u - 1}{\pi^l} \in H(i)$. Therefore l is the largest integer for which $\frac{g-1}{\pi^l} \in H(i)$. Thus, $A(\Xi(A(i, j, u))) = H(i, l)$.   □

8.4 Hopf Orders in KC_4, KC_9

In this section, we present an alternate approach to proving the valuation condition (Proposition 8.1.2) for the cases $p = 2, 3$. We show that if

$$R \to H(i) \to H \xrightarrow{s} H(j) \to R$$

is a short exact sequence where H is a Hopf order in KC_4 or KC_9, then either $pj \leq i$ or $pi' \leq j'$.

We begin with a lemma.

Lemma 8.4.1. *Let H be an R-Hopf order in KC_{p^2}. Suppose H^* can be written in the form*

$$H^* = R\left[\frac{\gamma^p - 1}{\pi^{j'}}, \frac{\gamma u - 1}{\pi^{i'}}\right],$$

where $u = \sum_{m=0}^{p-1} b_m \eta_m \in K\hat{C}_p$ and where η_m are the minimal idempotents in $K\langle \gamma^p \rangle$. Then H is of the form

$$H = R\left[\frac{g^p - 1}{\pi^i}, \frac{g a_v - 1}{\pi^j}\right],$$

where $a_v = \sum_{m=0}^{p-1} v^m f_m$, $v = \zeta_{p^2}^{-1} b_1^{-1}$, and where f_m are the minimal idempotents in $K\langle g^p \rangle$.

Proof. Let $v = \zeta_{p^2}^{-1} b_1^{-1}$. We claim that

$$\left\langle \frac{g a_v - 1}{\pi^j}, H^* \right\rangle \subseteq R,$$

which is equivalent to

$$\mathrm{ord}\left(\langle g a_v - 1, (\gamma^p - 1)^r (\gamma u - 1)^s \rangle\right) \geq j + rj' + si' \qquad (8.14)$$

for $r, s = 0, \ldots, p - 1, r + s > 0$. Now

$$\langle g a_v - 1, (\gamma^p - 1)^r (\gamma u - 1)^s \rangle = (\zeta_p - 1)^r \sum_{q=0}^{s} \binom{s}{q} (-1)^{s-q} \zeta_{p^2}^q v^q b_1^q$$

$$= (\zeta_p - 1)^r \sum_{q=0}^{s} \binom{s}{q} (-1)^{s-q}$$

since $\zeta_{p^2} v b_1 = 1$. Thus, for $s \geq 1$, (8.4) holds since the sum above is 0. For $s = 0$, one has

$$\langle g a_v - 1, (\gamma^p - 1)^r (\gamma u - 1)^s \rangle = (\zeta_p - 1)^r,$$

in which case (8.4) holds since $r \cdot \mathrm{ord}(\zeta_p - 1) = re' \geq j + rj'$. It follows that $\frac{g a_v - 1}{\pi^j} \in H$.

An application of Proposition 7.1.3 then shows that H is of the form claimed.
□

Proposition 8.4.1. *Let H be an R-Hopf order in KC_{p^2} that induces the short exact sequence $R \to H(i) \to H \to H(j) \to R$. Suppose H^* can be written in the form*

$$H^* = R\left[\frac{\gamma^p - 1}{\pi^{j'}}, \frac{\gamma u - 1}{\pi^{i'}}\right],$$

where $u = \sum_{m=0}^{p-1} b_m \eta_m \in K\langle \gamma^p \rangle$. *Then either* $i \geq pj$ *or* $j' \geq pi'$.

Proof. By Lemma 8.4.1, H has the form

$$H = R\left[\frac{g^p - 1}{\pi^i}, \frac{g a_v - 1}{\pi^j}\right]$$

for some unit v in R. Now, by [Ch00, 31.3],

$$\mathrm{ord}(\zeta_p v^p - 1) \geq pj + i' \quad \text{and} \quad \mathrm{ord}(v^p - 1) \geq j + pi',$$

and so, by [Ch00, 31.4], either $i \geq pj$ or $j' \geq pi'$. $\square$

So it remains to show that H^* can be written in the form of Lemma 8.4.1. We begin with the case $p = 2$. Let H be an R-Hopf order in KC_4, $\langle g \rangle = C_4$. Then $H = R[\frac{g^2-1}{\pi^i}, \Psi]$, where Ψ is an element of KC_4 for which $s(\Psi)$ is the generator $\frac{\overline{g}-1}{\pi^j}$ of the Larson order $H(j)$. Since $KC_4 = Ke_0 \otimes Ke_1 \otimes Ke_2 \otimes Ke_3$, Ψ has the form

$$\Psi = a_0 e_0 + a_1 e_1 + a_2 e_2 + a_3 e_3$$

for some elements $a_0, a_1, a_2, a_3 \in K$. Let e_0', e_1' be the idempotents in $K\langle \overline{g} \rangle$. Since $s(\Psi) = a_0 e_0' + a_2 e_2'$, $a_0 = 0$ and $a_2 = c = \frac{\zeta_4^2-1}{\pi^j}$. Thus $\Psi = a_1 e_1 + c e_2 + a_3 e_3$. Put $h = \frac{g^2-1}{\pi^i}$. Then an R-basis for H is

$$\{1, h, \Psi, h\Psi\}.$$

Now, let ι_m be the idempotents in $K\hat{C}_4$, and let

$$\Phi = b_0 \iota_0 + b_1 \iota_1 + b_2 \iota_2 + b_3 \iota_3$$

be an element in $K\hat{C}_4$. Let $b_0 = 0$ and $b_2 = d = \frac{\zeta_4^2-1}{\pi^{i'}}$. We seek conditions on Φ such that

$$\langle H, \Phi \rangle \subseteq R.$$

But these can be found by choosing b_1 and b_3 such that

$$\langle \Psi, \Phi \rangle = 0,$$
$$\langle (g^2 - 1)\Psi, \Phi \rangle = 0.$$

The system above corresponds to a system of equations

$$\langle a_1 e_1 + c e_2 + a_3 e_3, b_1 \iota_1 + d \iota_2 + b_3 \iota_3 \rangle = 0,$$

$$\langle a_1 e_1 + a_3 e_3, b_1 \iota_1 + d \iota_2 + b_3 \iota_3 \rangle = 0,$$

which yields the 2×2 linear system

$$\begin{pmatrix} a_1 \zeta_4^{-1} + c \zeta_4^{-2} + a_3 \zeta_4^{-3} & a_1 \zeta_4^{-3} + c \zeta_4^{-2} + a_3 \zeta_4^{-1} \\ a_1 \zeta_4^{-1} + a_3 \zeta_4^{-3} & a_1 \zeta_4^{-3} + a_3 \zeta_4^{-1} \end{pmatrix} \begin{pmatrix} b_1 \\ b_3 \end{pmatrix} = \begin{pmatrix} (a_1 - c + a_3) d \\ (a_1 + a_3) d \end{pmatrix},$$

whose solution is readily obtained as

$$b_1 = \frac{\zeta_4 x - 1}{\pi^{i'}}, \quad b_3 = \frac{\zeta_4^3 x - 1}{\pi^{i'}}, \quad \text{with } x = \frac{a_3 + a_1}{a_3 - a_1}.$$

Thus

$$\Phi = \frac{\zeta_4 x - 1}{\pi^{i'}} \iota_1 + \frac{\zeta_4^2 - 1}{\pi^{i'}} \iota_2 + \frac{\zeta_4^3 x - 1}{\pi^{i'}} \iota_3 = \frac{\gamma a_x - 1}{\pi^{i'}}.$$

Now, let $A = R[\frac{\gamma^2 - 1}{\pi^{j'}}]$ and let $J = A[\frac{\gamma a_x - 1}{\pi^{i'}}]$. Since $\langle H, \Phi \rangle \subseteq R$, $\langle H, J \rangle \subseteq R$. An application of Proposition 7.1.3 then shows that

$$H^* = R\left[\frac{\gamma^2 - 1}{\pi^{j'}}, \frac{\gamma a_x - 1}{\pi^{i'}} \right].$$

Hence H^* is in the form of Lemma 8.4.1.

We repeat this calculation for $p = 3$. Put $\zeta = \zeta_9$. Let H be an R-Hopf order in KC_9, $C_9 = \langle g \rangle$. Then $H = R[\frac{g^3 - 1}{\pi^i}, \Psi]$, where Ψ is an element of KC_9 for which $s(\Psi) = \frac{\overline{g} - 1}{\pi^j}$ generates the Larson order $H(j)$. We can assume that Ψ has the form

$$\Psi = c_0 e_0 + a_1 e_1 + a_2 e_2 + c_1 e_3 + a_4 e_4 + a_5 e_5 + c_2 e_6 + a_7 e_7 + a_8 e_8$$

for some elements $a_1, a_2, a_4, a_5, a_7, a_8 \in K$, and $c_0 = 0$, $c_1 = \frac{\zeta^3 - 1}{\pi^j}$, $c_2 = \frac{\zeta^6 - 1}{\pi^j}$. Put $h = \frac{g^3 - 1}{\pi^i}$. An R-basis for H is

$$\{1, h, h^2, \Psi, h\Psi, h^2\Psi, \Psi^2, h\Psi^2, h^2\Psi^2\}.$$

Now, let ι_m denote the idempotents in $K\hat{C}_9$, and let

$$\Phi = d_0 \iota_0 + b_1 \iota_1 + b_2 \iota_2 + d_1 \iota_3 + b_4 \iota_4 + b_5 \iota_5 + d_2 \iota_6 + b_7 \iota_7 + b_8 \iota_8$$

be an element of $K\hat{C}_9$ with $d_0 = 0$, $d_1 = \frac{\zeta^3 - 1}{\pi^i}$, and $d_2 = \frac{\zeta^6 - 1}{\pi^i}$. We seek conditions on Φ such that $\langle H, \Phi \rangle \subseteq R$. But these can be found by choosing $b_1, b_2, b_4, b_5, b_7, b_8$ such that

$$\langle \Psi, \Phi \rangle = 0,$$

$$\langle (g^3 - 1)\Psi, \Phi \rangle = 0,$$

$$\langle (g^3 - 1)^2 \Psi, \Phi \rangle = 0,$$

$$\langle \Psi^2, \Phi \rangle = 0,$$

$$\langle (g^3 - 1)\Psi^2, \Phi \rangle = 0,$$

$$\langle (g^3 - 1)^2 \Psi^2, \Phi \rangle = 0.$$

This system corresponds to the system of equations

$$\langle a_1 e_1 + a_2 e_2 + c_1 e_3 + a_4 e_4 + a_5 e_5 + c_2 e_6 + a_7 e_7 + a_8 e_8,$$

$$b_1 l_1 + b_2 l_2 + d_1 l_3 + b_4 l_4 + b_5 l_5 + d_2 l_6 + b_7 l_7 + b_8 l_8 \rangle = 0;$$

$$\langle (\zeta - 1)a_1 e_1 + (\zeta^2 - 1)a_2 e_2 + (\zeta - 1)a_4 e_4 + (\zeta^2 - 1)a_5 e_5$$

$$+ (\zeta - 1)a_7 e_7 + (\zeta^2 - 1)a_8 e_8,$$

$$b_1 l_1 + b_2 l_2 + d_1 l_3 + b_4 l_4 + b_5 l_5 + d_2 l_6 + b_7 l_7 + b_8 l_8 \rangle = 0;$$

$$\langle (\zeta - 1)^2 a_1 e_1 + (\zeta^2 - 1)^2 a_2 e_2 + (\zeta - 1)^2 a_4 e_4 + (\zeta^2 - 1)^2 a_5 e_5$$

$$+ (\zeta - 1)^2 a_7 e_7 + (\zeta^2 - 1)^2 a_8 e_8,$$

$$b_1 l_1 + b_2 l_2 + d_1 l_3 + b_4 l_4 + b_5 l_5 + d_2 l_6 + b_7 l_7 + b_8 l_8 \rangle = 0;$$

$$\langle a_1^2 e_1 + a_2^2 e_2 + c_1^2 e_3 + a_4^2 e_4 + a_5^2 e_5 + c_2^2 e_6 + a_7^2 e_7 + a_8^2 e_8,$$

$$b_1 l_1 + b_2 l_2 + d_1 l_3 + b_4 l_4 + b_5 l_5 + d_2 l_6 + b_7 l_7 + b_8 l_8 \rangle = 0;$$

$$\langle (\zeta - 1)a_1^2 e_1 + (\zeta^2 - 1)a_2^2 e_2 + (\zeta - 1)a_4^2 e_4 + (\zeta^2 - 1)a_5^2 e_5$$

$$+ (\zeta - 1)a_7^2 e_7 + (\zeta^2 - 1)a_8^2 e_8,$$

$$b_1 l_1 + b_2 l_2 + d_1 l_3 + b_4 l_4 + b_5 l_5 + d_2 l_6 + b_7 l_7 + b_8 l_8 \rangle = 0;$$

$$\langle (\zeta - 1)^2 a_1^2 e_1 + (\zeta^2 - 1)^2 a_2^2 e_2 + (\zeta - 1)^2 a_4^2 e_4 + (\zeta^2 - 1)^2 a_5^2 e_5$$

$$+ (\zeta - 1)^2 a_7^2 e_7 + (\zeta^2 - 1)^2 a_8^2 e_8,$$

$$b_1 l_1 + b_2 l_2 + d_1 l_3 + b_4 l_4 + b_5 l_5 + d_2 l_6 + b_7 l_7 + b_8 l_8 \rangle = 0.$$

Set

$$A = a_1\zeta^{-1} + a_4\zeta^{-4} + a_7\zeta^{-7}, \qquad B = a_1\zeta^{-2} + a_4\zeta^{-8} + a_7\zeta^{-5},$$
$$C = a_1 + a_4 + a_7, \qquad\qquad\qquad E = a_2\zeta^{-2} + a_5\zeta^{-5} + a_8\zeta^{-8},$$
$$F = a_2\zeta^{-4} + a_5\zeta^{-1} + a_8\zeta^{-7}, \qquad G = a_2 + a_5 + a_8,$$
$$A' = a_1^2\zeta^{-1} + a_4^2\zeta^{-4} + a_7^2\zeta^{-7}, \qquad B' = a_1^2\zeta^{-2} + a_4^2\zeta^{-8} + a_7^2\zeta^{-5},$$
$$C' = a_1^2 + a_4^2 + a_7^2, \qquad\qquad\quad E' = a_2^2\zeta^{-2} + a_5^2\zeta^{-5} + a_8^2\zeta^{-8},$$
$$F' = a_2^2\zeta^{-4} + a_5^2\zeta^{-1} + a_8^2\zeta^{-7}, \quad G' = a_2^2 + a_5^2 + a_8^2,$$
$$J = c_1\zeta^{-3} + c_2\zeta^{-6}, \qquad\qquad K = c_1\zeta^{-6} + c_2\zeta^{-3},$$
$$J' = c_1^2\zeta^{-3} + c_2^2\zeta^{-6}, \qquad\qquad K' = c_1^2\zeta^{-6} + c_2^2\zeta^{-3}.$$

Note that

$$K = c_1\zeta^{-6} + c_2\zeta^{-3}$$

$$= \zeta^{-2}\left(\frac{\zeta - 1}{\pi^j}\right) + \zeta^{-1}\left(\frac{\zeta^2 - 1}{\pi^j}\right)$$

$$= 0.$$

Using the relation $\langle e_m, \iota_n \rangle = \zeta^{-mn}/9$, one obtains the 6×6 linear system

$$M\begin{pmatrix} b_1 \\ b_2 \\ b_4 \\ b_5 \\ b_7 \\ b_8 \end{pmatrix} = \begin{pmatrix} (-\zeta^{-3}C - \zeta^{-6}G - c_1 - c_2)d_1 + (-\zeta^{-6}C - \zeta^{-3}G - c_1 - c_2)d_2 \\ (-\zeta^{-3}d_1 - \zeta^{-6}d_2)C \\ (-\zeta^{-6}d_1 - \zeta^{-3}d_2)G \\ (-\zeta^{-3}C' - \zeta^{-6}G' - c_1 - c_2)d_1 + (-\zeta^{-6}C' - \zeta^{-3}G' - c_1 - c_2)d_2 \\ (-\zeta^{-3}d_1 - \zeta^{-6}d_2)C' \\ (-\zeta^{-6}d_1 - \zeta^{-3}d_2)G' \end{pmatrix},$$

where M is the matrix

$$\begin{pmatrix} A + E + J & B + F & \zeta^{-3}A + \zeta^{-6}E + J & \zeta^{-3}B + \zeta^{-6}F \\ A & B & \zeta^{-3}A & \zeta^{-3}B \\ E & F & \zeta^{-6}E & \zeta^{-6}F \\ A' + E' + J' & B' + F' + K' & \zeta^{-3}A' + \zeta^{-6}E' + J' & \zeta^{-3}B' + \zeta^{-6}F' + K' \\ A' & B' & \zeta^{-3}A' & \zeta^{-3}B' \\ E' & F' & \zeta^{-6}E' & \zeta^{-6}F' \end{pmatrix}$$

$$\left.\begin{matrix} \zeta^{-6}A + \zeta^{-3}E + J & \zeta^{-6}B + \zeta^{-3}E \\ \zeta^{-6}A & \zeta^{-6}B \\ \zeta^{-3}E & \zeta^{-3}F \\ \zeta^{-6}A' + \zeta^{-3}E' + J' & \zeta^{-6}B' + \zeta^{-3}E' + K' \\ \zeta^{-6}A' & \zeta^{-6}B' \\ \zeta^{-3}E' & \zeta^{-3}F' \end{matrix}\right)$$

This system is equivalent to the system

$$\left(\begin{array}{cccccc|c} 1 & 0 & 1 & 0 & 1 & 0 & \frac{-3}{\pi^{i'}} \\ \frac{A}{B} & 1 & \zeta^{-3}\frac{A}{B} & \zeta^{-3} & \zeta^{-6}\frac{A}{B} & \zeta^{-6} & \frac{-3}{\pi^{i'}}\frac{C}{B} \\ 1 & \frac{F}{E} & \zeta^{-6} & \zeta^{-6}\frac{F}{E} & \zeta^{-3} & \zeta^{-3}\frac{F}{E} & 0 \\ J' & K' & J' & K' & J' & K' & \frac{3(c_1^2+c_2^2)}{\pi^{i'}} \\ \frac{A'}{B'} & 1 & \zeta^{-3}\frac{A'}{B'} & \zeta^{-3} & \zeta^{-6}\frac{A'}{B'} & \zeta^{-6} & \frac{-3}{\pi^{i'}}\frac{C'}{B'} \\ 1 & \frac{F'}{E'} & \zeta^{-6} & \zeta^{-6}\frac{F'}{E'} & \zeta^{-3} & \zeta^{-3}\frac{F'}{E'} & 0 \end{array}\right),$$

which reduces to the system

$$\left(\begin{matrix} 1 & 0 & 1 & 0 & 1 \\ \frac{A}{B}-\frac{A'}{B'} & 0 & \zeta^{-3}\left(\frac{A}{B}-\frac{A'}{B'}\right) & 0 & \zeta^{-6}\left(\frac{A}{B}-\frac{A'}{B'}\right) \\ 0 & \frac{F}{E}-\frac{F'}{E'} & 0 & \zeta^{-6}\left(\frac{F}{E}-\frac{F'}{E'}\right) & 0 \\ 0 & 1 & 0 & 1 & 0 \\ \frac{A'}{B'} & 1 & \zeta^{-3}\frac{A'}{B'} & \zeta^{-3} & \zeta^{-6}\frac{A'}{B'} \\ 1 & \frac{F'}{E'} & \zeta^{-6} & \zeta^{-6}\frac{F'}{E'} & \zeta^{-3} \end{matrix}\right.$$

$$\left.\begin{array}{c|c} 0 & \frac{-3}{\pi^{i'}} \\ 0 & \frac{-3}{\pi^{i'}}\left(\frac{C}{B}-\frac{C'}{B'}\right) \\ \zeta^{-3}\left(\frac{F}{E}-\frac{F'}{E'}\right) & 0 \\ 1 & \frac{-3}{\pi^{i'}} \\ \zeta^{-6} & \frac{-3}{\pi^{i'}}\frac{C'}{B'} \\ \zeta^{-3}\frac{F'}{E'} & 0 \end{array}\right),$$

and ultimately to the system

$$
\left(
\begin{array}{cccccc|c}
1 & 0 & 1 & 0 & 1 & 0 & \frac{-3}{\pi^{i'}} \\
1 & 0 & \zeta^{-3} & 0 & \zeta^{-6} & 0 & \frac{3}{\pi^{i'}}\left(\frac{BC'-CB'}{AB'-BA'}\right) \\
0 & 1 & 0 & \zeta^{-6} & 0 & \zeta^{-3} & 0 \\
0 & 1 & 0 & 1 & 0 & 1 & \frac{-3}{\pi^{i'}} \\
0 & 1 & 0 & \zeta^{-3} & 0 & \zeta^{-6} & \frac{3}{\pi^{i'}}\left(\frac{CA'-AC'}{AB'-BA'}\right) \\
1 & 0 & \zeta^{-6} & 0 & \zeta^{-3} & 0 & 0
\end{array}
\right).
$$

Now, the required values of $b_1, b_2, b_4, b_5, b_7, b_8$ satisfy

$$b_1 + b_4 + b_7 = \frac{-3}{\pi^{i'}},$$

$$b_1 + \zeta^{-3}b_4 + \zeta^{-6}b_7 = \frac{3}{\pi^{i'}}\left(\frac{BC' - CB'}{AB' - BA'}\right),$$

$$b_2 + \zeta^{-6}b_5 + \zeta^{-3}b_8 = 0,$$

$$b_2 + b_5 + b_8 = \frac{-3}{\pi^{i'}},$$

$$b_2 + \zeta^{-3}b_5 + \zeta^{-6}b_8 = \frac{3}{\pi^{i'}}\left(\frac{CA' - AC'}{AB' - BA'}\right),$$

$$b_1 + \zeta^{-6}b_4 + \zeta^{-3}b_7 = 0. \tag{8.15}$$

Let x and y be elements of K, and set $b_1 = \frac{\zeta x - 1}{\pi^{i'}}$, $b_2 = \frac{\zeta^2 y - 1}{\pi^{i'}}$, $b_4 = \frac{\zeta^4 x - 1}{\pi^{i'}}$, $b_5 = \frac{\zeta^5 y - 1}{\pi^{i'}}$, $b_7 = \frac{\zeta^7 x - 1}{\pi^{i'}}$, and $b_8 = \frac{\zeta^8 y - 1}{\pi^{i'}}$. Then, with

$$x = \zeta^{-1}\left(\frac{BC' - CB'}{AB' - BA'}\right),$$

$$y = \zeta^{-2}\left(\frac{CA' - C'A}{AB' - BA'}\right),$$

the equations in (8.15) are satisfied.
   Now

$$\Phi = \frac{\zeta x - 1}{\pi^{i'}}l_1 + \frac{\zeta^2 y - 1}{\pi^{i'}}l_2 + \frac{\zeta^3 - 1}{\pi^{i'}}l_3 + b_4\frac{\zeta^4 x - 1}{\pi^{i'}}l_4$$

$$+ \frac{\zeta^5 y - 1}{\pi^{i'}}l_5 + \frac{\zeta^6 - 1}{\pi^{i'}}l_6 + \frac{\zeta^7 x - 1}{\pi^{i'}}l_7 + \frac{\zeta^8 y - 1}{\pi^{i'}}l_8 = \frac{\gamma u - 1}{\pi^{i'}},$$

where $u = \eta_0 + x\eta_1 + y\eta_2 \in K\langle\gamma^3\rangle$, with η_m the idempotents in $K\langle\gamma^3\rangle$. With this definition of Φ, we have

$$\langle H, \Phi\rangle \subseteq R.$$

Let $A = R[\frac{\gamma^3-1}{\pi^{j'}}]$, and let $J = A[\frac{\gamma u-1}{\pi^{i'}}]$. Since $\langle H, \Phi\rangle \subseteq R$, $\langle H, J\rangle \subseteq R$. An application of Proposition 7.1.3 then yields

$$H^* = R\left[\frac{\gamma^3-1}{\pi^{j'}}, \frac{\gamma u-1}{\pi^{i'}}\right].$$

8.5 Chapter Exercises

Exercises for §8.1

1. Let K be a finite extension of $\mathbb{Q}_2$ with $\mathrm{ord}(2) = e$. Let H be an R-Hopf order in KC_4. Prove that there is no short exact sequence of R-Hopf orders of the form

$$R \to H(e/2) \to H \to H(e/2) \to R.$$

2. Suppose $\zeta_{p^2} \in K$, let H be an R-Hopf order in KC_{p^2}, and let

$$R \to H(i) \to H \to H(j) \to R$$

be a short exact sequence with $pj > i$. Prove that $\mathrm{ord}(1 - \zeta_{p^2}) \geq i' + (j/p)$.

Exercises for §8.2

3. Let $h : C_m \to G$ be a function with $h(1) = 1$. Let $f_h : C_m \times C_m \to G$ be defined as

$$f_h(\tau^i, \tau^j) = h(\tau^i)(h(\tau^{i+j}))^{-1}h(\tau^j)$$

for $i, j = 0, \ldots, m - 1$. Prove that f_h is a cocycle.
4. Prove Proposition 8.2.1.
5. Compute all of the non-equivalent extensions in $E(\mathbb{Z}, C_3)$.
6. Compute all of the non-equivalent extensions in $E(\mathbb{Z}/(p^2), C_p)$.

Exercises for §8.3

7. Prove Proposition 8.3.8.
8. Prove Lemma 8.3.1.
9. Prove that the R-Hopf orders $A(i, j, v)$ and $A(i, j, w)$ are equal if and only if $\mathrm{ord}(v - w) \geq i' + j$.
10. Let $A(i, j, v)$ be a Greither order in KC_{p^2}.

    (a) Show that $A(i, j, v^{-1})$ is a Greither order in KC_{p^2}.
    (b) Show that $A(i, j, v) = A(i, j, v^{-1})$ if and only if $p = 2$.

11. Let $H(i, j)$ be a Larson order in KC_{p^2}. Find conditions on i, j so that the linear dual $H(i, j)^*$ is a Larson order in KC_{p^2}.

12. Let $A(i, j, u)$ be an R-Hopf order in KC_{p^2}. Show that $\text{ord}(1 - u) \geq i' + (j/p)$.

13. Suppose $A(i, j, v)$ is a Greither order with $\text{ord}(1 - \zeta_{p^2} v) \geq i'$. Show that there exists an R-Hopf order of the form $H(a, b)^*$ for which $H(a, b)^* \subseteq A(i, j, v)$.

14. By Proposition 8.3.5, $C/B \cong U_{pi'+j}(R)/U^p_{i'+j}(R)$, and by Proposition 8.2.3, $E(U_j(R), C_p) \cong U_j(R)/U^p_j(R)$. Is C/B a subgroup of $E(U_j(R), C_p)$?

Exercises for §8.4

15. How would one generalize the results of §8.4 to $p > 3$?

Chapter 9
Hopf Orders in KC_{p^3}

In this chapter, we move on to the construction of Hopf orders in KC_{p^3}. We assume throughout this chapter that K is a finite extension of $\mathbb{Q}_p$ with $\zeta_{p^3} \in K$. Though all Hopf orders in KC_p and KC_{p^2} are known, this is not the case for Hopf orders in KC_{p^3}. L. Childs and R. Underwood have explored various ways to construct Hopf orders in KC_{p^3}; the reader is referred to the papers [CU03], [Un08b], [Un06], [UC05], and [Un96]. In the first section here, we briefly review the construction of "duality Hopf orders" of [UC05].

9.1 Duality Hopf Orders in KC_{p^3}

Let g be a generator for C_{p^3}, and let γ be a generator for $\hat{C}_{p^3}$. Let $\overline{g}$ denote the image of g under the mapping $KC_{p^3} \to KC_{p^2}$, $g^{p^2} \mapsto 1$, and let $\overline{\gamma}$ denote the image of γ under the mapping $K\hat{C}_{p^3} \to K\hat{C}_{p^2}$, $\gamma^{p^2} \mapsto 1$. For an integer m, $0 \le m \le e'$, set $m' = e' - m$, and, for a unit $u \in R$, set $\tilde{u} = \zeta_{p^2}^{-1} u^{-1}$. Let

$$A(i, j, u) = R\left[\frac{g^{p^2} - 1}{\pi^i}, \frac{g^p a_u - 1}{\pi^j} \right]$$

and

$$A(j, k, w) = R\left[\frac{\overline{g}^p - 1}{\pi^j}, \frac{\overline{g} a_w - 1}{\pi^k} \right]$$

be Greither orders in KC_{p^2}. Here u, w are units in R with

$$a_u = \sum_{l=0}^{p-1} u^l \frac{1}{p} \sum_{q=0}^{p-1} \zeta_p^{-lq} g^{p^2 q} \quad \text{and} \quad a_w = \sum_{l=0}^{p-1} w^l \frac{1}{p} \sum_{q=0}^{p-1} \zeta_p^{-lq} \overline{g}^{pq}.$$

Moreover, $pj \le i$ and $pk \le j$.

R.G. Underwood, *An Introduction to Hopf Algebras*, DOI 10.1007/978-0-387-72766-0_9, 181
© Springer Science+Business Media, LLC 2011

By Proposition 8.3.9, the linear duals of these Hopf orders are Hopf orders in KC_{p^2} of the form

$$A(i,j,u)^* = A(j',i',\tilde{u}) = R\left[\frac{\overline{\gamma}^p - 1}{\pi^{j'}}, \frac{\overline{\gamma}a_{\tilde{u}} - 1}{\pi^{i'}}\right], \quad a_{\tilde{u}} \in K\langle\overline{\gamma}^p\rangle,$$

$$A(j,k,w)^* = A(k',j',\tilde{w}) = R\left[\frac{\gamma^{p^2} - 1}{\pi^{k'}}, \frac{\gamma^p a_{\tilde{w}} - 1}{\pi^{j'}}\right], \quad a_{\tilde{w}} \in K\langle\gamma^{p^2}\rangle.$$

We shall extend the rank p^2 Hopf order $A(i,j,u)$ to obtain a Hopf order of rank p^3. To do this, we need to select a "correct" generator $\Psi \in KC_{p^3}$ that maps to $\dfrac{\overline{g}a_w - 1}{\pi^k}$ under the canonical surjection $KC_{p^3} \to KC_{p^2}$.

Let v be a unit of R, and let $a_v = \sum_{l=0}^{p-1} v^l \frac{1}{p} \sum_{q=0}^{p-1} \zeta_p^{-lq} g^{p^2 q}$. Let $\{e_{pm+n}\}$, $m,n = 0, \ldots, p-1$, denote the set of minimal idempotents in $K\langle g^p\rangle \cong KC_{p^2}$, and let

$$b_w = \sum_{m,n=0}^{p-1} w^m e_{pm+n}.$$

Note that

$$a_v b_w = \sum_{m,n=0}^{p-1} v^n w^m e_{pm+n}.$$

Let

$$H = A(i,j,k,u,v,w) = A(i,j,u)\left[\frac{ga_v b_w - 1}{\pi^k}\right].$$

Also, let x be a unit of R and let $a_x = \sum_{l=0}^{p-1} x^l \frac{1}{p} \sum_{q=0}^{p-1} \zeta_p^{-lq} \gamma^{p^2 q}$. Let $\{\iota_{pm+n}\}$, $m,n = 0, \ldots, p-1$, denote the set of minimal idempotents in $K\langle\gamma^p\rangle \cong K\hat{C}_{p^2}$, and let

$$b_{\tilde{u}} = \sum_{m,n=0}^{p-1} \tilde{u}^m \iota_{pm+n}$$

and

$$J = A(k',j',i',\tilde{w},x,\tilde{u}) = A(k',j',\tilde{w})\left[\frac{\gamma a_x b_{\tilde{u}} - 1}{\pi^{i'}}\right].$$

We wish to find conditions on v and x such that H and J are R-orders in KC_{p^3} and $\langle H, J\rangle \subseteq R$. Then one can show that $\text{disc}(J) = \text{disc}(H^*)$. Thus, by Proposition 4.4.10, $J = H^*$ and H, J are Hopf orders.

First, we find conditions for H to be an R-order.

Lemma 9.1.1. *The algebra*

$$H = A(i,j,k,u,v,w) = R\left[\frac{g^{p^2} - 1}{\pi^i}, \frac{g^p a_u - 1}{\pi^j}, \frac{ga_v b_w - 1}{\pi^k}\right]$$

is an R-order in KC_{p^3} if the following inequalities hold:

(i) $\operatorname{ord}(v^p \zeta_{p^2} - 1) \geq i' + pk;$

(ii) $\operatorname{ord}(\tilde{u} - 1) \geq i' > 0;$

(iii) $\operatorname{ord}(\tilde{u} - 1) + \operatorname{ord}(w^p \zeta_p - 1) \geq e' + i' + pk.$

Proof. The conditions (i), (ii), and (iii) imply that $\dfrac{ga_v b_w - 1}{\pi^k}$ satisfies a monic polynomial of degree p with coefficients in $A(i, j, u)$ (see [UC05, §3]). Thus H is finitely generated over R. Thus, by [Rot02, Theorem 9.3], H is free and of finite rank over R. Clearly, $KH = KC_{p^3}$, and so H is an R-order. □

Similarly, we have the following lemma.

Lemma 9.1.2. *The algebra*

$$J = A(k', j', i', \tilde{w}, x, \tilde{u}) = R\left[\frac{\gamma^{p^2} - 1}{\pi^{k'}}, \frac{\gamma^p a_{\tilde{w}} - 1}{\pi^{j'}}, \frac{\gamma a_x b_{\tilde{u}} - 1}{\pi^{i'}}\right]$$

is an R-order in $K\hat{C}_{p^3}$ if the following inequalities hold:

(i) $\operatorname{ord}(x^p \zeta_{p^2} - 1) \geq k + pi';$

(ii) $\operatorname{ord}(w - 1) \geq k > 0;$

(iii) $\operatorname{ord}(w - 1) + \operatorname{ord}(\tilde{u}^p \zeta_p - 1) \geq e' + k + pi'.$

Proof. The conditions (i), (ii), and (iii) imply that $\dfrac{\gamma a_x b_{\tilde{w}} - 1}{\pi^k}$ satisfies a monic polynomial with coefficients in $A(k', j', \tilde{w})$ (see [UC05, §3]). Thus J is free and of finite rank over R. Since $KJ = K\hat{C}_{p^3}$, J is an R-order in $K\hat{C}_{p^3}$. □

With some additional conditions, we can show that J and H are dual Hopf orders in KC_{p^3}. For $a, b \in R$, let $G(a, b)$ denote the **Gauss sum of a and b**, defined as

$$G(a, b) = \frac{1}{p} \sum_{m=0}^{p-1} \sum_{n=0}^{p-1} \zeta_p^{-mn} a^n b^m.$$

Proposition 9.1.1. *Let $H = A(i, j, k, u, v, w)$ and $J = A(k', j', i', \tilde{w}, x, \tilde{u})$ be R-algebras that satisfy the hypotheses of Lemmas 9.1.1 and 9.1.2, with the additional conditions*

(i) $e' > \operatorname{ord}(w - 1),$

(ii) $e' > \operatorname{ord}(\tilde{u} - 1),$

(iii) $\operatorname{ord}(w - 1) + \operatorname{ord}(\tilde{u} - 1) \geq e' + (\frac{p-1}{p})(i' + k + e'),$ *and*

(iv) $vx\zeta_{p^3} G(\tilde{u}, w) = 1.$

Then H and J are Hopf orders in KC_{p^3} with $J = H^*$.

Proof. By Lemma 9.1.1, H is an R-order in KC_{p^3}, and, by Lemma 9.1.2, J is an R-order in $K\hat{C}_{p^3}$. The conditions (i)–(iv) above imply that $\langle J, H \rangle \subseteq R$. Moreover, $\operatorname{disc}(H^*) = \operatorname{disc}(J)$. (For details of these calculations, see [UC05, §3].) Thus H and $J = H^*$ are Hopf orders by Proposition 4.4.10. □

The Hopf orders constructed in Proposition 9.1.1 are called **duality Hopf orders** in KC_{p^3}.

For p prime, the group ring KC_{p^3} is a K-Hopf algebra. There exists a Hopf inclusion $KC_{p^2} \cong K\langle g^p \rangle \to KC_{p^3}$, and a Hopf surjection $KC_{p^3} \overset{g^p \mapsto 1}{\to} K\langle \overline{g} \rangle \cong KC_p$, with

$$KC_{p^3}/i(K\langle g^p \rangle^+)KC_{p^3} \cong K\langle \overline{g} \rangle,$$

and thus there is a short exact sequence

$$K \to K\langle g^p \rangle \to KC_{p^3} \overset{g^p \mapsto 1}{\to} K\langle \overline{g} \rangle \to K. \tag{9.1}$$

At the same time, there exists a Hopf inclusion $KC_p \cong K\langle g^{p^2} \rangle \to KC_{p^3}$, and a Hopf surjection $KC_{p^3} \overset{g^{p^2} \mapsto 1}{\to} K\langle \overline{g} \rangle \cong KC_{p^2}$, with

$$KC_{p^3}/i(K\langle g^{p^2} \rangle^+)KC_{p^3} \cong K\langle \overline{g} \rangle,$$

and thus there is a short exact sequence

$$K \to K\langle g^{p^2} \rangle \to KC_{p^3} \overset{g^{p^2} \mapsto 1}{\to} K\langle \overline{g} \rangle \to K. \tag{9.2}$$

Proposition 9.1.2. *Let $A(i,j,k,u,v,w)$ be a duality Hopf order in KC_{p^3}. Then there exist short exact sequences of R-Hopf orders*

$$R \to A(i,j,u) \to A(i,j,k,u,v,w) \to H(k) \to R, \tag{9.3}$$

$$R \to H(i) \to A(i,j,k,u,v,w) \to A(j,k,w) \to R. \tag{9.4}$$

Proof. One shows that $A(i,j,k,u,v,w) \cap K\langle g^p \rangle = A(i,j,u)$ and that the image of $A(i,j,k,u,v,w)$ is $H(k)$ under the map given by $g^p \mapsto 1$. Thus (9.3) is a short exact sequence by §4.4 (4.10).

Moreover, $A(i,j,k,u,v,w) \cap K\langle g^{p^2} \rangle = H(i)$, and the image of $A(i,j,k,u,v,w)$ under the map given by $g^{p^2} \mapsto 1$ is $A(j,k,w)$, and so (9.4) is a short exact sequence by §4.4 (4.10). We leave the details to the reader as an exercise. $\square$

If H is an arbitrary R-Hopf order in KC_{p^3}, then the extensions (9.1) and (9.2) induce extensions of R-Hopf orders,

$$R \to A(i,j,u) \to H \to H(k) \to R \tag{9.5}$$

and

$$R \to H(i) \to H \to A(j,k,w) \to R, \tag{9.6}$$

for R-Hopf orders $A(i, j, u)$ and $A(j, k, w)$. The sequence (9.6) dualizes to yield the short exact sequence

$$R \to A(k', j', \tilde{w}) \to H^* \to H(i') \to R. \qquad (9.7)$$

In an effort to classify all Hopf orders in KC_{p^3}, R. Underwood has given the following analog of Definition 8.1.1 for Hopf orders in KC_{p^3}.

Definition 9.1.1. Let H be an R-Hopf order in KC_{p^3} inducing the short exact sequences (9.5) and (9.7) as above. Let $\Xi(A(i, j, u))$ denote the p-adic obgv determined by $A(i, j, u)$, and let $\Xi(A(k', j'\tilde{w}))$ denote the p-adic obgv given by $A(k', j', \tilde{w})$ (see Proposition 5.3.9). Then H satisfies the **valuation condition for** $n = 3$ if either

$$pk \le \Xi(A(i, j, u))(g^p)$$

or

$$pi' \le \Xi(A(k', j', \tilde{w}))(\gamma^p).$$

To see Definition 9.1.1 as an extension of Definition 8.1.1, let

$$R \to H(i) \to H \to H(j) \to R$$

and

$$R \to H(j') \to H^* \to H(i') \to R$$

be short exact sequences, where H is an R-Hopf order in KC_{p^2}. Then H satisfies the valuation condition for $n = 2$ if and only if either

$$pj \le \Xi(H(i))(\tau), \ \langle \tau \rangle = C_p,$$

or

$$pi' \le \Xi(H(j'))(\eta), \ \langle \eta \rangle = \hat{C}_p,$$

since $\Xi(H(i))(\tau) = i$ and $\Xi(H(j'))(\eta) = j'$.

In [UC05], the authors show that every duality Hopf order $A(i, j, k, u, v, w)$ satisfies the valuation condition for $n = 3$ (see [UC05, Theorem 3.8]), and it has been conjectured that an arbitrary Hopf order in KC_{p^3} satisfies the valuation condition. Indeed, there are no known examples where the condition fails.

9.2 Circulant Matrices and Hopf Orders in KC_{p^3}

In this section, we construct another collection of Hopf orders in KC_{p^3}. We keep the notation of the previous section: $\bar{g}$ denotes the image of g under the mapping $KC_{p^3} \to KC_{p^2}$, $g^{p^2} \mapsto 1$, and $\bar{\gamma}$ denotes the image of γ under the mapping

$K \hat{C}_{p^3} \rightarrow K \hat{C}_{p^2}$, $\gamma^{p^2} \mapsto 1$. For an integer $n \geq 1$, let $\langle \ , \rangle_n$ denote the duality map $K \hat{C}_{p^n} \times K C_{p^n} \rightarrow K$. For an integer m, $0 \leq m \leq e'$, let $m' = e' - m$, and for a unit $u \in R$, let $\tilde{u} = \zeta_{p^2}^{-1} u^{-1}$.

Let

$$A(i, j, u) = R \left[\frac{g^{p^2} - 1}{\pi^i}, \frac{g^p a_u - 1}{\pi^j} \right]$$

and

$$A(j, k, w) = R \left[\frac{\overline{g}^p - 1}{\pi^j}, \frac{\overline{g} a_w - 1}{\pi^k} \right]$$

be Hopf orders in $K C_{p^2}$ with linear duals

$$A(i, j, u)^* = A(j', i', \tilde{u}) = R \left[\frac{\overline{\gamma}^p - 1}{\pi^{j'}}, \frac{\overline{\gamma} a_{\tilde{u}} - 1}{\pi^{i'}} \right], \quad a_{\tilde{u}} \in K \langle \overline{\gamma}^p \rangle,$$

$$A(j, k, w)^* = A(k', j', \tilde{w}) = R \left[\frac{\gamma^{p^2} - 1}{\pi^{k'}}, \frac{\gamma^p a_{\tilde{w}} - 1}{\pi^{j'}} \right], \quad a_{\tilde{w}} \in K \langle \gamma^{p^2} \rangle.$$

We assume that $j' > p i'$, $j + i' > \text{ord}(1 - \tilde{u})$ and $j' + k > \text{ord}(1 - w)$.

We construct our collection of Hopf orders by choosing generators $\Psi \in K C_{p^3}$ and $\Phi \in K \hat{C}_{p^3}$ such that the R-modules $H = A(i, j, u)[\Psi]$ and $J = A(k', j', \tilde{w})[\Phi]$ are invariant under the comultiplications of $K C_{p^3}$ and $K \hat{C}_{p^3}$, respectively. We also require that $\langle J, H \rangle_3 \subseteq R$. Then, as we shall see, H and $J = H^*$ are Hopf orders in $K C_{p^3}$.

We begin with the construction of Ψ. Let

$$\iota_{pm+n} = \frac{1}{p^2} \sum_{a=0}^{p-1} \sum_{b=0}^{p-1} \zeta_{p^2}^{-(pm+n)(pa+b)} \gamma^{p(pa+b)}, \quad m, n = 0, \dots, p-1,$$

$$\rho_q = \frac{1}{p} \sum_{n=0}^{p-1} \zeta_p^{-qn} \overline{\gamma}^{pn}, \quad 0 \leq q \leq p-1,$$

and

$$e_{pm+n} = \frac{1}{p^2} \sum_{a=0}^{p-1} \sum_{b=0}^{p-1} \zeta_{p^2}^{-(pm+n)(pa+b)} g^{p(pa+b)}$$

denote the idempotents for $K \langle \gamma^p \rangle \cong K \hat{C}_{p^2}$, $K \langle \overline{\gamma}^p \rangle \cong K \hat{C}_p$, and $K \langle g^p \rangle \cong K C_{p^2}$, respectively. Let s_{pm+n}, $m, n = 0, \dots, p-1$, be units of R with $s_{pm} = \tilde{u}^m$, and set

$$\tau = \sum_{m=0}^{p-1} \sum_{n=0}^{p-1} s_{pm+n} \iota_{pm+n}, \quad \tau \in K \langle \gamma^p \rangle,$$

$$d = \sum_{q=0}^{p-1} s_1^{-1} s_{pq+1} \rho_q, \quad d \in K\langle \bar{\gamma}^p \rangle.$$

Let x_{pm+n}, $m, n = 0, \ldots, p-1$, be indeterminate, and set

$$x = \sum_{m=0}^{p-1} \sum_{n=0}^{p-1} x_{pm+n} e_{pm+n}, \quad x_{pm} = w^m.$$

We seek values for x_{pm+n}, $n > 0$, such that

$$\left\langle (\gamma^{p^2} - 1)^q (\gamma^p a_{\tilde{w}} - 1)^r (\gamma\tau - 1)^s, gx - 1 \right\rangle_3 = 0 \tag{9.8}$$

for $q, r, s = 0, \ldots, p-1$.

Lemma 9.2.1. *The solution to* (9.8) *is a vector* (x_{pm+n}), $0 \leq m, n \leq p-1$, *of elements of* R *defined as*

$$x_{pm+n} = \zeta_{p^3}^{-n} s_1^{-n} \langle \bar{\gamma}^{pm} d^{-n}, a_w \rangle_1.$$

Proof. One has

$$(\gamma^{p^2}-1)^q(\gamma^p a_{\tilde{w}}-1)^r(\gamma\tau-1)^s = \sum_{m,n=0}^{p-1} \left(\zeta_p^n - 1\right)^q \left(\zeta_{p^2}^{pm+n}\tilde{w}^n - 1\right)^r (\gamma s_{pm+n}-1)^s \iota_{pm+n}$$

and

$$\langle \gamma^t \iota_{pm+n}, g e_{pa+b} \rangle = \begin{cases} \dfrac{1}{p}\zeta_{p^3}^t \zeta_p^{-ma} & \text{if } n = 1, b = t \\ 0 & \text{otherwise,} \end{cases}$$

and thus (9.8) expands to

$$(\zeta_p - 1)^q \sum_{m=0}^{p-1} \left(\zeta_{p^2}^{pm+1}\tilde{w} - 1\right)^r \sum_{t=0}^{s} \binom{s}{t}(-1)^{s-t} s_{pm+1}^t \sum_{a=0}^{p-1} x_{pa+t}\left(\frac{1}{p}\zeta_{p^3}^t \zeta_p^{-ma}\right) = 0.$$

$$\tag{9.9}$$

For integers $l, n = 0, \ldots, p-1$, let

$$h_l^{(n)} = \frac{1}{p}\sum_{q=0}^{p-1} \zeta_p^{-lq} s_{pq+1}^n.$$

Then (9.9) can be rewritten as

$$\sum_{y=0}^{r} \binom{r}{y} (-1)^{r-y} \zeta_{p^2}^{y} \tilde{w}^{y} \sum_{n=0}^{s} \binom{s}{n} (-1)^{s-n} \zeta_{p^3}^{n} \sum_{l=0}^{p-1} x_{pl+n} h_{l-y}^{(n)} = 0, \qquad (9.10)$$

where the subscripts on $h_{l-y}^{(n)}$ are read modulo p.
Equation (9.10) is equivalent to the system

$$\begin{cases}
\zeta_{p^3}^{n} \left(h_0^{(n)} x_n + h_1^{(n)} x_{p+n} + \cdots + h_{p-1}^{(n)} x_{(p-1)p+n} \right) & = 1 \\[2mm]
\zeta_{p^3}^{n} \tilde{w} \zeta_{p^2} \left(h_{p-1}^{(n)} x_n + h_0^{(n)} x_{p+n} + \cdots + h_{p-2}^{(n)} x_{(p-1)p+n} \right) & = 1 \\[2mm]
\zeta_{p^3}^{n} \left(\tilde{w} \zeta_{p^2} \right)^2 (h_{p-2}^{(n)} x_n + h_{p-1}^{(n)} x_{p+n} + \cdots + h_{p-3}^{(n)} x_{(p-1)p+n}) & = 1 \qquad (9.11) \\[2mm]
\qquad\qquad \vdots \\[2mm]
\zeta_{p^3}^{n} \left(\tilde{w} \zeta_{p^2} \right)^{p-1} (h_1^{(n)} x_n + h_2^{(n)} x_{p+n} + \cdots + h_0^{(n)} x_{(p-1)p+n}) & = 1.
\end{cases}$$

In matrix form, (9.11) appears as

$$\begin{pmatrix}
h_0^{(n)} & h_1^{(n)} & h_2^{(n)} & \dots & h_{p-1}^{(n)} \\
h_{p-1}^{(n)} & h_0^{(n)} & h_1^{(n)} & \dots & h_{p-2}^{(n)} \\
h_{p-2}^{(n)} & h_{p-1}^{(n)} & h_0^{(n)} & \dots & h_{p-3}^{(n)} \\
\vdots & \vdots & & & \vdots \\
h_1^{(n)} & h_2^{(n)} & h_3^{(n)} & \dots & h_0^{(n)}
\end{pmatrix}
\begin{pmatrix}
x_n \\
x_{p+n} \\
x_{2p+n} \\
\vdots \\
x_{(p-1)p+n}
\end{pmatrix}
=
\begin{pmatrix}
\zeta_{p^3}^{-n} \\
w \zeta_{p^3}^{-n} \\
w^2 \zeta_{p^3}^{-n} \\
\vdots \\
w^{p-1} \zeta_{p^3}^{-n}
\end{pmatrix}. \qquad (9.12)$$

Here the coefficient matrix is the circulant matrix

$$M^{(n)} = \mathrm{circ} \left(h_0^{(n)}, h_1^{(n)}, h_2^{(n)}, \dots, h_{p-1}^{(n)} \right).$$

Note that the eigenvalues of $M^{(n)}$ are $s_{pq+1}^{n} \neq 0$, for $0 \leq q \leq p-1$, with corresponding eigenvectors (ζ_p^{lq}), $0 \leq l \leq p-1$. Thus $M^{(n)}$ is invertible with inverse $\Theta^{(n)} = (\theta_{m,l}^{(n)})$ for $m, l = 0, \dots, p-1$. Consequently, the matrix equations in (9.12) have unique solutions for $m, n = 0, \dots, p-1, n > 0$. These solutions are computed to be

$$x_{pm+n} = \zeta_{p^3}^{-n} s_1^{-n} \langle \bar{\gamma}^{pm} d^{-n}, a_w \rangle 1.$$

$\square$

Now let $u_{pm+n} = x_{pm+n}$, $n > 0$, and let

$$b = \sum_{m=0}^{p-1}\sum_{n=0}^{p-1} u_{pm+n}e_{pm+n}, \quad u_{pm} = w^m.$$

Put

$$\Psi = \frac{gb-1}{\pi^k},$$

and let

$$H = A(i,j,u)[\Psi] = A(i,j,u)\left[\frac{gb-1}{\pi^k}\right]$$

denote the R-module that is the $A(i,j,u)$-span of the set

$$\left\{1, \frac{gb-1}{\pi^k}, \left(\frac{gb-1}{\pi^k}\right)^2, \ldots, \left(\frac{gb-1}{\pi^k}\right)^{p-1}\right\}.$$

Lemma 9.2.2. *Let H be the R-module as above. Suppose*

(i) $A(j',i',\tilde{u}) = R\left[\dfrac{\overline{\gamma}^p - 1}{\pi^{j'}}, \dfrac{\overline{\gamma}d - 1}{\pi^{i'}}\right]$,

(ii) $\operatorname{ord}\left(\zeta_{p^2}s_1^p\langle\overline{\gamma}^p d^p, a_w\rangle_1 - 1\right) \geq pi' + k$, *and*

(iii) $\operatorname{ord}(\zeta_{p^3}s_1 - 1) \geq i' + \dfrac{k}{p^2}$.

Then H is invariant under the comultiplication of KC_{p^3}.

Proof. (Sketch.) We show that $\Delta_{KC_{p^3}}(H) \subseteq H \otimes_R H$. Since

$$\Delta_{KC_{p^3}}(\Psi) = \Psi \otimes 1 + 1 \otimes \Psi + \pi^k\Psi \otimes \Psi + \frac{\Delta(gb) - gb \otimes gb}{\pi^k},$$

H is invariant under $\Delta = \Delta_{KC_{p^3}}$ if and only if

$$\frac{\Delta(gb) - gb \otimes gb}{\pi^k} = \left(\frac{\Delta(b) - b \otimes b}{\pi^k}\right)(g \otimes g) \in H \otimes_R H.$$

The condition $j' > pi'$ guarantees that $A(i,j,u) \neq RC_{p^2}^*$, and so, by Chapter 4, Exercise 12, $A(i,j,u)$ is a local ring with maximal ideal $(\pi, A(i,j,u)^+)$. One has $b \in A(i,j,u)$ and $b \notin (\pi, A(i,j,u)^+)$, and so b is a unit in $A(i,j,u)$. Thus g is a unit in H. Therefore H is invariant under $\Delta_{KC_{p^3}}$ if and only if

$$\frac{\Delta(b) - b \otimes b}{\pi^k} \in A(i,j,u) \otimes A(i,j,u),$$

which follows from the conditions of the lemma. The reader is referred to [Un08b, Lemma 2.6] for the details of this computation.   □

Our next task is to construct Φ. Suppose that $H = A(i, j, u)[\dfrac{gb-1}{\pi^k}]$ is an R-module as constructed by Lemma 9.2.2, with $b = \sum_{m=0}^{p-1}\sum_{n=0}^{p-1} u_{pm+n}e_{pm+n}$. Set $c = \sum_{m=0}^{p-1} u_1^{-1} u_{pm+1} f_m$, where f_m are the minimal idempotents in $K\langle \overline{g}^p \rangle \cong KC_p$. The quantities c and f_m are the analogs in the dual situation for d and ρ_q. Let y_{pm+n}, $m, n = 0, \ldots, p-1$ be indeterminate, and set

$$y = \sum_{m=0}^{p-1}\sum_{n=0}^{p-1} y_{pm+n} l_{pm+n}, \qquad y_{pm} = \tilde{u}^m.$$

The y_{pm+n} are the analogs in the dual situation for x_{pm+n}. Then, following the construction of b as above, we find values for y_{pm+n}, $n > 0$, for which

$$\left\langle \gamma y - 1, \left(g^{p^2}-1\right)^l (g^p a_u - 1)^m (gb-1)^t \right\rangle = 0 \tag{9.13}$$

for $l, m, t = 0, \ldots, p-1, t > 0$.

Lemma 9.2.3. *The solution to (9.13) is a vector (y_{pm+n}), $0 \le m, n \le p-1$, of elements of R defined as*

$$y_{pm+n} = \zeta_{p^3}^{-n} u_1^{-n} \langle \overline{g}^{pm} c^{-n}, a_{\tilde{u}} \rangle$$

for $m, n = 0, \ldots, p-1, n > 0$.

Proof. We follow the method of Lemma 9.2.1. For integers $l, n = 0, \ldots, p-1$, define $\eta_l^{(n)}$ as

$$\eta_l^{(n)} = \frac{1}{p}\sum_{q=0}^{p-1} \zeta_p^{-lq} u_{pq+1}^n.$$

Then, finding quantities y_{pm+n} that satisfy (9.13) is equivalent to solving the matrix equations for $n = 1, 2, 3, \ldots, p-1$:

$$\begin{pmatrix} \eta_0^{(n)} & \eta_1^{(n)} & \eta_2^{(n)} & \cdots & \eta_{p-1}^{(n)} \\ \eta_{p-1}^{(n)} & \eta_0^{(n)} & \eta_1^{(n)} & \cdots & \eta_{p-2}^{(n)} \\ \eta_{p-2}^{(n)} & \eta_{p-1}^{(n)} & \eta_0^{(n)} & \cdots & \eta_{p-3}^{(n)} \\ \vdots & \vdots & \vdots & & \vdots \\ \eta_1^{(n)} & \eta_2^{(n)} & \eta_3^{(n)} & \cdots & \eta_0^{(n)} \end{pmatrix} \begin{pmatrix} y_n \\ y_{p+n} \\ y_{2p+n} \\ \vdots \\ y_{(p-1)p+n} \end{pmatrix} = \begin{pmatrix} \zeta_{p^3}^{-n} \\ \tilde{u}\zeta_{p^3}^{-n} \\ \tilde{u}^2\zeta_{p^3}^{-n} \\ \vdots \\ \tilde{u}^{p-1}\zeta_{p^3}^{-n} \end{pmatrix}. \tag{9.14}$$

Here the coefficient matrix is the circulant matrix

$$N^{(n)} = \text{circ}\left(\eta_0^{(n)}, \eta_1^{(n)}, \eta_2^{(n)}, \dots, \eta_{p-1}^{(n)}\right).$$

Let $\Phi^{(n)} = (\phi_{m,l}^{(n)})$, $m, l = 0, \dots, p-1$ denote the inverse of $N^{(n)}$. Then the matrix equations in (9.14) have unique solutions,

$$y_{pm+n} = v_{pm+n} = \zeta_{p^3}^{-n} \sum_{l=0}^{p-1} \phi_{m,l}^{(n)} \tilde{u}^l$$

$$= \zeta_{p^3}^{-n} u_1^{-n} \langle \overline{g}^{pm} c^{-n}, a_{\tilde{u}} \rangle,$$

for $m, n = 0, \dots, p-1, n > 0$. $\qquad\square$

Now, let

$$y = \beta = \sum_{m=0}^{p-1}\sum_{n=0}^{p-1} v_{pm+n} l_{pm+n}, \quad v_{pm} = \tilde{u}^m,$$

put

$$\Phi = \frac{\gamma\beta - 1}{\pi^{i'}},$$

and let

$$J = A(k', j', i')[\Phi] = A(k', j', \tilde{w})\left[\frac{\gamma\beta - 1}{\pi^{i'}}\right]$$

denote the R-module that is the $A(k', j', \tilde{w})$-span of the set

$$\left\{1, \frac{\gamma\beta - 1}{\pi^{i'}}, \left(\frac{\gamma\beta - 1}{\pi^{i'}}\right)^2, \dots, \left(\frac{\gamma\beta - 1}{\pi^{i'}}\right)^{p-1}\right\}.$$

Lemma 9.2.4. *Suppose H satisfies the hypothesis of Lemma 9.2.2. Suppose $i \geq pj$, $k' \geq p^2 i'$, $j \geq p^2 k > pk$, and $e' \geq i + j + k$. Then the R-module J is invariant under the comultiplication of $K\hat{C}_{p^3}$.*

Proof. We use the criteria of Lemma 9.2.2 to show that $\Delta_{K\hat{C}_{p^3}}(J) \subseteq J \otimes_R J$. In this case, J is invariant if $j > pk$,

$$A(j, k, w) = R\left[\frac{\overline{g}^p - 1}{\pi^j}, \frac{\overline{g}c - 1}{\pi^k}\right],$$

$$\text{ord}\left(\zeta_{p^2} u_1^p \langle \overline{g}^p c^p, a_{\tilde{u}} \rangle - 1\right) \geq pk + i',$$

and

$$\operatorname{ord}(\zeta_{p^3} u_1 - 1) \geq k + \frac{i'}{p^2},$$

which follow from the conditions of the lemma. Details of the computations are found in [Un08b, Lemma 2.7].

Let $H = A(i, j, u)\left[\frac{gb-1}{\pi^k}\right]$ and $J = A(k', j', \tilde{w})[\frac{\gamma\beta - 1}{\pi^{i'}}]$ be R-modules as constructed above by Lemma 9.2.2 and Lemma 9.2.4. We need several more lemmas before we can show that H and J are Hopf orders.

Let

$$H^* = \{\alpha \in K\hat{C}_{p^3} : \langle \alpha, H \rangle_3 \subseteq R\}$$

denote the linear dual of the R-module H.

Lemma 9.2.5. H^* *is an R-algebra.*

Proof. By Lemma 9.2.2, $\Delta_{KC_{p^3}} : H \to H \otimes H$. Thus there is a map of linear duals $\Delta^*_{KC_{p^3}} : (H \otimes_R H)^* \to H^*$. Since $H^* \otimes_R H^* \subseteq (H \otimes_R H)^*$, $\Delta^*_{KC_{p^3}}$ serves as multiplication on H^*. $\square$

Lemma 9.2.6. *Let $\overline{H}$ be the image of H under the mapping $KC_{p^3} \to KC_{p^2}$, $g^{p^2} \mapsto 1$. Then*

$$K\hat{C}_{p^2} \cap H^* = \overline{H}^* = A(j, k, w)^* = A(k', j', \tilde{w}).$$

Proof. We show that $\overline{H}^* = K\hat{C}_{p^2} \cap H^*$. Let $\alpha \in \overline{H}^*$. Then $\alpha \in K\hat{C}_{p^2}$ and $\langle \alpha, \overline{H} \rangle_2 \subseteq R$. Let $f \in H$, and let $\overline{f}$ be the image of f under the mapping $KC_{p^3} \to KC_{p^2}$. Then $\langle \alpha, f \rangle_3 = \langle \alpha, \overline{f} \rangle_2$. Hence

$$\langle \alpha, H \rangle_3 = \langle \alpha, \overline{H} \rangle_2 \subseteq R,$$

so that $\alpha \in H^*$. Hence $\alpha \in K\hat{C}_{p^2} \cap H^*$.

Now suppose $\alpha \in K\hat{C}_{p^2} \cap H^*$. Then $\alpha \in K\hat{C}_{p^2}$ and $\langle \alpha, H \rangle_3 \subseteq R$. Consequently, $\alpha \in \overline{H}^*$, which shows that $\overline{H}^* = K\hat{C}_{p^2} \cap H^*$.

Since $\overline{b} = a_w$, $\overline{H} = A(j, k, w)$, which completes the proof of the lemma. $\square$

Lemma 9.2.7. $J \subseteq H^*$.

Proof. By Lemma 9.2.5, H^* is an algebra. Thus it suffices to show that $A(k', j', \tilde{w}) \subseteq H^*$ and $\frac{\gamma\beta - 1}{\pi^{i'}} \in H^*$. By Lemma 9.2.6, $H^* \cap K\hat{C}_{p^2} = A(k', j', \tilde{w})$, and thus $A(k', j', \tilde{w}) \subseteq H^*$. We claim that $\frac{\gamma\beta - 1}{\pi^{i'}} \in H^*$, but this amounts to showing that

$$\langle \gamma\beta - 1, H \rangle_3 \subseteq \pi^{i'} R.$$

Since $\dfrac{\gamma\beta - 1}{\pi^{i'}}$ acts on $A(i, j, u)$ as $\dfrac{\overline{\gamma}a_{\tilde{u}} - 1}{\pi^{i'}}$, it suffices to show that

$$\mathrm{ord}\left(\left\langle \gamma\beta - 1, \left(g^{p^2} - 1\right)^l (g^p a_u - 1)^m (gb - 1)^t \right\rangle\right) \geq i' + li + mj + tk \quad (9.15)$$

for $l, m, t = 0, \ldots, p - 1,\, t > 0$. But (9.15) is satisfied since (9.13) holds with $y = \beta$. $\square$

Proposition 9.2.1. *Let $A(i, j, u)$ and $A(j, k, w)$ be Hopf orders in KC_{p^2} with linear duals $A(j', i', \tilde{u})$ and $A(k', j', \tilde{w})$, respectively. Let s be a unit of R, and put $s_{pm+n} = s^n \tilde{u}^m$ for $m, n = 0, \ldots, p - 1$. Let*

$$\tau = \sum_{m=0}^{p-1}\sum_{n=0}^{p-1} s_{pm+n} \iota_{pm+n} = \sum_{m=0}^{p-1}\sum_{n=0}^{p-1} s^n \tilde{u}^m \iota_{pm+n}.$$

Let

$$b = \sum_{m=0}^{p-1}\sum_{n=0}^{p-1} u_{pm+n} e_{pm+n},$$

where $u_{pm+n} = \zeta_{p^3}^{-n} s^{-n} G(\zeta_p^m \tilde{u}^{-n}, w)$.

Suppose

$$\beta = \sum_{m=0}^{p-1}\sum_{n=0}^{p-1} v_{pm+n} \iota_{pm+n}, \quad v_{pm} = \tilde{u}^m,$$

satisfies

$$\left\langle \gamma\beta - 1, \left(g^{p^2} - 1\right)^l (g^p a_u - 1)^m (gb - 1)^t \right\rangle = 0$$

for $l, m, t = 0, \ldots, p - 1,\, t > 0$.

Additionally, suppose the following conditions are satisfied:

(i) $\mathrm{ord}\left(\zeta_{p^2} s^p G(u^{-p}, w) - 1\right) \geq pi' + k$;

(ii) $\mathrm{ord}(\zeta_{p^3} s - 1) \geq i' + \dfrac{k}{p^2}$;

(iii) $i \geq pj$;

(iv) $j' > pi'$;

(v) $k' \geq p^2 i'$;

(vi) $j \geq p^2 k > pk$;

(vii) $e' \geq i + j + k$.

Then $H = A(i, j, u)\left[\dfrac{gb-1}{\pi^k}\right]$ is a Hopf order in KC_{p^3} with linear dual $J = A(k', j', \tilde{w})\left[\dfrac{\gamma\beta-1}{\pi^{i'}}\right]$.

Proof. Conditions (i)–(vii) show that $\Delta_{KC_{p^3}}(H) \subseteq H \otimes_R H$ and $\Delta_{K\hat{C}_{p^3}}(J) \subseteq J \otimes_R J$.

We show that J is an R-Hopf order. By Lemmas 9.2.5 and 9.2.6, H^* is an R-algebra with $H^* \cap K\hat{C}_{p^2} = A(k', j', \tilde{w})$. By Lemma 9.2.7, $\frac{\gamma\beta-1}{\pi^{i'}} \in H^*$. Thus $\frac{\gamma\beta-1}{\pi^{i'}}$ satisfies a monic polynomial of degree p with coefficients in $A(k', j', \tilde{w})$. Thus J is an R-algebra, and consequently J is an R-Hopf order.

By Lemma 9.2.7, $H \subseteq J^*$. Since $J^* \cap KC_{p^2} = A(i, j, u)$, $\frac{gb-1}{\pi^k}$ satisfies a monic polynomial of degree p with coefficients in $A(i, j, u)$. Thus H is an R-Hopf order. An application of Proposition 7.1.3 then shows that $H^* = J$.

In light of the fact that circulant matrices play a key role in their construction, we call the Hopf orders constructed in Proposition 9.2.1 **circulant matrix Hopf orders** in KC_{p^3}.

9.3 Chapter Exercises

Exercises for §9.1

1. Let K be a finite extension of $\mathbb{Q}_p$. Show that there are Larson orders in KC_{p^3} that are not duality Hopf orders.
2. Let K be a finite extension of $\mathbb{Q}_2$. Construct an example of a duality Hopf order in KC_8.
3. In the construction of the duality Hopf orders, prove that either $A(j, k, w)$ or $A(j', i', \tilde{u})$ is a Larson order.
4. Compute the p-adic obgv determined by an arbitrary duality Hopf order.
5. Give the details in the proof of Proposition 9.1.2.

Exercises for §9.2

6. Prove that every Larson order in KC_{p^3} is a circulant matrix Hopf order.
7. Construct an example of a circulant matrix Hopf order in KC_8.
8. Prove that there exist circulant matrix Hopf orders that are not duality.
9. Prove that there exist duality Hopf orders that are not circulant matrix.
10. Does the valuation condition for $n = 3$ hold for the collection of circulant matrix Hopf orders?

Chapter 10
Hopf Orders and Galois Module Theory

For this chapter, we return to the global situation where K is a finite extension of $\mathbb{Q}$, R is the integral closure of $\mathbb{Z}$ in K, and L is a Galois extension of K with group G and ring of integers S. In this chapter, we study applications of Hopf orders to Galois module theory. Galois module theory is the branch of number theory that seeks to describe S as a module over the group ring RG. We begin with a review of some Galois theory.

10.1 Some Galois Theory

Proposition 10.1.1. *(The Fundamental Theorem of Algebra) Let $f(x)$ be a non-constant polynomial in $\mathbb{C}[x]$. Then there is a zero of $f(x)$ in $\mathbb{C}$.*

Proof. Various proofs can be found (see, for example, [Rot02, Theorem 4.49]); a familiar proof that employs Liouville's Theorem can be found in [Fr03, Theorem 31.17]. □

Proposition 10.1.2. *Let $p(x)$ be an irreducible monic polynomial of degree $m \geq 1$ in $K[x]$. Then the zeros of $p(x)$ are distinct.*

Proof. By Proposition 10.1.1, there exists a zero α of $p(x)$ in $\mathbb{C}$. Let $\phi_\alpha : K[x] \to \mathbb{C}$ be the evaluation homomorphism defined as $f(x) \mapsto f(\alpha)$. Then $\ker(\phi_\alpha) = (p(x))$, and so $p(x)$ is the monic polynomial of smallest degree for which α is a zero.

By way of contradiction, assume that α has multiplicity ≥ 2. Since K has characteristic 0, $p'(x)$ is a non-constant polynomial of degree $m - 1 < m$ for which α is a root. This contradicts that $p(x)$ is a polynomial of smallest degree for which $p(\alpha) = 0$, and thus the zeros of $p(x)$ are distinct. □

R.G. Underwood, *An Introduction to Hopf Algebras*, DOI 10.1007/978-0-387-72766-0_10, 195
© Springer Science+Business Media, LLC 2011

Let $p(x)$ be an irreducible monic polynomial of degree m in $K[x]$. By repeated uses of Proposition 10.1.1 and the Factor Theorem, $f(x)$ factors linearly over $\mathbb{C}$ as

$$p(x) = (x - \alpha_1)(x - \alpha_2) \cdots (x - \alpha_m),$$

where $\alpha_1, \alpha_2, \ldots, \alpha_m$ are the distinct zeros of $p(x)$.

Clearly, $\mathbb{C}$ is a field extension of K that contains all of the zeros of $p(x)$. There is a smallest field extension L/K that contains all of the zeros of $p(x)$. This is the **splitting field of the irreducible polynomial** $p(x)$. The splitting field of $p(x)$ is necessarily a finite extension of K; that is, the splitting field is a finite-dimensional vector space over K. Finite extensions of K are simple extensions.

Proposition 10.1.3. *Let L/K be a finite extension of fields. Then there exists an element $\eta \in L$ for which $L = K(\eta)$.*

Proof. It suffices to prove the proposition in the case $L = K(\alpha, \beta)$, where $\alpha, \beta \in \mathbb{C}$. Let $q(x)$ be the irreducible polynomial of α. By Proposition 10.1.2, the roots of $q(x)$ are distinct, and we may list them as $\alpha = \alpha_1, \alpha_2, \ldots, \alpha_l$. Let $r(x)$ be the irreducible polynomial of β. Again, the roots of $r(x)$ are distinct, and we list them as $\beta = \beta_1, \beta_2, \ldots, \beta_k$.

Since K is an infinite field, there exists an element $t \in K$ for which $t \neq (\alpha_i - \alpha)/ (\beta - \beta_j)$ for all i, j, $1 \leq i \leq l$, $2 \leq j \leq k$. Set $\eta = \alpha + t\beta$. Then, as the reader can verify, $L = K(\alpha, \beta) = K(\eta)$ (see [Fr03, Theorem 51.15]). $\square$

Isomorphisms of simple extensions of K that fix K can be characterized as follows.

Proposition 10.1.4. *Let L/K be a finite extension of fields. Let $\alpha, \beta \in L$, and let $m = [K(\alpha) : K]$. The map $\phi_\alpha^\beta : K(\alpha) \to K(\beta)$, defined as*

$$a_0 + a_1\alpha + \cdots + a_{m-1}\alpha^{m-1} \mapsto a_0 + a_1\beta + \cdots + a_{m-1}\beta^{m-1}, \quad a_i \in K,$$

is an isomorphism fixing K if and only if α and β are zeros of the same irreducible polynomial in $K[x]$.

Proof. Suppose $\phi_\alpha^\beta : K(\alpha) \to K(\beta)$ is an isomorphism that fixes K, and let $p(x) = x^m + b_{m-1}x^{m-1} + \cdots + b_1 x + b_0$ be the irreducible polynomial for α. Then

$$0 = \phi_\alpha^\beta(0) = \phi_\alpha^\beta(p(\alpha)) = p\left(\phi_\alpha^\beta(\alpha)\right) = p(\beta),$$

and so β is a root of $p(x)$.

Conversely, suppose $p(x)$ is the irreducible polynomial for both α and β. Let $\phi_\alpha^\beta : K(\alpha) \to K(\beta)$ be the map defined as

$$a_0 + a_1\alpha + \cdots + a_{m-1}\alpha^{m-1} \mapsto a_0 + a_1\beta + \cdots + a_{m-1}\beta^{m-1}, \quad a_i \in K.$$

Then ϕ_α^β is an isomorphism that fixes K. $\square$

Let L/K be a finite extension of fields. The collection of all automorphisms of L is a group under function composition, denoted by $\text{Aut}(L)$. The **group of L/K,** denoted by $G(L/K)$, is the subgroup of $\text{Aut}(L)$ consisting of elements of $\text{Aut}(L)$ that fix K.

Definition 10.1.1. A finite extension of number fields L/K is a **Galois extension** if L is the splitting field of an irreducible polynomial $p(x) \in K[x]$. If L/K is a Galois extension, then the group $G(L/K)$ is the **Galois group** *of L/K*, denoted as $\text{Gal}(L/K)$.

Proposition 10.1.5. *Let L/K be a Galois extension with Galois group $G = \text{Gal}(L/K)$. Then $[L : K] = |\text{Gal}(L/K)|$.*

Proof. By definition, L/K is the splitting field of an irreducible monic polynomial $p(x) \in \mathbb{Q}[x]$ of degree m. Let β_1 be a zero of $p(x)$, and put $E_1 = K(\beta_1)$. By Proposition 10.1.4, the set of isomorphisms of E_1 that fix K contributes $n_1 = m$ elements to $\text{Gal}(L/K)$ (in the form of elements of the permutation group S_m).

Next, let β_2 be a zero of $p(x)$ that is not in E_1 (if indeed there are any such zeros). Form the extension $E_2 = E_1(\beta_2)$, and let $n_2 = [E_2 : E_1]$. By Proposition 10.1.4, this contributes n_2 elements of S_m to $\text{Gal}(L/K)$ in the form of n_2 permutations that fix all of the roots of $p(x)$ that are in E_1 and move β_2 to roots of $p(x)$ not in E_1.

Next, let β_3 be a zero of $p(x)$ that is not in E_2. Let $E_3 = E_2(\beta_3)$, and let $n_3 = [E_3 : E_2]$. This contributes n_3 elements of S_m to $\text{Gal}(L/K)$. These n_3 permutations must fix all of the roots of $p(x)$ that are in E_2 and must move β_3 to roots of $p(x)$ not in E_2. We repeat this process until $E_k = L$ for some k. Then

$$[L : K] = n_1 n_2 \cdots n_k = |\text{Gal}(L/K)|. \qquad \square$$

We state the fundamental theorem of Galois theory.

Proposition 10.1.6. *Suppose L/K is a Galois extension of number fields with group G.*

(i) If H is a subgroup of G, then

$$L' = \{x \in L : h(x) = x, \forall h \in H\}$$

is a subfield of L. Moreover, L is a Galois extension of L' with $\text{Gal}(L/L') \cong H$. If H is a normal subgroup of G, then L' is a Galois extension of K with $\text{Gal}(L'/K) \cong G/H$.

(ii) Let L', $K \subseteq L' \subseteq L$ be an intermediate field. Then

$$H = \{g \in G : g(x) = x, \forall x \in L'\}$$

is a subgroup of G, and L is a Galois extension of L' with group H. If L' is a Galois extension of K, then H is a normal subgroup of G.

Proof. We prove (i). Various proofs of (ii) can be found in [Fr03, Chapter X, §53], [Rot02, Chapter 4, §4.2], or [La84, Chapter VIII, §1].

Suppose $H \leq G$. Clearly, $K \subseteq L'$. Let $a, b \in L'$, and let $h \in H$. Then $h(a + b) = h(a) + h(b) = a + b$ and $h(ab) = h(a)h(b) = ab$, and so L' is a subring of L, which implies that L' is a subfield of L. It follows that L is a Galois extension of L' with group $\mathrm{Gal}(L/L') = H$.

Now assume that $H \lhd G$. By Proposition 10.1.3, $L' = K(\eta)$ for some $\eta \in L'$. Observe that $g^{-1}hg \in H$ for all $h \in H$, $g \in G$. Thus, for $a \in L'$, $(g^{-1}hg)(a) = a$, and so $h(g(a)) = g(a)$ for all $h \in H$. Thus $g(a) \in L'$ for all $g \in G$, and so $g(\eta) \in L'$ for all $g \in G$. It follows that L' is the splitting field of the irreducible polynomial of η. Thus L'/K is a Galois extension of fields. Now, let gH be an element in the factor group G/H, and let $gh \in gH$. Now, for $a \in L'$, $(gh)(a) = g(h(a)) = g(a)$, and so the Galois action of elements in the coset gH depends only on the action of the representative g. It follows that $\mathrm{Gal}(L'/K) = G/H$. □

Proposition 10.1.7. *Let L/K be a Galois extension with group G, $n = |G|$. Let $a \in L/K$, and let $f(x)$ be the irreducible polynomial of a over K. Then $s = n/\deg(f(x))$ is an integer with $f(x)^s = \prod_{g \in G}(x - g(a))$.*

Proof. We have $K \subseteq K(a) \subseteq L$. By Proposition 10.1.6(ii), there is a subgroup H of G consisting of elements that fix $K(a)$. Let $\{a = a_1, a_2, \dots, a_l\}$ be the zeros of $f(x)$. Now, with $m = |H|$, $t = [G : H]$,

$$\prod_{g \in G}(x - g(a)) = \prod_{i=1}^{l}(x - a_i)^m = f(x)^m = f(x)^{n/t}$$

since $m = n/t$

Since L/K is Galois, $[L : K] = n$, and since $L/K(a)$ is Galois with group H, $[L : K(a)] = m$. Thus, by the index formula, $[K(a) : K] = \deg(f(x)) = n/m$. Thus $n/t = m = n/\deg(f(x))$, and so $t = \deg(f(x))$. □

Expanding $\prod_{g \in G}(x - g(a))$ in powers of x, one has $\prod_{g \in G} g(a) \in K$ (why?). Thus there is a map $N_{L/K} : L \to K$ defined as $a \mapsto \prod_{g \in G} g(a)$, which is the **norm map of L/K**. One also has $\sum_{g \in G} g(a) \in K$, and so there is a map $\mathrm{tr}_{L/K} : L \to K$ defined as $a \mapsto \sum_{g \in G} g(a)$, which is the **trace map**. Note that $\mathrm{tr}_{L/K}(S) \subseteq R$ and $N_{L/K}(S) \subseteq R$.

We give some calculations of the Galois group of the splitting field of various polynomials.

Let $p > 2$ be prime and let $p(x) = x^{p-1} + x^{p-2} + \cdots + x^2 + x + 1$. Then, by the Eisenstein criterion, $p(x)$ is irreducible over $\mathbb{Q}$. Let $\zeta = \zeta_p$ denote a primitive pth root of unity. Then the roots of $p(x)$ are $\{\zeta, \zeta^2, \cdots, \zeta^{p-1}\}$, and the splitting field of $p(x)$ is $K = \mathbb{Q}(\zeta)$.

Proposition 10.1.8. *With the notation above, $\mathrm{Gal}(K/\mathbb{Q}) \cong C_{p-1}$, where C_{p-1} denotes the cyclic group of order $p - 1$.*

Proof. Since $[\mathbb{Q}(\zeta) : \mathbb{Q}] = p - 1$, $\mathrm{Gal}(K/\mathbb{Q})$ is a group of order $p - 1$. By [Fr03, Corollary 23.6], the group of units $U(\mathbb{F}_p)$ is cyclic, generated by the primitive element $a \in \mathbb{F}_p$. If we number the roots of $p(x)$ according to the rule $r_i = \zeta^i$ for $i = 1, \ldots, p - 1$, then there is a set of relations among the roots of $p(x)$,

$$r_1^a = r_a, \; r_2^a = r_{2a}, \; r_3^a = r_{3a}, \; \ldots, \; r_{p-1}^a = r_{(p-1)a}.$$

(Here, the subscripts are assumed to be the least positive residue modulo p.) These relations correspond to a permutation of the subscripts $1, 2, \ldots, p - 1$, which can be written as

$$g = \begin{pmatrix} 1 & 2 & \cdots & k & \cdots & p-1 \\ a & 2a & \cdots & ka & \cdots & (p-1)a \end{pmatrix}.$$

(Again, the subscripts in the bottom row are the least positive residue modulo p.)

The powers $\{g^i\}$ for $i = 1, \ldots, p - 1$ determine $p - 1$ automorphisms of K that fix $\mathbb{Q}$, and thus $\mathrm{Gal}(K/\mathbb{Q}) = C_{p-1}$.   □

For a less well-known example, we consider the collection of polynomials defined as follows. Let $p \geq 5$ be prime, let $\zeta = \zeta_p$ denote a primitive pth root of unity, and let a, b be complex numbers that satisfy the relations

$$a^p + b^p = 1 \text{ and } ab = 1. \tag{10.1}$$

Let x be indeterminate, and let

$$A = \begin{pmatrix} a & x & b & 0 & 0 & \cdots & 0 \\ 0 & a & x & b & 0 & \cdots & 0 \\ 0 & 0 & a & x & b & \cdots & 0 \\ \vdots & & & & & & \vdots \\ x & b & 0 & 0 & 0 & \cdots & a \end{pmatrix}$$

denote the $p \times p$ circulant matrix $\mathrm{circ}(a, x, b, 0, 0, \ldots, 0)$. Then $\det(A)$ defines a monic degree p polynomial in x,

$$f_p(x) = x^p + \sum_{i=1}^{(p-1)/2} (-1)^i \frac{p}{p-i} \binom{p-i}{i} x^{p-2i} + 1, \tag{10.2}$$

which factors as

$$f_p(x) = \prod_{i=0}^{p-1} (x\zeta^i + a + b\zeta^{2i})$$

(see [FLSU08, §2]).

We find specific values for a and b that satisfy the relations (10.1). Let $\eta = \frac{1+i\sqrt{3}}{2}$ with conjugate $\overline{\eta} = \frac{1-i\sqrt{3}}{2}$.

Lemma 10.1.1. η^p *is a pth root of* η.

Proof. Since $\eta = \cos(2\pi/6) + i\sin(2\pi/6)$, $\eta^6 = 1$. Since $p \geq 5$, $p^2 - 1 = (p+1)(p-1) \equiv 0 \mod 6$. Thus $\eta^{p^2-1} = \eta^{6m} = 1$ for some integer m, and hence $\eta^{p^2} = \eta$. $\qquad\square$

Put $\theta = \eta^p$. Then $\theta, \overline{\theta}$ satisfy (10.1). Indeed, by Lemma 10.1.1, $\theta^p + \overline{\theta}^p = \eta + \overline{\eta} = 1$ and $\theta\overline{\theta} = 1$, and thus the polynomial $f_p(x)$ factors as

$$f_p(x) = \prod_{i=0}^{p-1}(x\zeta^i + \theta + \overline{\theta}\zeta^{2i}).$$

Consequently, the roots of $f_p(x)$ are

$$r_0 = -\theta - \overline{\theta}, \; r_1 = -\theta\zeta^{-1} - \overline{\theta}\zeta, \; r_2 = -\theta\zeta^{-2} - \overline{\theta}\zeta^2,$$

$$r_3 = -\theta\zeta^{-3} - \overline{\theta}\zeta^3, \ldots, r_{p-1} = -\theta\zeta^{-(p-1)} - \overline{\theta}\zeta^{p-1}.$$

Lemma 10.1.2. $r_0 = -\theta - \overline{\theta} = -1$.

Proof. We have $\eta^p + \overline{\eta}^p = 2\cos(2\pi p/6)$. Now, since $p \geq 5$, $p = 6m \pm 1$ for some integer m. Thus

$$2\cos(2\pi p/6) = 2\cos(2\pi(6m \pm 1)/6)$$

$$= 2\cos(2\pi m \pm 2\pi/6)$$

$$= 2\cos(\pm 2\pi/6) = 1.$$

$\qquad\square$

By Lemma 10.1.2, $f_p(x)$ has a rational zero $r_0 = -1$, and so $f_p(x)$ is not irreducible over $\mathbb{Q}$. However, the polynomial $f_p(x)/(x+1)$ of degree $p - 1$ is irreducible over $\mathbb{Q}$, and its roots are $r_1, r_2, \ldots, r_{p-1}$.

Proposition 10.1.9. $f_p(x)/(x+1)$ *is irreducible over* $\mathbb{Q}$.

Proof. The reader is referred to [FLSU08, Lemma 5] for a proof. $\qquad\square$

Proposition 10.1.10. *Let* $p \geq 5$, *and let* K *be the splitting field of* $f_p(x)/(x+1)$ *over* $\mathbb{Q}$. *Then* $\mathrm{Gal}(K/\mathbb{Q})$ *is cyclic of order* $p - 1$.

Proof. We prove the proposition assuming that -2 generates $U(\mathbb{F}_p)$. For a complete proof, the reader is directed to [FLSU08, Theorem 6].

For $1 \leq j \leq p - 1$, one has

$$
\begin{aligned}
r_j^2 = (-\theta\zeta^{-j} - \overline{\theta}\zeta^j)^2 &= \theta^2\zeta^{-2j} + \overline{\theta}^2\zeta^{2j} + 2 \\
&= -\overline{\theta}\zeta^{p-2j} - \theta\zeta^{-(p-2j)} + 2 \\
&= -\theta\zeta^{-(p-2j)} - \overline{\theta}\zeta^{p-2j} + 2 \\
&= r_{p-2j} + 2.
\end{aligned}
\tag{10.3}
$$

These relations yield a permutation on $p - 1$ letters,

$$
g = \begin{pmatrix} 1 & 2 & 3 & \cdots & \frac{p-1}{2} & \frac{p+1}{2} & \cdots & p-2 & p-1 \\ p-2 & p-4 & p-6 & \cdots & 1 & p-1 & \cdots & 4 & 2 \end{pmatrix}.
$$

Let $\rho : K \to K$ be the map defined as

$$
\rho(r_2) = r_{g(2)} = r_2^2 - 2.
\tag{10.4}
$$

Then ρ is an automorphism of K that fixes $\mathbb{Q}$. We claim that

$$
\rho(r_{g^i(2)}) = r_{g^{i+1}(2)}, \forall i \geq 0.
$$

To prove this assertion, we proceed by induction on i, with (10.4) being the trivial case. Assume that $\rho(r_{g^{i-1}(2)}) = r_{g^i(2)}$. By (10.3), $r_{g^i(2)} = r_{g^{i-1}(2)}^2 - 2$, and thus

$$
\rho(r_{g^i(2)}) = (\rho(r_{g^{i-1}(2)}))^2 - 2 = r_{g^i(2)}^2 - 2 = r_{g^{i+1}(2)},
$$

which completes the induction proof.

Now, since g has order $p - 1$ if and only if $\langle -2 \rangle = U(\mathbb{F}_p)$, ρ has order $p - 1$. Thus $\mathrm{Gal}(K/\mathbb{Q}) = C_{p-1}$. $\square$

For another example, we consider the polynomial $p(x) = x^p - a \in \mathbb{Q}[x]$.

Proposition 10.1.11. *Either the polynomial $p(x)$ is irreducible over $\mathbb{Q}$ or a is a pth power in $\mathbb{Q}$; that is, there exists $\alpha \in \mathbb{Q}$ for which $\alpha^p = a$.*

Proof. It is known that the zeros of $x^p - a$ consist of $\{a^{\frac{1}{p}}\zeta_p^i\}_{i=0}^{p-1}$. If $p(x)$ is reducible over $\mathbb{Q}$, then $p(x) = q(x)s(x)$, where the zeros of $s(x)$ consist of $\{a^{\frac{1}{p}}\zeta_p^i\}_{i=1}^{p-1}$. Thus $q(x)$ is linear over $\mathbb{Q}$ with zero $a^{\frac{1}{p}} \in \mathbb{Q}$. $\square$

Assume that a is not a pth power in $\mathbb{Q}$, so that $x^p - a$ is irreducible by Proposition 10.1.11. Let $\zeta = \zeta_p$. Since the roots of $x^p - a$ are the p elements

$$
r_1 = \alpha, r_2 = \alpha\zeta, r_3 = \alpha\zeta^2, \ldots, r_p = \alpha\zeta^{p-1}, \alpha = a^{\frac{1}{p}},
$$

the splitting field K of $x^p - a$ is $\mathbb{Q}(\zeta)(\alpha) = \mathbb{Q}(\zeta, \alpha)$. We compute the Galois group of K over the base field $\mathbb{Q}(\zeta)$.

Proposition 10.1.12. $\mathrm{Gal}(K/\mathbb{Q}(\zeta)) = C_p$.

Proof. Since a is not a pth power in $\mathbb{Q}(\zeta)$, $p(x) = x^p - a$ is irreducible over $\mathbb{Q}(\zeta)$. Thus $[K : \mathbb{Q}(\zeta)] = p$. Since K is the splitting field of $x^p - a$, $\mathrm{Gal}(K/\mathbb{Q}(\zeta))$ has order p and thus is isomorphic to C_p.

The p mappings $g_j(\alpha) = \alpha\zeta^j$, $0 \le j \le p - 1$, determine p automorphisms of K that fix $\mathbb{Q}(\zeta)$, and so $\mathrm{Gal}(K/\mathbb{Q}(\zeta))$ is generated by g_1. As a permutation of the subscripts of the r_i, g_1 appears as

$$\begin{pmatrix} 1 & 2 & 3 & \cdots & p-1 & p \\ 2 & 3 & 4 & \cdots & p & 1 \end{pmatrix},$$

which is a cycle of length p.   □

10.2   Ramification

Let K be a finite extension of $\mathbb{Q}$ with ring of integers R, and let p be a rational prime. By §4.2, Exercises 9 and 10, the principal ideal $(p) \subseteq R$ factors uniquely into a product of prime ideals,

$$(p) = P_1^{e_1} P_2^{e_2} \cdots P_g^{e_g}. \tag{10.5}$$

This factorization takes on a nicer form in the case where K is a Galois extension.

Proposition 10.2.1. *Let $K/\mathbb{Q}$ be a Galois extension with group G. Let p be a rational prime, and let P_i be a prime ideal in the factorization (10.5). For each $h \in G$, $h(P_i) = P_j$ for some j, $1 \le j \le g$. Moreover, for each prime ideal P_j in the factorization (10.5), there exists an element $h \in G$ for which $h(P_i) = P_j$.*

Proof. Let $h \in G$. It is easy to show that $h(P_i)$ is a prime ideal of R. From the factorization (10.5), we obtain

$$\begin{aligned} h((p)) &= h(P_1^{e_1} P_2^{e_2} \cdots P_g^{e_g}) \\ &= h(P_1)^{e_1} h(P_2)^{e_2} \cdots h(P_g)^{e_g}, \end{aligned}$$

which equals (p) since h fixes $\mathbb{Q}$. Since the factorization (10.5) is unique, one concludes that $h(P_i) = P_j$ for some j, $1 \le j \le g$.

For the second statement of the proposition, let $G = \{h_1, h_2, \ldots, h_n\}$. Suppose P_j is a prime factor of (p) for which $P_j \ne h_l(P_i)$ for all $1 \le l \le n$. Renumbering the elements of G if necessary, let

$$\{h_1(P_i), h_2(P_i), \ldots, h_\alpha(P_i)\}$$

be the collection of distinct images of P_i under elements of G.

Let l, m be integers with $l \neq m$, and suppose that $h_l(P_i) + h_m(P_i) \subseteq Q$ for some prime Q of R. Then, $h_l(P_i) \subseteq Q$ and $h_m(P_i) \subseteq Q$. Since R is a Dedekind domain, $h_l(P_i)$ and $h_m(P_i)$ are maximal, and so $h_l(P_i) = Q = h_m(P_i)$, which is a contradiction. Thus, by Proposition 1.1.1, $h_l(P_i) + h_m(P_i) = R$ for $l \neq m$, $1 \leq l, m \leq \alpha$.

Put $I = h_1(P_i)h_2(P_i) \cdots h_\alpha(P_i)$. By the Chinese Remainder Theorem for Rings [IR90, Chapter 12, Proposition 12.3.1],

$$R/I \cong R/h_1(P_i) \times R/h_2(P_i) \times \cdots \times R/h_\alpha(P_i). \tag{10.6}$$

One has $P_j + I = R$, and so there exists an element $a \in P_j$ for which $a \equiv 1$ (mod I) and, by the isomorphism in (10.6), $a \equiv 1$ (mod $h_l(P_i)$) for $1 \leq l \leq n$. Thus $N_{K/\mathbb{Q}}(a) = \prod_{m=1}^{n} h_m(a) \in P_j$. Necessarily, $N_{K/\mathbb{Q}}(a) \in P_j \cap \mathbb{Z} = p\mathbb{Z}$, and so $N_{K/\mathbb{Q}}(a) \in P_i$, which says that $h_l(a) \in P_i$ for some l. Thus $a \in h_m(P_i)$ for some m, which is a contradiction. Thus no factor P_j exists with the property that $P_j \neq h_l(P_i)$ for all $1 \leq l \leq n$.   □

Proposition 10.2.2. *Let $K/\mathbb{Q}$ be a Galois extension with group G, and let p be a prime of $\mathbb{Z}$. Then*

$$(p) = (P_1 P_2 \cdots P_g)^e$$

for an integer $e \geq 1$.

Proof. Let

$$(p) = P_1^{e_1} P_2^{e_2} \cdots P_i^{e_i} \cdots P_j^{e_j} \cdots P_g^{e_g}$$

be the factorization of p. By Proposition 10.2.1, there exists $h \in G$ for which $h(P_i) = P_j$. Thus,

$$(p) = h(P_1)^{e_1} h(P_2)^{e_2} \cdots P_i^{e_i} \cdots h(P_j)^{e_j} \cdots h(P_g)^{e_g},$$

and so, by the uniqueness of factorization, $e_i = e_j$ for all $1 \leq i, j \leq g$.   □

The integer e in Proposition 10.2.2 is the **ramification index of p in R**.

Let K be a finite Galois extension of $\mathbb{Q}$ with ring of integers R. By Proposition 10.1.3, we can write $K = \mathbb{Q}(\alpha)$ for some $\alpha \in K$.

Let p be a prime of $\mathbb{Z}$. By Proposition 10.2.2, $(p) = (P_1 P_2 \cdots P_g)^e$ for some integer $e \geq 1$. Let $\mathbb{Z}_p$ denote the completion of $\mathbb{Z}$ with respect to $| \ |_p$, with fraction field $\mathbb{Q}_p$. By Proposition 5.1.5, each prime P_j, $1 \leq j \leq g$, corresponds to a discrete absolute value $[\ ]_{p,j}$ that extends $| \ |_p$.

Let K_{P_j} denote the completion of K with respect to $[\ ]_{pj}$. By Proposition 5.3.11, K_{P_j} is a finite extension of $\mathbb{Q}_p$. Moreover, by [FT91, Chapter III, §1, Theorem 17], there is an isomorphism

$$\mathbb{Q}_p \otimes_{\mathbb{Q}} K \to \bigoplus_{j=1}^{g} K_{P_j}, \tag{10.7}$$

where the primitive idempotent η_j of $\mathbb{Q}_p \otimes_{\mathbb{Q}} K$ corresponds to the prime ideal P_j.

Proposition 10.2.3. *With the notation above,*

$$[K : \mathbb{Q}] = [(\mathbb{Q}_p \otimes_{\mathbb{Q}} K) : \mathbb{Q}_p] = \sum_{i=1}^{g} [K_{P_i} : \mathbb{Q}_p].$$

Proof. Let $n = [K : \mathbb{Q}]$. Since $\{\alpha^m\}_{m=0}^{n-1}$ is a basis for K over $\mathbb{Q}$, $\{1 \otimes \alpha^m\}_{m=0}^{n-1}$ is a spanning set for $\mathbb{Q}_p \otimes_{\mathbb{Q}} K$ over $\mathbb{Q}_p$; $\mathbb{Q}_p \otimes_{\mathbb{Q}} K$ is a finite-dimensional $\mathbb{Q}_p$-vector space. By Lemma 5.1.1,

$$[(\mathbb{Q}_p \otimes_{\mathbb{Q}} K) : \mathbb{Q}_p] = \sum_{i=1}^{l} [\mathbb{Q}_p(\alpha_i) : \mathbb{Q}_p].$$

Here α_i is a zero of the irreducible factor $r_i(x)$ in the factorization $q(x) = r_1(x) \cdots r_l(x)$ over $\mathbb{Q}_p$. Thus $n = \sum_{i=1}^{l} [\mathbb{Q}_p(\alpha_i) : \mathbb{Q}_p]$, and so $n = [(\mathbb{Q}_p \otimes_{\mathbb{Q}} K) : \mathbb{Q}_p]$. Now, by (10.7), $[K : \mathbb{Q}] = \sum_{j=1}^{g} [K_{P_j} : \mathbb{Q}_p]$. $\square$

For i, $1 \le i \le g$, let R_{P_i} denote the completion of R at P_i. Then $R_{P_i}/P_i R_{P_i}$ is a field extension of $\mathbb{Z}_p/p\mathbb{Z}_p \cong \mathbb{F}_p$. Put $f_i = [R_{P_i}/P_i R_{P_i} : \mathbb{F}_p]$. Let $n_i = [K_{P_i} : \mathbb{Q}_p]$.

Proposition 10.2.4. *With the notation above, $n_i = ef_i$ for $1 \le i \le g$.*

Proof. Put $N_i = P_i R_{P_i}$. There is an isomorphism of $\mathbb{F}_p$-vector spaces

$$R_{P_i}/pR_{P_i} \cong R_{P_i}/N_i \oplus N_i/N_i^2 \oplus N_i^2/N_i^3 \oplus \cdots \oplus N_i^{e-1}/N_i^e. \qquad (10.8)$$

Now, since R_{P_i} is free and of rank n_i over $\mathbb{Z}_p$, the dimension of the vector space R_{P_i}/pR_{P_i} over $\mathbb{F}_p$ is n_i. Now, as vector spaces over $\mathbb{F}_p$,

$$R_{P_i}/N_i \cong N_i^j/N_i^{j+1}$$

for $j = 1, \ldots, e-1$. Thus, the right-hand side in (10.8) is the direct sum of e vector spaces, each of dimension f_i over $\mathbb{F}_p$, and thus the direct sum has dimension ef_i over $\mathbb{F}_p$. Consequently, $n_i = ef_i$. $\square$

Proposition 10.2.5. *Let $K/\mathbb{Q}$ be a Galois extension with group G, $n = |G|$. Let $(p) = (P_1 \cdots P_g)^e$, and let $f_i = [R_{P_j}/P_j R_{P_j} : \mathbb{F}_p]$, $1 \le j \le g$. Then $f_1 = f_2 = \cdots = f_g$ and $n = efg$, where $f = f_1$.*

Proof. By Proposition 10.2.3, $n = \sum_{i=1}^{g} [K_{P_i} : \mathbb{Q}_p]$, and, by Proposition 10.2.4, $\sum_{i=1}^{g} [K_{P_i} : \mathbb{Q}_p] = \sum_{i=1}^{g} ef_i$. As in the proof of Proposition 10.2.4,

$$R_{P_i}/N_i \cong N_i^j/N_i^{j+1} \cong N_k^j/N_k^{j+1} \cong R_{P_k}/N_k$$

for all i, k. Thus all of the f_i are equal to a common value, say f. Thus $n = \sum_{i=1}^{g} ef = efg$. $\square$

To illustrate the usefulness of the formula $n = efg$, take $K = \mathbb{Q}(i)$, where $i = \sqrt{-1}$. We know that $\mathbb{Q}(i)$ is a Galois extension of $\mathbb{Q}$ with group C_2. The ring of integers of K is $R = \mathbb{Z}[i]$, the ring of Gaussian integers. We have already shown how prime ideals of $\mathbb{Z}$ factor in $\mathbb{Z}[i]$ (see §1.2). For example, $(2) = P_1^2$ with $P_1 = (1+i)$, and thus $e = 2$, $f = g = 1$. Thus (10.7) is

$$\mathbb{Q}_2 \otimes_{\mathbb{Q}} \mathbb{Q}(i) = \mathbb{Q}(i)_{(1+i)},$$

and (10.8) is

$$\mathbb{Z}[i]_{(1+i)}/2\mathbb{Z}[i]_{(1+i)} = \mathbb{Z}[i]_{(1+i)}/(1+i)\mathbb{Z}[i]_{(1+i)} \oplus (1+i)\mathbb{Z}[i]_{(1+i)}/(1+i)^2\mathbb{Z}[i]_{(1+i)}.$$

Moreover, the ideal (5) factors as $(5) = P_1 P_2$ with $P_1 = (2+i)$, $P_2 = (2-i)$. In this case, $e = 1$, $f = 1$, and $g = 2$, and therefore

$$\mathbb{Q}_5 \otimes_{\mathbb{Q}} \mathbb{Q}(i) = \mathbb{Q}(i)_{(2+i)} \oplus \mathbb{Q}(i)_{(2-i)} = \mathbb{Q}_5 \oplus \mathbb{Q}_5,$$

and

$$\mathbb{Z}[i]_{(2+i)}/5\mathbb{Z}[i]_{(2+i)} = \mathbb{Z}[i]_{(2+i)}/(2+i)\mathbb{Z}[i]_{(2+i)} = \mathbb{Z}_5/5\mathbb{Z}_5 \cong \mathbb{Z}/5\mathbb{Z}$$

since $i \in \mathbb{Z}_5$ by Hensel's Lemma.

Let $K/\mathbb{Q}$ be a Galois extension with group G and ring of integers R. Suppose $(p) = (P_1 \cdots P_g)^e$, and let P be one of the prime ideals P_i.

Proposition 10.2.6. K_P is a Galois extension of $\mathbb{Q}_p$.

Proof. Let $K = \mathbb{Q}(\alpha)$, and let $q(x)$ be the irreducible polynomial of α. Let $r_1(x)r_2(x)\cdots r_l(x)$ be the factorization of $q(x)$ over $\mathbb{Q}_p$ into irreducible polynomials, and let α_i be a zero of $r_i(x)$. Since there is an isomorphism $\bigoplus_{i=1}^{l} \mathbb{Q}_p(\alpha_i) \cong \bigoplus_{i=1}^{l} K_{P_i}$ (see Proposition 5.1.5), $K_P \cong \mathbb{Q}_p(\alpha_i)$ for some i. Necessarily, K_P is the splitting field of the polynomial $r_i(x)$. Thus, $K_P/\mathbb{Q}_p$ is a Galois extension. $\square$

The Galois group of $K_P/\mathbb{Q}_p$ has an elegant characterization. First, a definition: the **decomposition group of** P is the subgroup of G defined as

$$G_P = \{g \in G : g(P) = P\}.$$

Proposition 10.2.7. $G_P \cong \mathrm{Gal}(K_P/\mathbb{Q}_p)$.

Proof. Let p be the prime of $\mathbb{Z}$ lying below P, and let $[\ ]_p$ be the extension of $|\ |_p$ corresponding to the prime P. Let $h \in G_P$. The function $f_h : K \to K$ defined by $f_h(x) = h(x)$ is continuous with respect to the $[\ ]_p$-topology on K. Thus f_h extends to a function $\tilde{f}_h : K_P \to K_P$ that is defined as

$$\tilde{f}_h \left(\lim_{n\to\infty} a_n \right) = \lim_{n\to\infty} (f_h(a_n)) = \lim_{n\to\infty} (h(a_n)),$$

where $\{a_n\}$ is a $[\ ]_p$-Cauchy sequence in K.

Since $\tilde{f}_h$ is an automorphism of K_P that fixes $\mathbb{Q}_p$, the Galois group of $K_P/\mathbb{Q}_p$ is the group of functions $\{\tilde{f}_h : h \in G_P\}$, which is isomorphic to G_P. $\square$

With respect to the Galois group G_P, we define the **trace map** $\text{tr}_{K_P/\mathbb{Q}_p}$: $K_P \to \mathbb{Q}_p$ by the rule

$$\text{tr}_{K_P/\mathbb{Q}_p}(a) = \sum_{h \in G_P} h(a)$$

for $a \in K_P$.

Let $f = [R_P/PR_P : \mathbb{F}_p]$. The finite field extension R_P/PR_P over $\mathbb{F}_p$ has p^f elements that are precisely the p^f distinct zeros of the polynomial $x^{p^f} - x$ in $\mathbb{F}_p[x]$. Moreover, R_P/PR_P is the splitting field of an irreducible polynomial $q(x) \in \mathbb{F}_p[x]$ of degree f. Thus $(R_P/PR_P)/\mathbb{F}_p$ is a Galois extension whose group is computed as follows.

The group

$$G^P = \{g \in G : g(x) \equiv x \pmod{P}, \forall x \in R\}$$

is the **inertia group** of P. Observe that G^P is a normal subgroup of G_P.

Proposition 10.2.8. $J_P = G_P/G^P \cong \text{Gal}((R_P/PR_P)/\mathbb{F}_p)$.

Proof. Exercise. $\square$

The group J_P is cyclic of order f and is generated by the automorphism ϕ of R_P/PR_P, defined as $\phi(\overline{a}) = \overline{a}^p$, where $\overline{a}$ is the image of $a \in R_P$ under the canonical surjection $R_P \to R_P/PR_P$. Note that $\phi^f(\overline{a}) = \overline{a}$ for all $\overline{a} \in R_P/PR_P$.

By Proposition 10.2.4, $[K_P : \mathbb{Q}_p] = |G_P| = ef$, and so, since $|J_P| = f$, $|G^P| = e$.

Set $E = R_P/PR_P$. Since E has Galois group $\langle \phi \rangle$ over $\mathbb{F}_p$, one can define the trace function $\text{tr}_{E/(\mathbb{F}_p)} : E \to \mathbb{F}_p$ as

$$\text{tr}_{E/(\mathbb{F}_p)}(\overline{a}) = \sum_{j=1}^{f} \phi^j(\overline{a}) = \overline{a} + \overline{a}^p + \overline{a}^{p^2} + \cdots + \overline{a}^{p^{f-1}}.$$

Proposition 10.2.9. *Let $\varrho : \mathbb{Z}_p \to \mathbb{Z}_p/p\mathbb{Z}_p \cong \mathbb{F}_p$ denote the canonical surjection. Then*

$$\varrho(\text{tr}_{K_P/\mathbb{Q}_p}(a)) = e \cdot \text{tr}_{E/(\mathbb{F}_p)}(\overline{a}) \tag{10.9}$$

for $a \in R_P$.

Proof. This follows from the decomposition (10.8) with $P = P_i$. $\square$

Proposition 10.2.10. *With the notation above,*

$$\text{tr}_{E/(\mathbb{F}_p)}(E) = \mathbb{F}_p. \tag{10.10}$$

Proof. Let $b \in \mathbb{F}_p$. Since $\mathrm{tr}_{E/(\mathbb{F}_p)}(0) = 0$, we assume that $b \neq 0$. The polynomial $x + x^p + x^{p^2} + \cdots + x^{p^{f-1}}$ has at most p^{f-1} zeros in R_P / PR_P, but since E has $p^f > p^{f-1}$ elements, there must be an element $d \in R_P$ for which $\mathrm{tr}_{E/(\mathbb{F}_p)}(d) = c \neq 0$. Now, $\mathrm{tr}_{E/(\mathbb{Z}_p\mathbb{Z})}$ is surjective since

$$\mathrm{tr}_{E/(\mathbb{F}_p)}(bd/c) = (b/c)\mathrm{tr}_{(E/\mathbb{F}_p)}(d) = (b/c)c = b.$$ □

We next discuss the notion of relative ramification. Let K be a finite extension of $\mathbb{Q}$ with ring of integers R, and let L/K be a Galois extension of fields with group G, $n = |G|$. Let S denote the ring of integers in L. Let Q be a prime ideal of R lying above the prime $p \in \mathbb{Z}$. Analogous to Proposition 10.2.2, the ideal $QS \subseteq S$ factors uniquely as

$$QS = (P_1 P_2 \cdots P_g)^{e_{L/K}}, \tag{10.11}$$

where P_i are distinct prime ideals of S and $e_{L/K}$ is a positive integer. We call the integer $e_{L/K}$ the **relative ramification index of Q in S.**

We say that Q **ramifies in L** if $e_{L/K} > 1$, and that Q is **unramified in L** if $e_{L/K} = 1$. The prime Q is **tamely ramified in L** if p does not divide $e_{L/K}$; Q is **totally ramified** in L if $e_{L/K} = [L : K]$.

Recall that the trace map $\mathrm{tr}_{L/K} : L \to K$ is defined as $\mathrm{tr}_{L/K}(x) = \sum_{g \in G} g(x)$, $x \in L$.

Analogous to Lemma 4.2.3, there exists a basis $\{x_1, x_2, \ldots, x_n\}$ for L over K that is contained in S. Let $(\mathrm{tr}_{L/K}(x_i x_j))$ denote the $n \times n$ matrix whose i, jth entry is $\mathrm{tr}_{L/K}(x_i x_j)$.

Definition 10.2.1. The **discriminant of S over R** is the unique ideal disc(S/R) of R that is generated by the set

$$\{\det(\mathrm{tr}_{L/K}(x_i x_j))\},$$

where $\{x_i\}$ runs through all of the bases for L over K that are contained in S.

Let

$$S^D = \{x \in L : \mathrm{tr}_{L/K}(xS) \subseteq R\}$$

denote the dual module of S. Since S^D is a finitely generated R-module with $S \subseteq S^D$, S^D is a fractional ideal of L, and hence

$$(S^D)^{-1} = \{x \in L : xS^D \subseteq S\}$$

is an integral ideal of S, which we call the **different**. The different is denoted as $\mathcal{D}$.

Suppose that P is one of the primes in the decomposition (10.11). Let $[\ ]_{p,P}$ be the extension of $|\ |_p$ to L that corresponds to the prime P. Since p is also the unique prime lying below Q, there is an extension $[\ ]_{p,Q}$ that extends $|\ |_p$ to K. In fact, the restriction of $[\ ]_{p,P}$ to K is $[\ ]_{p,Q}$, and so $[\ ]_{p,P}$ is an extension of $[\ ]_{p,Q}$ to L.

Let L_P denote the completion of L with respect to $[\ ]_{p,P}$, and let K_Q denote the completion of K with respect to $[\ ]_{p,Q}$. Analogous to Proposition 10.2.7, $\mathrm{Gal}(L_P/K_Q) \cong G_P$, where G_P is the decomposition group

$$G_P = \{h \in G : h(P) = P\}.$$

The **norm of** L_P/K_Q is the map $N_{L_P/K_Q} : L_P \to K_Q$ defined as

$$N_{L_P/K_Q}(s) = \prod_{g \in G_P} g(s)$$

for $s \in L_P$, and the **trace of** L_P/K_Q is the map $\mathrm{tr}_{L_P/K_Q} : L_P \to K_Q$ defined as

$$\mathrm{tr}_{L_P/K_Q}(s) = \sum_{g \in G_P} g(s).$$

Let $f = [S_P/PS_P : R_Q/QR_Q]$, $q = [R_Q/QR_Q : \mathbb{F}_p]$. The finite field extension R_P/PR_P over $\mathbb{F}_p$ has p^q elements, and the extension S_P/PS_P over R_Q/QR_Q has $(p^q)^f = p^{qf}$ elements that are precisely the p^{qf} distinct zeros of the polynomial $x^{p^{qf}} - x$ in $\mathbb{F}_p[x]$. Moreover, S_P/PS_P is the splitting field of an irreducible polynomial $q(x) \in (R_Q/QR_Q)[x]$ of degree f with Galois group

$$J_P = \mathrm{Gal}((S_P/PS_P)/(R_Q/QR_Q)) = G_P/G^P,$$

where
$$G^P = \{g \in G : g(x) \equiv x \pmod{P}, \forall x \in S\}$$

is the **inertia group**.

The group J_P is cyclic of order f and is generated by the automorphism ϕ of S_P/PS_P defined as $\phi(\overline{a}) = \overline{a}^{p^q}$, where $\overline{a}$ is the image of $a \in S_P$ under the canonical surjection $S_P \to S_P/PS_P$. Note that $\phi^f(\overline{a}) = \overline{a}$ for all $\overline{a} \in S_P/PS_P$.

The element ϕ lifts to an element $g \in G_P$ for which

$$g(a) \equiv a^{p^q} \mod P$$

for all $a \in S$.

Let $E = S_P/PS_P$, $F = R_Q/QR_Q$. Since E has Galois group $\langle \phi \rangle$ over F, one can define the trace function $\mathrm{tr}_{E/F} : E \to F$ as

$$\mathrm{tr}_{E/F}(\overline{a}) = \sum_{j=1}^{f} \phi^j(\overline{a}) = \overline{a} + \overline{a}^p + \overline{a}^{p^2} + \cdots + \overline{a}^{p^{f-1}}.$$

Let $\varrho : R_Q \to F$ denote the canonical surjection. Then, analogous to (10.9),

$$\varrho(\mathrm{tr}_{L_P/K_Q}(a)) = e_{L/K} \cdot \mathrm{tr}_{E/F}(\overline{a}) \tag{10.12}$$

for $a \in S_P$. Moreover, analogous to (10.10),

$$\mathrm{tr}_{E/F}(E) = F. \tag{10.13}$$

Note that S_P is a finitely generated torsion-free module over the PID R_Q. Thus S_P is free over R_Q and of finite rank $m = [L_P : K_Q]$.

Let $\{x_1, x_2, \ldots, x_m\}$ be an R_Q-basis for S_P. The **discriminant** of S_P/R_Q is the R_Q-ideal $\det(\mathrm{tr}_{L_P/K_Q}(x_i x_j)) R_Q$.

Let S_P^D denote the dual module of S_P; that is,

$$S_P^D = \{x \in L_P : \mathrm{tr}_{L_P/K_Q}(x S_P) \subseteq R_Q\}.$$

As in the global case, the **different** is the S_P-ideal

$$(S_P^D)^{-1} = \{x \in L : x S_P^D \subseteq S_P\},$$

which we denote as $\mathcal{D}_{S_P}$. Observe that $\mathcal{D}_{S_P} = S_P \otimes_R \mathcal{D} = (S_P^D)^{-1}$.

The discriminant and the different are related in the following way.

Proposition 10.2.11. $\mathrm{disc}(S_P/R_Q) = N_{L_P/K_Q}(\mathcal{D}_{S_P})$.

Proof. For a proof, see [Se79, Lemma I.6.3].   □

A lower bound for the P-order of $\mathcal{D}$ can now be computed.

Proposition 10.2.12. *Let L be a Galois extension of K with ring of integers S, R, respectively. Let P be a prime ideal of S that lies above the prime ideal Q in R, and let e denote the ramification index of Q in S. Then $\mathrm{ord}_P(\mathcal{D}) \geq e - 1$.*

Proof. We have the quotients S_P/PS_P, S_P/QS_P, and

$$PS_P/QS_P = PS_P/(PS_P)^e.$$

For $x \in S_P$, let $\tilde{x}$ denote the image of x under the canonical surjection $S_P \to S_P/QS_P$. Let $f = [S_P/PS_P : R_Q/QR_Q]$. There exists an R_Q/QR_Q-basis $\{a_i\}_{i=1}^{ef}$ for S_P/QS_P, for which $\{a_i\}$, $1 \leq i \leq (e-1)f$, is an R_Q/QR_Q-basis for PS_P/QS_P. By Nakayama's Lemma, there exists an R_Q-basis $\{x_i\}$ for S_P for which $x_i \equiv a_i \mod QS_P$ for $1 \leq i \leq ef$.

Now, for $i = 1, 2, \ldots, (e-1)f$, and $j = 1, 2, \ldots, ef$, $x_i x_j \in PS_P$, and so $\mathrm{tr}_{L_P/K_Q}(x_i x_j) \in QR_Q$. Thus,

$$\mathrm{ord}_Q(\mathrm{disc}(S_P/R_Q)) \geq (e-1)f,$$

and so, by Proposition 10.2.11,

$$\frac{1}{f} \mathrm{ord}_Q(N_{L_P/K_Q}(\mathcal{D}_{S_P})) \geq (e-1).$$

As an integral ideal of S_P, $\mathcal{D}_{S_P}$ factors as $\mathcal{D}_{S_P} = (PS_P)^l$, and hence $\mathrm{ord}_P(\mathcal{D}_{S_P}) = l$. But since $N_{L_P/K_Q}(PS_P) = (QR_Q)^f$, $\mathrm{ord}_Q(N_{L_P/K_Q}(\mathcal{D}_{S_P})) = lf$, and so

$$\mathrm{ord}_P(\mathcal{D}_{S_P}) = l = \frac{1}{f}\mathrm{ord}_Q(N_{L_P/K_Q}(\mathcal{D}_{S_P})) \geq (e - 1).$$

Of course, $\mathrm{ord}_P(\mathcal{D}_{S_P}) = \mathrm{ord}_P(\mathcal{D})$, which completes the proof. $\square$

We can now give the two main results in this section.

Proposition 10.2.13. *Let L/K be a Galois extension of number fields with group G. Let P be a prime ideal of S that lies above the prime ideal Q in R, and let $e = e_{L/K}$ denote the (relative) ramification index of Q in S. Then the following conditions are equivalent:*

(i) $\mathrm{ord}_P(\mathcal{D}) = e - 1$;
(ii) $\mathrm{tr}_{L_P/K_Q}(S_P) = R_Q$ (that is, the trace map is surjective);
(iii) L_P is tamely ramified over K_Q (that is, p does not divide e).

Proof. (i) $\Leftrightarrow$ (ii) Put $r = \mathrm{ord}_Q(\mathrm{tr}_{L_P/K_Q}(S_P))$, $v = \mathrm{ord}_P(\mathcal{D})$. By definition,

$$S_P^D = \{x \in L_P : \mathrm{tr}_{L_P/K_Q}(xS_P) \subseteq R_Q\},$$

and so

$$\begin{aligned} S_P^D \cap K_Q &= \{x \in K_Q : \mathrm{tr}_{L_P/K_Q}(xS_P) \subseteq R_Q\} \\ &= \{x \in K_Q : \mathrm{tr}_{L_P/K_Q}(S_P)x \subseteq R_Q\} \\ &= (\mathrm{tr}_{L_P/K_Q}(S_P))^{-1}, \end{aligned}$$

so that

$$\mathcal{D}_{S_P} \cap K_Q = \mathrm{tr}_{L_P/K_Q}(S_P).$$

Thus $\frac{v}{e} \geq r$ or

$$v \geq re.$$

Now, if $\mathrm{ord}_P(\mathcal{D}) = e - 1$, then $e - 1 \geq re$, so that $r = 0$ and $\mathrm{tr}_{L_P/K_Q}(S_P) = R_Q$.

Conversely, if $\mathrm{tr}_{L_P/K_Q}(S_P) = R_Q$, then $r = 0$ and $\mathcal{D}_{S_P} = S_P$. It follows that $QR_Q \subset S_P^D \cap K_Q$, and so $e > v$. Now, by Proposition 10.2.12, $e > v \geq e - 1$, which yields $v = e - 1$.

(ii) $\Leftrightarrow$ (iii) Let $E = S_P/PS_P$, $F = R_Q/QR_Q$. Suppose that $\mathrm{tr}_{L_P/K_Q}(S_P) = R_Q$. Then, by (10.12), $F = e \cdot \mathrm{tr}_{E/F}(E)$, and so, by (10.13), $F = e \cdot F$. Consequently, $(p, e) = 1$, and thus L_P/K_Q is tamely ramified.

On the other hand, if $(p, e) = 1$, then e is a unit in F, and

$$\varrho(\mathrm{tr}_{L_P/K_Q}(S_P)) = \mathrm{tr}_{E/F}(E) = F.$$

Thus, $\mathrm{tr}_{L_P/K_Q}(S_P) = R_Q$. $\square$

Proposition 10.2.14. *Let L/K be a Galois extension of number fields with group G. Let P be a prime ideal of S that lies above the prime ideal Q in R, and let e denote the ramification index of Q in S. Then the prime Q ramifies in S if and only if Q divides* $\mathrm{disc}(S/R)$.

Proof. Suppose Q does not divide $\mathrm{disc}(S/R)$. Then $\mathrm{disc}(S_P/R_Q) = R_Q$, and so, by Proposition 10.2.11, $0 = \mathrm{ord}_Q(N_{L_P/K_Q}(\mathcal{D}_{S_P}))$, which says that $0 = \mathrm{ord}_P(\mathcal{D}) \ge e - 1$ by Proposition 10.2.12. Thus $e = 1$.

Conversely, suppose that $e = 1$, and let $\{x_i\}$ be an R_Q-basis for S_P. Then formula (10.12) yields

$$\varrho(\mathrm{tr}_{L_P/K_Q}(x_i x_j)) = \mathrm{tr}_{E/F}(\overline{x_i x_j}).$$

Since $\mathrm{tr}_{E/F}(\overline{x_i x_j}) \ne 0$, $\varrho(\mathrm{tr}_{L_P/K_Q}(x_i x_j))$, and consequently $\varrho(\mathrm{disc}(S_P/R_Q))$ is not zero in $F = R_Q/QR_Q$ and therefore Q does not divide $\mathrm{disc}(S/R)$. $\qquad\square$

Remark 10.2.1. If K is a Galois extension of $\mathbb{Q}$, then $\mathrm{disc}(R/\mathbb{Z})$ is always a nontrivial proper ideal of $\mathbb{Z}$. Thus, at least one prime $p \in \mathbb{Z}$ ramifies in K. This deep result is due to H. Minkowski [Ne99, Theorem III.2.17].

Using the preceding propositions, we can calculate the ring of integers of several Galois extensions.

Proposition 10.2.15. *Let $K = \mathbb{Q}(\zeta_p)$. Then $R = \mathbb{Z}[\zeta_p]$.*

Proof. As a $\mathbb{Q}$-vector space,

$$\mathbb{Q}(\zeta_p) = \mathbb{Q} \oplus \mathbb{Q}(1 - \zeta_p) \oplus \mathbb{Q}(1 - \zeta_p)^2 \oplus \cdots \oplus \mathbb{Q}(1 - \zeta_p)^{p-2}.$$

Let P denote the principal ideal of R generated by $1 - \zeta_p$. Let $\alpha \in R$. Then

$$\alpha^l = a_0 + a_1 \alpha + \cdots + a_{l-1} \alpha^{l-1}$$

for integers $a_i \in \mathbb{Z}$ and some integer $l \ge 0$. Let $\overline{\alpha} = \alpha \pmod P$. Then

$$\overline{\alpha}^l = a_0 + a_1 \overline{\alpha} + \cdots + a_{l-1} \overline{\alpha}^{l-1},$$

and thus $\overline{\alpha} \in R/P$ is integral over $\mathbb{Z}$.

Since R is finitely generated and torsion-free over $\mathbb{Z}$, a PID, R is free over $\mathbb{Z}$ of rank $p - 1$. Let $\{b_1, b_2, \ldots, b_{p-1}\}$ be a $\mathbb{Z}$-basis for $R \subseteq \mathbb{Q}(\zeta_p)$. One has $R/P \subseteq \mathbb{Q}$. Thus $R/P \subseteq \mathbb{Z}$ since the integral closure of $\mathbb{Z}$ in $\mathbb{Q}$ is $\mathbb{Z}$. Evidently, $\mathbb{Z} \subseteq R/P$. Thus $R/P = \mathbb{Z}$, so that P is a prime ideal of R.

The ideal P^{p-1} factors as $P^{p-1} = (p)(z)$, where $z \in R$ with $(1 - \zeta_p, z) = 1$. Thus, $(p) = P^{p-1}$ is the prime factorization of (p), and so $e = p - 1$ is the ramification index of p in R.

Since $(p, p - 1) = 1$, K_P is tamely ramified over $\mathbb{Q}_p$. Let $\mathbb{Z}[\zeta_p]$ denote the free $\mathbb{Z}$-module on the basis $B = \{1, \zeta_p, \zeta_p^2, \ldots, \zeta_p^{p-2}\}$. Then $\mathbb{Z}[\zeta_p] \subseteq R$; the problem is to show the reverse containment.

Relative to the basis B, one computes $\mathrm{disc}(\mathbb{Z}_p[\zeta_p]/\mathbb{Z}_p) = (p^{e-1})$. On the other hand, we have $\mathcal{D}_{R_P} = P^{e-1} R_P$ by Proposition 10.2.13, and so

$$N_{K_P/\mathbb{Q}_p}(\mathcal{D}_{R_P}) = N_{K_P/\mathbb{Q}_p}(P^{e-1}R_P)$$

$$= (N_{K_P/\mathbb{Q}_p}(PR_P))^{e-1}$$

$$= \left(\prod_{i=1}^{p-1}\left(1 - \zeta_p^i\right)\right)^{e-1} = p^{e-1}.$$

But now, by Proposition 10.2.11, $\mathrm{disc}(R_P/\mathbb{Z}_p) = p^{e-1}$, and so $R_P = \mathbb{Z}_p[\zeta_p]$. Consequently, $R = \mathbb{Z}[\zeta_p]$. □

For our next example, we consider the polynomial $f_3(x) = x^3 - 3x + 1$, which is obtained from formula (10.2) with $p = 3$. Clearly, f_3 is irreducible over $\mathbb{Q}$. Moreover, the zeros of f_3 are $r_1 = -\theta - \overline{\theta}$, $r_2 = -\theta\zeta^{-1} - \overline{\theta}\zeta$, and $r_3 - \theta\zeta^{-2} - \overline{\theta}\zeta^2$, with $\theta^3 = \eta$, $\eta = (1 + i\sqrt{3})/2$, $\zeta = \zeta_3$. These roots are related by the equations $r_1^2 = r_2 + 2$, $r_2^2 = r_3 + 2$, and $r_3^2 = r_1 + 2$. Thus, the splitting field of f_3 is $K = \mathbb{Q}(r_1)$. We have $[K : \mathbb{Q}] = 3$, so that $\mathrm{Gal}(K/\mathbb{Q}) = C_3$.

To compute a generator for the Galois group, we proceed in a manner analogous to Proposition 10.1.10. The relations among the roots yield the permutation on three letters

$$g = \begin{pmatrix} 1 & 2 & 3 \\ 2 & 3 & 1 \end{pmatrix}.$$

Let $\rho : K \to K$ be the map defined as $\rho(r_1) = r_{g(1)} = r_2 = r_1^2 - 2$. Then ρ is an automorphism of K fixing $\mathbb{Q}$ with $\rho(r_{g^i(1)}) = r_{g^{i+1}(1)}$, $\forall i \geq 0$. Consequently, ρ has order 3 since g has order 3 in S_3.

Proposition 10.2.16. *Let K be the splitting field of $f_3(x) = x^3 - 3x + 1$. Then $R = \mathbb{Z}[r_1]$.*

Proof. Let $\mathbb{Z}[r_1]$ denote the free $\mathbb{Z}$-module on the basis $\{1, r_1, r_1^2\}$. Then $\mathrm{disc}(\mathbb{Z}[r_1]/\mathbb{Z}) = \det(M)\mathbb{Z}$, where M is the 3×3 matrix

$$M = \begin{pmatrix} \mathrm{tr}(1) & \mathrm{tr}(r_1) & \mathrm{tr}(r_1^2) \\ \mathrm{tr}(r_1) & \mathrm{tr}(r_1^2) & \mathrm{tr}(r_1^3) \\ \mathrm{tr}(r_1^2) & \mathrm{tr}(r_1^3) & \mathrm{tr}(r_1^4) \end{pmatrix}.$$

Now, in view of the relations among the zeros of f_3,

$$M = \begin{pmatrix} \mathrm{tr}(1) & \mathrm{tr}(r_1) & \mathrm{tr}(r_2 + 2) \\ \mathrm{tr}(r_1) & \mathrm{tr}(r_2 + 2) & \mathrm{tr}(3r_1 - 1) \\ \mathrm{tr}(r_2 + 2) & \mathrm{tr}(3r_1 - 1) & \mathrm{tr}(r_3 + 4r_2 + 6) \end{pmatrix} = \begin{pmatrix} 3 & 0 & 6 \\ 0 & 6 & -3 \\ 6 & -3 & 18 \end{pmatrix},$$

and thus $\mathrm{disc}(\mathbb{Z}[r_1]/\mathbb{Z}) = \det(M)\mathbb{Z} = (81)$.

Since $\mathbb{Z}[r_1] \subseteq R$, disc$(R/\mathbb{Z})$ divides disc$(\mathbb{Z}[r_1]/\mathbb{Z})$, and so, by Proposition 10.2.14 and Remark 10.2.1, 3 is the only prime that ramifies in R. Since $3 = [K : \mathbb{Q}]$, 3 is totally ramified in R; that is, the ramification index of 3 in R is $e = 3$. We have $(3) = Q^3$ for some prime ideal of Q of R. Now

$$\text{disc}(\mathbb{Z}_3[r_1]/\mathbb{Z}_3) = [R_Q : \mathbb{Z}_3[r_1]]^2 \text{disc}(R_Q/\mathbb{Z}_3),$$

where $[R_Q : \mathbb{Z}_3[r_1]]$ is the module index of $\mathbb{Z}_3[r_1]$ in R_Q. Now $[R_Q : \mathbb{Z}_3[r_1]] = (3^a)$ for some $a \geq 0$. Since disc$(\mathbb{Z}_3[r_1]/\mathbb{Z}_3) = (81) = (3^4)$, $0 \leq a \leq 2$.

If $a = 2$, then disc$(R_Q/\mathbb{Z}_3) = \mathbb{Z}_3$, which is a contradiction.

If $a = 1$, then disc$(R_Q/\mathbb{Z}_3) = (9)$, which yields ord$_3(\text{disc}(R_Q/\mathbb{Z}_3)) = 2$, and so, by Proposition 10.2.11,

$$2 = \text{ord}_3(\text{disc}(R_Q/\mathbb{Z}_3)) = \text{ord}_3(N_{K_Q/\mathbb{Q}_3}(\mathcal{D}_{R_Q})).$$

By Proposition 12.2.12, $\mathcal{D}_{R_Q} = Q^b$, where $b \geq e - 1 = 2$. Thus

$$\text{ord}_Q(\mathcal{D}_{R_Q}) = b = 2,$$

and so, by Proposition 10.2.13, $K_Q/\mathbb{Q}_3$ is tamely ramified, which is a contradiction. Consequently, the only possibility is $a = 0$, and in this case disc$(\mathbb{Z}_3[r_1]/\mathbb{Z}_3) = $ disc$(R_Q/\mathbb{Z}_3)$, which yields $R_Q = \mathbb{Z}_3[r_1]$. It follows that $R = \mathbb{Z}[r_1]$. $\square$

10.3 Galois Extensions of Rings

Let R be an integral domain with field of fractions K, and let S be an R-algebra that is finitely generated and projective as an R-module. We assume that S has unity 1_S and that $\lambda(1_R) = 1_S$, where $\lambda : R \to S$ is the R-algebra structure map.

The collection Hom$_{R\text{-alg}}(S, S)$ is a group under function composition. Let G be a finite subgroup of Hom$_{R\text{-alg}}(S, S)$, and let $D(S, G)$ denote the collection of sums $\sum_{g \in G} a_g g$, $a_g \in S$. On $D(S, G)$, endow an S-module structure that is exactly the S-module structure of the group ring SG. Define a multiplication on $D(S, G)$ as follows. For $\sum_{g \in G} a_g g$, $\sum_{h \in G} b_h h \in D(S, G)$, put

$$\left(\sum_{g \in G} a_g g \right) \left(\sum_{h \in G} b_h h \right) = \sum_{g,h \in G} a_g g(b_h) gh,$$

where gh is the group product in G. The resulting S-algebra $D(S, G)$ is the **crossed product algebra of S by G**.

Let End$_R(S)$ denote the collection of R-linear maps $\phi : S \to S$. Then End$_R(S)$ is an R-module with scalar multiplication defined as $(r\phi)(t) = r\phi(t)$, for $r \in R$,

$t \in S$, and addition defined pointwise: for $\phi, \psi \in \mathrm{End}_R(S)$, we have $(\phi + \psi)(t) = \phi(t) + \psi(t)$. The R-module $\mathrm{End}_R(S)$ is an R-algebra with multiplication defined by function composition: $(\phi\psi)(t) = \phi(\psi(t))$.

Definition 10.3.1. The R-algebra S is a **Galois extension of R with group G** if the R-module map

$$j : D(S, G) \to \mathrm{End}_R(S), \tag{10.14}$$

defined as $j(\sum_{g \in G} a_g g)(t) = \sum_{g \in G} a_g g(t)$ for all $t \in S$, is a bijection.

Every Galois extension of fields is a Galois extension of rings. Indeed, let L/K be a Galois extension of fields with group G. Then $G = \mathrm{Gal}(L/K) \leq \mathrm{Hom}_{K\text{-alg}}(L, L)$, and the map

$$j : D(L, \mathrm{Gal}(L/K)) \to \mathrm{End}_K(L),$$

defined as $j\left(\sum_{g \in G} a_g g\right)(t) = \sum_{g \in G} a_g g(t)$ for all $t \in L$, is a bijection.

For the next three propositions, we specialize to the situation where L/K is a Galois extension with group G, R is the ring of integers in K, and S is the ring of integers in L.

Proposition 10.3.1. *(S. Chase, D. Harrison, and A. Rosenberg) S is a Galois extension of R with group G if and only if for each maximal ideal M of S, and for each non-trivial $g \in G$, there exists an element $t \in S$ for which $g(t) \not\equiv t$ (mod M).*

Proof. See [CHR65, Theorem 1.3].   □

Proposition 10.3.2. *S is a Galois extension of R if and only if every prime ideal Q of R is unramified in L.*

Proof. Suppose S is a Galois extension of R. Since S is a Dedekind domain, every non-zero prime ideal of S is maximal. Thus Proposition 10.3.1 shows that, for each prime P of S, the inertia group

$$G^P = \{g \in G : g(t) \equiv t \ (\mathrm{mod}\ P), \forall t \in S\}$$

is trivial. Thus the fixed field of G^P is all of L, and so, by [CF67, Chapter I, §7, Theorem 2], each prime Q of R is unramified in L.

Now suppose every prime ideal Q of R is unramified in L. Then, for each prime P lying above Q, the inertia group G^P is trivial by [CF67, Chapter I, §7, Theorem 2]. Thus, by Proposition 10.3.1, S is a Galois extension of R.   □

Proposition 10.3.3. *S is a Galois extension of R if and only if $\mathrm{disc}(S/R) = R$.*

Proof. Suppose $\mathrm{disc}(S/R) = R$. Then, by Proposition 10.2.14, every prime Q of R is unramified in L, and so by Proposition 10.3.2, S is a Galois extension.

Conversely, if S is a Galois extension, then each prime Q is unramified, and so $\text{disc}(S/R) = R$. $\qquad\qquad\qquad\qquad\qquad\qquad\qquad\qquad\qquad\qquad\qquad\qquad\qquad$ □

We revert to the general situation where R is an integral domain with field of fractions K. If S is a Galois extension of R, then the group G, and consequently the group ring RG, acts on S. The fact that RG is an R-Hopf algebra suggests that we may be able to generalize the notion of a Galois extension of rings to Hopf algebras.

Let H be an R-Hopf algebra, and let S be an R-algebra that is finitely generated and projective as an R-module and is also an H-module with action denoted by $h(s)$ for $h \in H, s \in S$. Then S is an H-**module algebra** if, for all $s, t \in S$,

$$h(st) = \sum_{(h)} h_{(1)}(s) h_{(2)}(t) \qquad\qquad (10.15)$$

and

$$h(1_S) = \epsilon_H(h) 1_S.$$

The **fixed ring** of the H-module algebra S is defined as

$$S^H = \{s \in S : h(s) = \epsilon_H(h)s, \forall h \in H\}.$$

The H-module algebra S is an H-**extension of** R if $S^H = R$.

Let S be an H-extension of R. Then there exists a map $j : S \otimes_R H \to \text{End}_R(S)$ defined as $j(s \otimes h)(t) = sh(t)$ for $s, t \in S, h \in H$.

Definition 10.3.2. Let S be an H-extension. Then S is a **Galois H-extension of** R if the map

$$j : S \otimes_R H \to \text{End}_R(S)$$

defined as $j(s \otimes h)(t) = sh(t)$ is an isomorphism of R-modules.

The notion of a Galois H-extension generalizes the notion of a Galois extension: if $H = RG$, then S is a Galois RG-extension of R if and only if S is a Galois extension of R with group G.

Let S be an H-module algebra. We define a multiplication on the R-module $S \otimes_R H$ as follows: for $a \otimes b, c \otimes d \in S \otimes_R H$,

$$(a \otimes b)(c \otimes d) = \sum_{(b)} ab_{(1)}(c) \otimes b_{(2)}d. \qquad\qquad (10.16)$$

Then $S \otimes_R H$ together with the multiplication (10.14) is an R-algebra, which is the **smash product** $S \sharp H$ of S by H. An element of the smash product is written as $\sum a \sharp b$.

Proposition 10.3.4. *Let S be an H-extension. Then S is a Galois H-extension of R if and only if the map*

$$j : S \sharp H \to \text{End}_R(S)$$

defined as $j(s \sharp h)(t) = sh(t)$ is an isomorphism of R-algebras.

Proof. Exercise.   □

Let S be an H-extension of R. The image of $S \sharp 1$ under the map j consists of endomorphisms of the form $t \mapsto st$ for $s \in S$ these are endomorphisms defined by "left multiplication by s" and, for this reason, we denote the image $j(S \sharp 1)$ by S_{left}.

Our goal is to give a characterization of Galois H-extensions in terms of S_{left}. Let $\Lambda \in \int_H, t \in S, h \in H$. Then

$$h(\Lambda(t)) = (h\Lambda)(t) = (\epsilon_H(h)\Lambda)(t) = \epsilon_H(h)\Lambda(t),$$

and so $\int_H S \subseteq S^H$. Now, since $S^H = R$, $\int_H S \subseteq R$.

Observe that $\int_H$ acts on $j(S \sharp H)$ to produce an element of $\text{End}_R(S)$. Indeed, for $\Lambda \in \int_H, h \in H, s, t \in S$,

$$(\Lambda \cdot j(s \sharp h))(t) \overset{\text{def}}{=} \Lambda(sh(t))$$

$$= \sum_{(\Lambda)} \Lambda_{(1)}(s)\Lambda_{(2)}(h(t)) \quad \text{by (10.13)}$$

$$= \left(\sum_{(\Lambda)} \Lambda_{(1)}(s)\Lambda_{(2)}h \right)(t).$$

Thus $\Lambda \cdot j(s \sharp h) \in \text{End}_R(S)$.

With $\hat{s} = j(s \sharp 1) \in S_{\text{left}}$, one has

$$(\Lambda \cdot \hat{s})(t) = \Lambda(st),$$

which is in R since $\int_H S \subseteq R$.   ❙

Lemma 10.3.1. $\int_H \cdot j(S \sharp H) = \int_H \cdot S_{\text{left}}$.

Proof. Clearly, $\int_H \cdot S_{\text{left}} \subseteq \int_H \cdot j(S \sharp H)$, so it remains to show the reverse containment. Let $\Lambda \in \int_H, s, t \in S, h \in H$. Then

$$(\Lambda \cdot j(s \sharp h))(t) = \sum_{(\Lambda)} \Lambda_{(1)}(s)\Lambda_{(2)}(h(t)) \quad \text{by the module algebra property of } S$$

$$= \sum_{(\Lambda, h)} \Lambda_{(1)}(s)\Lambda_{(2)}(\epsilon(h_{(1)})h_{(2)}(t)) \quad \text{by the counit property}$$

$$= \sum_{(\Lambda, h)} \Lambda_{(1)}(s)\epsilon(\sigma(h_{(1)}))\Lambda_{(2)}(h_{(2)}(t)) \quad \text{since } \epsilon(\sigma(h)) = \epsilon(h)$$

$$= \sum_{(\Lambda, h)} \Lambda_{(1)}(\epsilon(\sigma(h_{(1)}))s)\Lambda_{(2)}(h_{(2)}(t))$$

$$= \sum_{(h)} \Lambda(\epsilon(\sigma(h_{(1)}))sh_{(2)}(t))$$

$$= \sum_{(h)} \epsilon(\sigma(h_{(1)}))\Lambda(sh_{(2)}(t))$$

$$= \sum_{(h)} (\Lambda\sigma(h_{(1)}))(sh_{(2)}(t)) \quad \text{since } \Lambda \text{ is an integral}$$

$$= \sum_{(h,\Lambda)} (\Lambda_{(1)}\sigma(h_{(1)}))(s)(\Lambda_{(2)}\sigma(h_{(2)})h_{(3)})(t)$$

$$= \sum_{(h,\Lambda)} (\Lambda_{(1)}\sigma(h_{(1)}))(s)(\Lambda_{(2)}\epsilon(h_{(2)})1_H)(t)$$

by the coinverse property

$$= \sum_{(h,\Lambda)} (\Lambda_{(1)}\epsilon(\sigma(h_{(2)}))\sigma(h_{(1)}))(s)\Lambda_{(2)}(t)$$

$$= \sum_{(\Lambda)} \Lambda_{(1)}(\sigma(h)(s))\Lambda_{(2)}(t) \quad \text{by the counit property}$$

$$= \Lambda(\sigma(h)(s)t)$$

$$= (\Lambda \cdot \widehat{\sigma(h)(s)})(t),$$

which is an element of $\Lambda \cdot S_{\text{left}}$.   □

Proposition 10.3.5. *Let H be an R-Hopf algebra that is finitely generated and projective as an R-module, and let S be an H-extension of R. Then S is a Galois H-extension of R if and only if*

$$\int_H \cdot S_{\text{left}} = S^*, \quad S^* = \text{Hom}_R(S, R).$$

Proof. Suppose that $\int_H \cdot S_{\text{left}} = S^*$. There is an isomorphism $\alpha : S \otimes S^* \to \text{End}_R(S)$ defined as $\alpha(s \otimes f)(t) = sf(t)$. Moreover, one has the map

$$\beta : S \otimes \int_H \cdot S_{\text{left}} \to j(S\sharp H),$$

defined as

$$\beta(q \otimes (\Lambda \cdot \hat{s}))(t) = q(\Lambda \cdot \hat{s})(t)$$

$$= q\Lambda(st)$$

$$= q \sum_{(\Lambda)} \Lambda_{(1)}(s)\Lambda_{(2)}(t)$$

$$= \sum_{(\Lambda)} q\Lambda_{(1)}(s)\Lambda_{(2)}(t)$$

$$= j\left(\sum_{(\Lambda)} q\Lambda_{(1)}(s)\sharp\Lambda_{(2)}\right)(t)$$

$$\in j(S\sharp H),$$

for $q, t \in S$, $\hat{s} = j(s\sharp 1)$, $\Lambda \in \int_H$.

The map $j : S\sharp H \to \text{End}_R(S)$, $s\sharp h \mapsto \phi$, where $\phi(t) = sh(t)$, is clearly an inclusion. So, there is a diagram

$$S \otimes \int_H \cdot S_{\text{left}} = S \otimes S^*$$

$$\downarrow \beta \qquad\qquad \downarrow \alpha$$

$$j(S\sharp H) \quad \subseteq \text{End}_R(R),$$

which commutes since

$$\beta(q \otimes (\Lambda \cdot \hat{s}))(t) = q(\Lambda \cdot \hat{s})(t) = \alpha(q \otimes (\Lambda \cdot \hat{s}))(t).$$

Thus $j(S\sharp H) = \text{End}_R(S)$, so that S is a Galois H-extension.

For the converse, assume that S is a Galois H-extension of R. One has an isomorphism

$$\text{End}_R(S) \cong S \otimes \int_H \cdot j(S\sharp H),$$

and so, by Lemma 10.3.1, there is an isomorphism

$$\text{End}_R(S) \cong S \otimes \int_H \cdot S_{\text{left}},$$

and thus $S \otimes S^* = S \otimes \int_H \cdot S_{\text{left}}$. Now, since the R-algebra map $R \to S$ is faithfully flat, $S^* = \int_H \cdot S_{\text{left}}$. $\square$

Proposition 10.3.6. *Let R be a Dedekind domain, and let H be an R-Hopf algebra that is finitely generated and projective as an R-module. Let S be an H-extension of R. Then S is a Galois H-extension of R if and only if the map*

$$\varphi : \int_H \otimes_R S \to S^*$$

defined as $\varphi(\Lambda \otimes s)(t) = \Lambda(st)$, for $s, t \in S$, $\Lambda \in \int_H$, is an isomorphism. (This map makes sense since $\int_H S \subseteq R$.)

Proof. Let S be a Galois H-extension of R. We first show that φ is surjective. Let $\alpha \in S^*$. Then, by Proposition 10.3.5, there exist elements $\hat{s} \in S_{\text{left}}$ and $\Lambda \in \int_H$ for which $\sum_\Lambda \cdot \hat{s} = \alpha$. Since

$$(\Lambda \cdot \hat{s})(t) = \Lambda(st) = \varphi(\Lambda \otimes s)(t),$$

φ is surjective.

Let P be a prime ideal of R. By Lemma 4.3.5, $\int_{H_P}$ is a free R_P-module of finite rank and, by Proposition 4.3.2, $\text{rank}(\int_{H_P}) = 1$. Thus $\int_{H_P} \otimes_{R_P} S_P \cong S_P^*$ as R_P-modules. It follows that φ is an isomorphism of R-modules.

We leave the converse to the reader.   □

Proposition 10.3.6 can be applied to the case $S = H^*$.

Proposition 10.3.7. H^* *is a Galois H-extension of R.*

Proof. We know that H^* is a finitely generated and projective R-algebra and an H-module algebra with $(H^*)^H = R$. With H playing the role of H^*, the map

$$\varphi : H^* \otimes_R \int_H \to H, \quad \alpha \otimes \Lambda \mapsto \alpha \cdot \Lambda,$$

of Proposition 4.3.2 is now the required isomorphism of 10.3.6.   □

Let R be a Dedekind domain, let S be an H-extension of R, and let P be a prime ideal of R. Put $S_P = R_P \otimes_R S$. For each prime P of R, S_P is R_P-free; let $\{x_i\}_{i=1}^l$ denote a basis for S_P over R_P. Since R_P is a PID, there exists a generating integral Λ for $\int_{H_P}$. Define

$$\text{disc}_{H_P}(S_P/R_P) = \det(\Lambda(x_i x_j)) R_P.$$

Definition 10.3.3. The **discriminant of S over R with respect to** H is the unique ideal $\text{disc}_H(S/R)$ of R for which

$$\text{disc}_H(S/R) R_P = \text{disc}_{H_P}(S_P/R_P)$$

for all primes P in R. Equivalently, $\text{disc}_H(S/R)$ is the ideal of R that is generated by the set

$$\{\det(\Lambda(x_i x_j))\},$$

where $\{x_i\}$ runs through all of the bases for L/K that are contained in S and Λ runs through all elements of $\int_H$. These bases are precisely the minimal generating sets of S over R that become bases for S_P over R_P.

The next proposition generalizes Proposition 10.3.3 to Hopf algebras.

Proposition 10.3.8. *Let R be a Dedekind domain with field of fractions K, and let H be an R-Hopf algebra. Let S be an R-algebra that is finitely generated and projective as an R-module and is also an H-extension of R. Suppose $L = K \otimes_R S$ is a Galois $(K \otimes_R H)$-extension of K. Then S is a Galois H-extension of R if and only if $\operatorname{disc}_H(S/R) = R$.*

Proof. Let P be a prime ideal of R. It suffices to show that S_P is a Galois H_P-extension of R_P if and only if $\operatorname{disc}_{H_P}(S_P/R_P) = R_P$.

Let Λ be a generating integral for H_P, and suppose that S_P is a Galois H_P-extension of R_P. Then, by Proposition 10.3.5, $\int_{H_P} \cdot (S_P)_{\text{left}} = S_P^*$. Let $\{x_1, x_2, \ldots, x_m\}$ be a basis for S_P over R_P, and let $\{f_1, f_2, \ldots, f_m\}$ be the dual basis for S_P^*. Let y_i, $i = 1, \ldots, m$, be elements of S_P for which $\Lambda \cdot \hat{y}_i = f_i$ for $1 \le i \le m$. Write $y_i = \sum_{l=1}^m a_{i,l} x_l$ for $a_{i,l} \in R_P$. Then

$$\delta_{ij} = f_i(x_j) = (\Lambda \cdot \hat{y}_i) x_j = \Lambda(y_i x_j)$$

$$= \Lambda \left(\sum_{l=1}^m a_{i,l} x_l x_j \right) = \sum_{l=1}^m a_{i,l} \Lambda(x_l x_j),$$

with $\Lambda(x_l x_i) \in R_P$. Since each $a_{i,l} \in R_P$, the $m \times m$ matrix $(\Lambda(x_l x_j))$ is invertible over R_P. Thus, $\operatorname{disc}_{H_P}(S_P/R_P) = \det(\Lambda(x_l x_j)) R_P = R_P$.

For the converse, we suppose that $\operatorname{disc}_{H_P}(S_P/R_P) = R_P$. The R_P-basis $\{x_i\}$ for S_P is actually a K_P-basis for $L_P = K_P \otimes_K L$. As above, one obtains $y_i = \sum_{l=1}^m a_{i,l} x_l$ and

$$\delta_{ij} = \sum_{l=1}^m a_{i,l} \Lambda(x_l x_j),$$

with $a_{i,l} \in K_P$. Now, since $\det(\Lambda(x_l x_j))$ is a unit of R_P, the matrix $((a_{i,l})) = (\Lambda(x_l x_j))^{-1}$ has entries in R_P. Thus $y_i \in S_P$ for all i, and therefore $\int_{H_P} \cdot (S_P)_{\text{left}} = S_P^*$. $\qquad\square$

10.4 Hopf-Galois Extensions of a Local Ring

We return to the complete, local situation. Let p be a prime of $\mathbb{Z}$, and let K be a finite extension of $\mathbb{Q}_p$ of degree n, with $\zeta_p \in K$. K is endowed with the discrete valuation ord. The ring of integers R in K is a local ring with maximal ideal Q. By Proposition 5.3.12, there exists an element $\pi \in R$ with $(\pi) = Q$; π is a uniformizing parameter for R. One has $\operatorname{ord}((p)) = e$, and since $(p) = (1 - \zeta_p)^{p-1}$, $e' = e/(p-1)$ is an integer.

Let C_p denote the cyclic group of order p, generated by g. For each integer i, $0 \le i \le e'$, there exists a Larson order $H(i)$ in KC_p with linear dual isomorphic to $H(i') = R[\frac{g-1}{\pi^{i'}}]$, where $i' = e' - i$.

In what follows, we apply the results of §10.3 to construct a Galois $H(i')$-extension of R.

Choose integers i and r with $0 \leq i < e'$ and $1 \leq r \leq p - 1$. Set $w = 1 + \pi^{pi+r}$. Then w cannot be a pth power in R and the polynomial $x^p - w \in R[x]$ is irreducible by Proposition 10.1.11. Set $L = K(z)$, with $z = w^{\frac{1}{p}}$. Then L/K is a Galois extension with group C_p. Let S denote the ring of integers in L; S is a local PID with maximal ideal $P = (\lambda)$ and is endowed with the discrete valuation ord_λ. One has $(\pi) = (\lambda)^{e_{L/K}}$ for some positive integer $e_{L/K}$. If $e_{L/K} = 1$, then L/K is unramified, and if $e_{L/K} = p$, then L/K is totally ramified.

Proposition 10.4.1. $L = K(z)$ is a totally ramified extension of K.

Proof. We have $[L : \mathbb{Q}_p] = pn$, and so, by Proposition 10.2.4,

$$pn = (e \cdot e_{L/K})(f \cdot f_{L/K}),$$

where $f = [R/(\pi) : \mathbb{F}_p]$ and $f_{L/K} = [S/(\lambda) : R/(\pi)]$. By Proposition 10.2.4, $n = e \cdot f$, and so

$$p = e_{L/K} \cdot f_{L/K}.$$

Thus either $e_{L/K} = 1$ or $e_{L/K} = p$.

Assume that $i < e'$. We have

$$(z-1)^p = w - 1 + \binom{p}{1}(-1)^{p-1}z^{p-1} + \binom{p}{2}(-1)^{p-2}z^{p-2} + \cdots + \binom{p}{p-1}z$$

$$= w - 1 + up(z-1), \tag{10.17}$$

where u is a unit of S, and so

$$p \cdot \mathrm{ord}_\lambda(z - 1) \geq \min\{\mathrm{ord}_\lambda(w - 1), \mathrm{ord}_\lambda(p) + \mathrm{ord}_\lambda(z - 1)\}.$$

Now, if L/K is unramified and if

$$\mathrm{ord}_\lambda(1 - w) \geq \mathrm{ord}_\lambda(p) + \mathrm{ord}_\lambda(z - 1),$$

then, in view of (10.17), $\mathrm{ord}_\lambda(z - 1) \geq e'$, with

$$pi + r \geq e + e',$$

which is a contradiction since $i < e'$ and $r < p$. So, either L/K is totally ramified or

$$\mathrm{ord}_\lambda(1 - w) < \mathrm{ord}_\lambda(p) + \mathrm{ord}_\lambda(z - 1).$$

But, in the latter case, $p \cdot \mathrm{ord}_\lambda(z-1) = \mathrm{ord}_\lambda(w-1)$, and so $\mathrm{ord}_\lambda(1-w)$ is an integer divisible by p, which says that L/K is totally ramified. $\square$

Proposition 10.4.2. *Let* $x = (z - 1)/\pi^i$. *Then* $R[x]$ *is a Galois* $H(i')$-*extension of* R.

Proof. We use Proposition 10.3.8 to show that $R[x]$ is a Galois $H(i')$-extension of R. We first check that the conditions of Proposition 10.3.8 are satisfied: $R[x]$ is finitely generated and projective as an R-module, $R[x]$ is an $H(i')$-extension of R, and $L = K \otimes_R R[x]$ is a Galois KC_p-extension of K. (The verification of these statements is left to the reader.) So it is a matter of checking that $\mathrm{disc}_{H(i')}(R[x]/R) = R$.

By Proposition 4.4.5 and Proposition 5.3.3, $\Lambda = \pi^{-(p-1)i'} \sum_{g \in C_p} g$ is a generating integral for $H(i')$. Note that $\Lambda = \mathrm{tr}_{L/K}/\pi^{(p-1)i'}$.

We now compute $\mathrm{disc}_{H(i')}(R[x]/R)$. Note that $R[\pi^i x] = R[z] \subseteq R\left[\frac{z-1}{\pi^i}\right]$. We compute $\mathrm{disc}(R[z]/R)$. With respect to the basis $\{1, z, z^2, \ldots, z^{p-1}\}$, we have, by [La84, p. 348],

$$\mathrm{disc}(R[z]/R) = (-1)^{p(p-1)/2} N_{L/K}(pz^{p-1})R = p^p R.$$

Now, the matrix that multiplies the basis $\{(z - 1)^m/\pi^{mi}\}_{m=0}^{p-1}$ of $R[x]$ to give the basis $\{(z - 1)^m\}$ of $R[z]$ is

$$\begin{pmatrix} 1 & 0 & 0 & \cdots & 0 \\ 0 & \pi^i & 0 & \cdots & 0 \\ 0 & 0 & \pi^{2i} & \cdots & 0 \\ \vdots & & & & \vdots \\ 0 & 0 & 0 & \cdots & \pi^{(p-1)i} \end{pmatrix},$$

and so, by a familiar property of discriminants,

$$p^p R = \pi^{p(p-1)i} R \cdot \mathrm{disc}(R[x]/R).$$

Thus, $\mathrm{disc}(R[x]/R) = \frac{p^p}{\pi^{p(p-1)i}} R$, and so

$$\frac{p^p}{\pi^{p(p-1)i}} R = \det(\mathrm{tr}(x^a x^b))R$$

$$= \pi^{p(p-1)i'} \det(\mathrm{tr}(x^a x^b)/\pi^{(p-1)i'})R$$

$$= \pi^{p(p-1)i'} \det(\Lambda(x^a x^b))R.$$

Thus,

$$\det(\Lambda(x^a x^b))R = \pi^{pe - p(p-1)i - p(p-1)i'} R$$

$$= \pi^{pe - p(p-1)(i+i')} R$$

$$= \pi^{pe - p(p-1)e'} R$$

$$= R.$$

$\square$

We know that $R[x]$ is contained in the ring of integers of L. But when precisely is $R[x]$ the ring of integers of L?

Proposition 10.4.3. *Let $R[x]$ be defined as above with $0 \le i < e'$. Then $R[x]$ is the ring of integers of $L = K(z)$, $z^p = w$, $w = 1 + \pi^{pi+r}$ if and only if $r = 1$.*

Proof. Clearly, $R[x]$ is contained in the ring of integers in L. Since $i < e'$, [Ch87, Lemma 13.1] applies to show that $\text{ord}_\lambda(x) = r$, and so $R[x]$ is the ring of integers of L if and only if $r = 1$.   □

We summarize the discussion above. For a given R-Hopf order $H(i')$ in KC_p, we have found a Galois extension L/K whose ring of integers S is a Galois $H(i')$-extension. We say that $H(i')$ is "realizable as a Galois group." The formal definition is as follows.

Definition 10.4.1. Let H be an R-Hopf order in KG. Then H is **realizable as a Galois group** if there exists a Galois extension L/K with group G for which the ring of integers S in L is a Galois H-extension.

N. Byott [By04, Theorem 6.1] has provided the following convenient criterion for realizability.

Proposition 10.4.4. *(N. Byott) Let $n \ge 1$ be an integer, and let K be a finite extension of $\mathbb{Q}_p$ with ring of integers R containing $\zeta_{p^n} \in K$. Let (π) be the maximal ideal R. Let H be an R-Hopf order in KG, and suppose both H and H^* are local rings. Then H is realizable as a Galois group if and only if H^* is a monogenic R-algebra.*

Proof. (Sketch.) We prove the "if" assertion. Suppose H^* is monogenic of rank p^n over R. Then there exists an isogeny $\phi : F \to G$ of formal groups of dimension $m = p^n - 1$ with $H^* \cong R[[\overline{x}]]_F/\phi(\overline{x})$. Let $c \in (\pi)\backslash(\pi)^2$, and let $S = R[[\overline{x}]]/J$, where J is the ideal

$$(\phi_1(\overline{x}) - c, \phi_2(\overline{x}), \dots, \phi_m(\overline{x})).$$

Then S is a Galois H-extension of R.
Write

$$S \cong R[[x_1]]/(\psi(x_1) - c)$$

with $\psi(x_1) = \phi_1((c, 0, 0, \dots, 0))$. Then there exists an irreducible polynomial $q(x_1)$ for which

$$S \cong R[[x_1]]/(q(x_1)) \cong R[y]/q(y)$$

($c \in (\pi)\backslash(\pi)^2$ guarantees the irreducibility of q by the Eisenstein criterion). Let α be a zero of $q(y)$ in some extension field L/K. Then $S \cong R[\alpha]$ is the ring of integers of $K(\alpha)$. It follows that H is realizable as a Galois group.   □

The following are some applications of Byott's Theorem.

Proposition 10.4.5. *Assume that $\zeta_p \in K$. Let $H(i)$ be an R-Hopf order in KC_p with $0 < i < e'$. Then $H(i)$ is realizable as a Galois group.*

Proof. Exercise.   □

Proposition 10.4.6. *Assume that* $\zeta_{p^2} \in K$. *Let* $H = A(i, j, u)$ *be a Greither order in* KC_{p^2} *with* $e' > \mathrm{ord}(u-1) = i' + (j/p)$. *Then* H *is realizable as a Galois group.*

Proof. By Proposition 8.3.9, the linear dual of H is the R-Hopf order

$$A(i, j, u)^* = R\left[\frac{\gamma^p - 1}{\pi^{j'}}, \frac{\gamma a_{\tilde{u}} - 1}{\pi^{i'}}\right].$$

Observe that $e' > i' + (j/p) \geq 0$ guarantees that $A(i, j, u)$ and $A(j', i', \tilde{u})$ are local rings, and so Byott's Theorem will apply.

We claim that

$$R\left[\frac{\gamma a_{\tilde{u}} - 1}{\pi^{i'}}\right] = H^*.$$

Certainly $R\left[\frac{\gamma a_{\tilde{u}}-1}{\pi^{i'}}\right] \subseteq H^*$, so it suffices to show that

$$\mathrm{disc}\left(R\left[\frac{\gamma a_{\tilde{u}} - 1}{\pi^{i'}}\right]\right) = \mathrm{disc}\left(R\left[\frac{\gamma^p - 1}{\pi^{j'}}, \frac{\gamma a_{\tilde{u}} - 1}{\pi^{i'}}\right]\right).$$

By Proposition 8.1.1 (applied to H^*) and Proposition 4.4.13,

$$\mathrm{disc}\left(R\left[\frac{\gamma^p - 1}{\pi^{j'}}, \frac{\gamma a_{\tilde{u}} - 1}{\pi^{i'}}\right]\right) = (\pi^{p^2/(p-1)(j'+i')})^{p^2} R.$$

On the other hand, $\mathrm{disc}\left(R\left[\frac{\gamma a_{\tilde{u}}-1}{\pi^{i'}}\right]\right)$

$$= \frac{1}{\pi^{p^2(p^2-1)i'}}\mathrm{disc}\left(1, \gamma a_{\tilde{u}} - 1, (\gamma a_{\tilde{u}} - 1)^2, \ldots, (\gamma a_{\tilde{u}} - 1)^{p^2-1}\right)$$

$$= \frac{1}{\pi^{p^2(p^2-1)i'}}\mathrm{disc}\left(1, \gamma a_{\tilde{u}}, (\gamma a_{\tilde{u}})^2, \ldots, (\gamma a_{\tilde{u}})^{p^2-1}\right).$$

Now

$$(\gamma a_{\tilde{u}})^k = \sum_{pm+n=0}^{p^2-1} (\zeta_p^m u^{-n})^k e_{pm+n}$$

for $0 \leq k \leq p^2 - 1$, and so

$$\begin{pmatrix} 1 \\ \gamma a_{\tilde{u}} \\ (\gamma a_{\tilde{u}})^2 \\ \vdots \\ (\gamma a_{\tilde{u}})^{p^2-1} \end{pmatrix} = M \begin{pmatrix} e_0 \\ e_1 \\ e_2 \\ \vdots \\ e_{p^2-1} \end{pmatrix},$$

where M is the $p^2 \times p^2$ matrix whose $(pm + n + 1)$st column, $0 \leq m, n \leq p - 1$, is

$$\begin{pmatrix} 1 \\ \zeta_p^m u^{-n} \\ (\zeta_p^m u^{-n})^2 \\ (\zeta_p^m u^{-n})^3 \\ \vdots \\ (\zeta_p^m u^{-n})^{p^2 - 1} \end{pmatrix}.$$

Since

$$\operatorname{disc}(e_0, e_1, \ldots, e_{p^2 - 1}) = R,$$

it suffices to compute $(\det(M))^2$. Since M is Vandermonde,

$$\det(M) = \prod_{0 \leq pm + n < pm' + n' \leq p^2 - 1} (\zeta_p^{m'} u^{-n'} - \zeta_p^m u^{-n}).$$

But

$$\operatorname{ord}(\zeta_p^{m'} u^{-n'} - \zeta_p^m u^{-n}) = \operatorname{ord}(\zeta_p - 1) = e'$$

if $n = n'$, and

$$\operatorname{ord}(\zeta_p^{m'} u^{-n'} - \zeta_p^m u^{-n}) = \operatorname{ord}(u - 1) = i' + (k/p)$$

for all other cases. Thus,

$$\operatorname{ord}(\det(M)) = \left(\frac{p^2(p-1)}{2} \right) e' + \left(\frac{p^2(p^2-1)}{2} - \frac{p^2(p-1)}{2} \right) (i' + (k/p)),$$

and so

$$\operatorname{ord}\left(\operatorname{disc}\left(R\left[\frac{\gamma a_{\tilde{u}} - 1}{\pi^{i'}} \right] \right) \right) = p^2(p-1)e' + (p^2(p^2-1)$$

$$-p^2(p-1))(i' + (j/p)) - p^2(p^2-1)i'$$

$$= p^2(p-1)i + p^2(p-1)j$$

$$= \operatorname{ord}\left(\operatorname{disc}\left(R\left[\frac{\gamma^{p^2} - 1}{\pi^{j'}}, \frac{\gamma a_{\tilde{u}} - 1}{\pi^{i'}} \right] \right) \right),$$

which completes the proof. $\qquad \square$

A realizability result for Hopf orders in KC_{p^3} follows.

Proposition 10.4.7. *Let* $H = A(i, j, u)\left[\frac{gb-1}{\pi^k}\right]$ *be a Hopf order in* KC_{p^3} *with linear dual* $J = A(k', j', \tilde{w})\left[\frac{\gamma\beta-1}{\pi^{i'}}\right]$, *as constructed in Proposition 9.2.1. If*

(i) $\mathrm{ord}(1-u) = i' + \frac{i}{p}$ *and*

(ii) $\mathrm{ord}(\zeta_{p^3}s - 1) = i' + \frac{k}{p^2}$, *then* H *is realizable as a Galois group.*

Proof. The method here is similar to the proof of Proposition 10.4.6. For details, the reader is referred to [Un08b, §3].   □

10.5   The Normal Basis Theorem

In this section, we remain in the local situation where K is a finite extension of $\mathbb{Q}_p$, R is its ring of integers, π is a uniformizing parameter for R, and L is a Galois extension of K with group G and ring of integers S with uniformizing parameter λ. Set $n = [L : K]$. The integer $e = e_{L/K}$ denotes the ramification index of π in S; that is, $(\pi) = (\lambda)^e$.

We say that the extension L/K is **tame** if p does not divide e, and L/K is **wild** if p divides e. If $e = 1$, then L/K is unramified, and if $e = [L/K]$, then L/K is totally ramified.

Example 10.5.1. Let $K = \mathbb{Q}_p$, $L = \mathbb{Q}_p(\zeta_p)$. Then $(p) = (\zeta_p - 1)^{p-1}$, so that $e = p - 1$. Thus L/K is tame and totally ramified. On the other hand, if $L = \mathbb{Q}_p(\zeta_{p^2})$, then $e = p(p - 1)$, and so L/K is wild and totally ramified.

By Proposition 10.1.3, $L = K(\alpha)$, where α is a zero of an irreducible polynomial $q(x) \in K[x]$ of degree n. The set $\{1, \alpha, \alpha^2, \dots, \alpha^{n-1}\}$ is a K-basis for L, though obviously there are many other bases for L/K.

Let $G = \{1 = g_0, g_1, g_2, \dots, g_{n-1}\}$. Can we find a basis for L/K that is of the form $\{w, g_1(w), g_2(w), \dots, g_{n-1}(w)\}$ for an element $w \in L$? The answer is "yes," and this allows us to identify L with KG.

Proposition 10.5.1. *Let* L/K *be a Galois extension with group* G. *Then there exists an element* $w \in L$ *for which* $\{w, g_1(w), g_2(w), \dots, g_{n-1}(w)\}$ *is a basis for* L *over* K.

Proof. Let $1 = g_0, g_1, \dots, g_{n-1}$ denote the elements of G. Let $\{X_{g_i^{-1}g_j}\}$ be a collection of indeterminates where $0 \le i, j \le n - 1$. Note that there are only n distinct indeterminates $X_{g_0}, X_{g_1}, \dots, X_{g_{n-1}}$ in this collection. Let $(X_{g_i^{-1}g_j})$ denote the $n \times n$ matrix whose i, jth entry is $X_{g_i^{-1}g_j}$, and put

$$f(X_{g_0}, X_{g_1}, \dots, X_{g_{n-1}}) = \det((X_{g_i^{-1}g_j})).$$

Then f is not identically 0. Now define a subset B of $\underbrace{L \times L \times \cdots \times L}_{n}$ as

$$B = \{(g_0(x), g_1(x), \ldots, g_{n-1}(x)) : x \in L\}.$$

Since each g_i is an automorphism of L, $B = \underbrace{L \times L \times L \times \cdots \times L}_{n}$. Now, by [Wat79, Chapter 4, p. 29, Theorem], f cannot vanish on all of B. Consequently, there exists an element $w \in L$ for which

$$f(g_0(w), g_1(w), \ldots, g_{n-1}(w)) = \det((g_i^{-1} g_j(w))) \neq 0.$$

Next, let $a_0, a_1, \ldots, a_{n-1}$ be elements of K for which

$$g_0(w)a_0 + g_1(w)a_1 + \cdots + g_{n-1}(w)a_{n-1} = 0.$$

Then, for each i, $0 \leq i \leq n-1$,

$$g_i^{-1} g_0(w)a_0 + g_i^{-1} g_1(w)a_1 + \cdots + g_i^{-1} g_{n-1}(w)a_{n-1} = 0,$$

and so the homogeneous system

$$(g_i^{-1} g_j(w)) \begin{pmatrix} a_0 \\ a_1 \\ \vdots \\ a_{n-1} \end{pmatrix} = \begin{pmatrix} 0 \\ 0 \\ \vdots \\ 0 \end{pmatrix}$$

has only the trivial solution. It follows that $\{g_i(w)\}_{i=0}^{n-1}$ is a K-basis for L.   □

Proposition 10.5.1 is the classical **Normal Basis Theorem**. For example, we suppose that $K = \mathbb{Q}_p(\zeta_{p^3})$ and let $L = K(2^{\frac{1}{3}})$. Now, $E = \{1, 2^{\frac{1}{3}}, 2^{\frac{2}{3}}\}$ is a K-basis for L. Moreover, L/K is a Galois extension with group $C_3 = \{1, g, g^2\}$, where $g^i(2^{\frac{1}{3}}) = \zeta_{p^3}^i \cdot 2^{\frac{1}{3}}$ for $i = 0, 1, 2$. The Normal Basis Theorem says that there exists an element $w \in L$ for which $\{w, g(w), g^2(w)\}$ is a basis for L/K. Let us find such an element.

An element in L is of the form $b = a_0 + a_1 \cdot 2^{\frac{1}{3}} + a_2 \cdot 2^{\frac{2}{3}}$ for $a_0, a_1, a_2 \in K$. Now $1(b) = b$, $g(b) = a_0 + a_1 \zeta_{p^3} \cdot 2^{\frac{1}{3}} + a_2 \zeta_{p^3}^2 \cdot 2^{\frac{2}{3}}$, and $g^2(b) = a_0 + a_1 \zeta_{p^3}^2 \cdot 2^{\frac{1}{3}} + a_2 \zeta_{p^3} \cdot 2^{\frac{2}{3}}$. We want to find conditions on a_i such that the set $B = \{b, g(b), g^2(b)\}$ is linearly independent. With respect to the basis E, the coordinate matrix for B is

$$\begin{pmatrix} a_0 & a_0 & a_0 \\ a_1 & a_1 \zeta_{p^3} & a_1 \zeta_{p^3}^2 \\ a_2 & a_2 \zeta_{p^3}^2 & a_2 \zeta_{p^3} \end{pmatrix},$$

which has determinant

$$a_0(a_1a_2\zeta_{p^3}^2 - a_1a_2\zeta_{p^3}) - a_0(a_1a_2\zeta_{p^3} - a_1a_2\zeta_{p^3}^2) + a_0(a_1a_2\zeta_{p^3}^2 - a_1a_2\zeta_{p^3})$$

$$= a_0a_1a_2(\zeta_{p^3}^2 - \zeta_{p^3}),$$

and so B is linearly independent if and only if $a_0a_1a_2 \neq 0$; that is, if and only if $a_i \neq 0$ for $i = 0, 1, 2$. So we choose $a_0 = a_1 = a_2 = 1$, and see that $w = 1 + 2^{\frac{1}{3}} + 2^{\frac{2}{3}}$ is the required element.

The Normal Basis Theorem connects the structure of L to the structure of the group ring KG. The extension L/K is a module over KG with module action defined as

$$ax = \left(\sum_{g \in G} a_g g\right) x = \sum_{g \in G} a_g g(x) \tag{10.18}$$

for $x \in L$, $a = \sum_{g \in G} a_g g$, $a_g \in K$, so the Normal Basis Theorem can be restated as follows.

Proposition 10.5.2. *Let L/K be a Galois extension with group G. Then $L \cong KG$ as KG-modules.*

Now let S be the ring of integers of L and let R be the ring of integers of K. The ring extension S/R is an RG-module with the module structure given by restriction of (10.18) to S. But do we have an analog for the Normal Basis Theorem? That is, does there exist an element $w \in S$ for which $\{g_i(w)\}$ is an R-basis for S? In other words, when are $S \cong RG$ as RG-modules?

We have the following criterion, a classical result due to E. Noether.

Proposition 10.5.3. *(Noether's Theorem) Suppose L/K is Galois with group G. Then $S \cong RG$ as RG-modules if and only if L/K is tame.*

For example, if $L = \mathbb{Q}_p(\zeta_p)$, then $\text{Gal}(L/\mathbb{Q}_p) = C_{p-1}$ and $L/\mathbb{Q}_p$ is tame. Thus $S \cong \mathbb{Z}_p C_{p-1}$ as $\mathbb{Z}_p C_{p-1}$-modules.

Noether's Theorem can be restated as follows.

Proposition 10.5.4. *Suppose L/K is a Galois extension with group G. Then $S \cong RG$ as RG-modules if and only if $\int_{RG} S = R$, where $\int_{RG}$ is the ideal of integrals of the R-Hopf order RG in KG.*

Proof. Suppose $S \cong RG$ as RG-modules. Then, by Noether's Theorem, L/K is tame. Thus, by Proposition 10.2.13, the trace map tr : $S \to R$ is surjective. By the definition of the trace map, this is equivalent to $\Lambda S = R$, where $\Lambda = \sum_{g \in G} g$. Now, since $\int_{RG} = R\Lambda$, one has $\int_{RG} S = R$. The converse is left to the reader. $\square$

If L/K is wild, then Noether's Theorem does not apply: $S \not\cong RG$ as RG-modules. But the question remains, can we still characterize S as a Galois module?

The answer is "yes," but we will have to replace RG with a bigger order in KG, namely the **associated order**, which is defined as

$$A_{L/K} = \{x \in KG : xS \subseteq S\}.$$

To handle the wild case, we assume, as in (10.4), that K contains ζ_p, so that $e' = e/(p-1)$ is an integer. Choose an integer $0 \le i < e'$, and set $w = 1 + \pi^{pi+1}$. Then $L = K(z)$, $z = w^{\frac{1}{p}}$, is a cyclic extension with group $C_p = \langle g \rangle$, which is wild by Proposition 10.4.1.

Let S be the ring of integers in L. By Proposition 10.4.3, $S = R[x]$ with $x = \pi^{-i}(z-1)$, and $H(i') = R[\frac{g-1}{\pi^{i'}}]$ is an R-Hopf order in KC_p with $i' = e' - i$.

Proposition 10.5.5. $A_{L/K} = H(i')$ and $S \cong A_{L/K}$ as $A_{L/K}$-modules.

Proof. By Proposition 10.4.2, S is a Galois $H(i')$-extension of R, so there exists a bijection

$$j : S \otimes_R H(i') \to \text{End}_R(S),$$

defined as $j(s \otimes h)(t) = sh(t)$, for $s, t \in S$, $h \in H(i')$. Let ι be the inverse of j, and let $f \in A_{L/K}$. Then $f \in \text{End}_R(S)$, and $\iota(f) = \sum_{l=1}^{a} s_l \otimes h_l$, for $s_l \in S$, $h_l \in H(i')$, and so $f = \sum_{l=1}^{a} s_l h_l(t)$.

Since $f \in KG \cap SH(i') = H(i')$, $s_l \in R$, for all l, so that $f \in H(i')$. Thus $A_{L/K} \subseteq H(i')$. Clearly, $H(i') \subseteq A_{L/K}$, and so $A_{L/K} = H(i')$.

Since S is a left $H(i')$-module, S is a right $H(i)$-comodule with structure map $\alpha : S \to S \otimes_R H(i)$ given as

$$\alpha(x) = \sum_{i=1}^{p} h_i(x) \otimes f_i,$$

where $\{h_i\}$ is a basis for $H(i')$, with dual basis $\{f_i\}$. Now, from the map j, we obtain an R-module isomorphism

$$\eta : S \otimes_R S \to S \otimes_R H(i),$$

given as $\eta(s \otimes x) = (s \otimes 1)\alpha(x)$ for $s, x \in S$ (see [Ch00, Chapter 1, §2, (2.9) Proposition]). In fact, η is an isomorphism of $H(i')$-modules with $H(i')$ acting on the right factors.

Now, since S is a free R-module of rank p, $S \cong \underbrace{R \oplus R \oplus \cdots \oplus R}_{p}$, and so

$$\underbrace{(R \oplus R \oplus \cdots \oplus R)}_{p} \otimes_R S \cong \underbrace{(R \oplus R \oplus \cdots \oplus R)}_{p} \otimes_R H(i),$$

$$\underbrace{S \oplus S \oplus \cdots \oplus S}_{p} \cong \underbrace{H(i) \oplus H(i) \oplus \cdots \oplus H(i)}_{p},$$

as $H(i')$-modules. An application of the Krull-Schmidt Theorem [CR81, Introduction, §6B] shows that $S \cong H(i)$ as $H(i')$-modules.

Moreover, $H(i) \cong H(i')$ as $H(i')$-modules by Proposition 4.3.4, and so $S \cong H(i')$ as $H(i')$-modules. Since $0 \leq i < e'$ implies $0 < i' \leq e'$, $RC_p \subset H(i')$. $\square$

An R-basis for $H(i')$ is

$$\left\{ 1, \frac{g-1}{\pi^{i'}}, \left(\frac{g-1}{\pi^{i'}}\right)^2, \ldots, \left(\frac{g-1}{\pi^{i'}}\right)^{p-1} \right\},$$

and so there exists an element $w \in S = R[x]$ for which

$$\left\{ 1, \frac{g(w)-w}{\pi^{i'}}, \left(\frac{g(w)-w}{\pi^{i'}}\right)^2, \ldots, \left(\frac{g(w)-w}{\pi^{i'}}\right)^{p-1} \right\}$$

is a basis for S over R.

So we have seen that even though $S \not\cong RG$ as RG-modules, $S \cong H$ as H-modules, where H is an R-Hopf order in KG. A natural question arises: Is there an analog of Proposition 10.5.4 for Hopf orders?

Proposition 10.5.6. *Suppose L/K is a Galois extension with group G. Let H be an R-Hopf order in KG, and suppose that S is an H-extension. Then $S \cong H$ as H-modules if and only if $\int_H S = R$.*

Proof. (Sketch.) By [CH86, Theorem 5.4], $S \cong H^*$ as H-modules if and only if $\int_H S = R$. Now, by Proposition 4.3.4, $S \cong H$ as H-modules. $\square$

10.6 Chapter Exercises

Exercises for §10.1

1. Prove that $f(x) = x^{p-1} + x^{p-2} + \cdots + x^2 + x + 1$ is irreducible over $\mathbb{Q}$.
2. Let $f_7(x)$ denote the polynomial defined by (10.2) with $p = 7$. Compute the zeros of $f_7(x)$.
3. Let $f_p(x)$ be the polynomial defined by (10.2). Prove that the zeros of $f_p(x)$ are real numbers in the interval $[-2, 2]$.
4. Compute the Galois group of the splitting field of the polynomial $x^5 - 3$ over $\mathbb{Q}$.
5. Suppose that L/K is a Galois extension of degree p. Prove that $K(\alpha) = L$ for all $\alpha \in L \backslash K$.

Exercises for §10.2

6. Let p be a prime and let $K = \mathbb{Q}(\sqrt{p})$.

(a) For $p \equiv 1 \mod 4$, show that $R = \mathbb{Z}[\frac{1+\sqrt{p}}{2}]$.

(b) For $p \equiv 3 \mod 4$, show that $R = \mathbb{Z}[\sqrt{p}]$.

(c) For $q \neq p$, $q \geq 3$, suppose that Q is a prime of R that lies above q. Prove that $K(\sqrt{p})_Q$ is an unramified extension of $\mathbb{Q}_q$.

7. Let K be a finite extension of $\mathbb{Q}$.

(a) Prove that at most a finite number of primes ramify in K.

(b) Find a proof of Minkowski's result; that is, prove that at least one prime ramifies in $K \neq \mathbb{Q}$.

Exercises for §10.3

8. Let $K/\mathbb{Q}$ be a Galois extension with group G and ring of integers R. Prove that $RG \cong RG^*$ as RG-modules.

9. Prove Proposition 10.3.4.

10. Finish the proof of Proposition 10.3.5.

11. Prove the converse of Proposition 10.3.6.

Exercises for §10.4

12. Prove Proposition 10.4.5.

Exercise for §10.5

13. Prove the converse of Proposition 10.5.4.

Chapter 11
The Class Group of a Hopf Order

11.1 The Class Group of a Number Field

Let K be a finite extension of $\mathbb{Q}$ with ring of integers R. Let I be a fractional ideal in K, and let

$$I^{-1} = \{x \in K : xI \subseteq R\}.$$

By the discussion preceding Proposition 5.3.1, I^{-1} is a fractional ideal of R. Certainly, $I^{-1}I \subseteq R$. If $I^{-1}I = R$, then I is **invertible**.

Proposition 11.1.1. *Let I be a fractional ideal of R. Then I is invertible.*

Proof. Let P be a non-zero prime ideal of R. Then, by Corollary 4.2.2, R_P is a PID. Thus IR_P is a principal ideal of R_P, which can be written as R_{Pa} for some $a \in R_P$. We have $I^{-1}R_P = (IR_P)^{-1} = (R_{Pa})^{-1}$. Observe that R_{Pa} is invertible; that is, $(R_{Pa})^{-1} = R_{Pa}^{-1}$ satisfies

$$(R_{Pa})^{-1}R_{Pa} = R_P. \tag{11.1}$$

Since (11.1) holds for all non-zero P, it follows that $I^{-1}I = R$. $\qquad\square$

Let $\mathcal{F}(R)$ denote the collection of all fractional ideals of R. We define a binary operation on $\mathcal{F}(R)$ as follows. For $I, J \in \mathcal{F}(R)$, let

$$I * J = IJ,$$

where IJ is the collection of all finite sums $\sum ab$ with $a \in I, b \in J$. Then $\mathcal{F}(R)$ together with $*$ is an Abelian monoid with R playing the role of the identity element. In fact, as a result of Proposition 11.1, $\mathcal{F}(R)$ is an Abelian group; the inverse of I is I^{-1}.

R.G. Underwood, *An Introduction to Hopf Algebras*, DOI 10.1007/978-0-387-72766-0_11,   233
© Springer Science+Business Media, LLC 2011

Every principal ideal Ra is a fractional ideal with "principal" fractional inverse Ra^{-1}. The collection of all principal fractional ideals is a subgroup of $\mathcal{F}(R)$ denoted by $\mathcal{PF}(R)$. The quotient group

$$\mathcal{F}(R)/\mathcal{PF}(R)$$

is the **class group** of R, denoted by $\mathcal{C}(R)$.

For a given finite extension $K/\mathbb{Q}$, the class group $\mathcal{C}(R)$ is a finite Abelian group. In this section, we shall prove this important fact using topology. We begin with some preliminaries. We know that up to equivalence classes the absolute values on $\mathbb{Q}$ consist of

$$\{|\ |_2, |\ |_3, |\ |_5, \cdots, |\ |_\infty\}. \tag{11.2}$$

By Propositions 5.1.3 and 5.1.6, each absolute value on $\mathbb{Q}$ extends to a finite number of absolute values on K. For a prime $p \in \mathbb{Z}$, let $(p) = P_1^{e_1} P_2^{e_2} \cdots P_g^{e_g}$ be the factorization of (p), and let $[\ ]_{p,i}$ be the extension of $|\ |_p$ corresponding to the prime P_i. Let K_{P_i} denote the completion of K with respect to the $[\ ]_{p,i}$-topology. The ring of integers in K_{P_i} is $\hat{R}_{P_i}$. If π_i is a uniformizing parameter for $\hat{R}_{P_i}$, then

$$[\pi_i^{e_i}]_{p,i} = [p]_{p,i} = |p|_p = \frac{1}{p},$$

and so

$$[\pi_i]_{p,i} = \frac{1}{p^{1/e_i}}.$$

We want to normalize $[\ ]_{p,i}$ so that the absolute value of π_i is $\frac{1}{p^{f_i}}$, where $f_i = [R_{P_i}/P_i R_{P_i} : \mathbb{F}_p]$. To do this, we define an absolute value on K by the rule

$$|\ |_{p,i} = [\ ]_{p,i}^{n_i}, \quad n_i = e_i f_i,$$

where n_i is the local degree $[K_{P_i} : \mathbb{Q}_p]$. Then $|\ |_{p,i}$ is equivalent to $[\ ]_{p,i}$ with $|\pi_i|_{p,i} = \frac{1}{p^{f_i}}$; $|\ |_{p,i}$ is the **normalized discrete absolute value** on K corresponding to the prime P_i.

To normalize the extensions of the ordinary absolute value $|\ |_\infty$, we also employ the local degree. But in this case the completion of K with respect to an extension $\|\ \|_{\infty,i}$ of $|\ |_\infty$ is either $\mathbb{R}$ or $\mathbb{C}$. Since the completion of $\mathbb{Q}$ with respect to $|\ |_\infty$ is $\mathbb{R}$, the local degree is either 1 or 2. Therefore the **normalized Archimedean absolute value** on K is defined for $b \in K$ as

$$|b|_{\infty,i} = \begin{cases} \|b\|_{\infty,i} = |\lambda_i(b)| & \text{if } \lambda_i \text{ is a real embedding} \\ \|b\|_{\infty,i}^2 = |\lambda_i(b)|^2 & \text{if } \lambda_i \text{ is a complex embedding}, \end{cases}$$

where $|\ |$ is the ordinary absolute value on $\mathbb{C}$.

Let $\{| \; |_v\}_{v \in S}$ be the set of normalized absolute values on K, and let $\{| \; |_v\}_{v \in D}$ denote the subset of normalized discrete absolute values on K.

Definition 11.1.1. The **ideal group** I_K is the free Abelian group on the collection $\{| \; |_v\}_{v \in D}$ of normalized discrete absolute values on K.

Identifying $| \; |_v$ with v, I_K consists of all sums of the form

$$\sum_{v \in D} n_v v,$$

where $n_v = 0$ for all but a finite number of v.

The following is a critical proposition that describes fractional ideals in K as elements of I_K.

Proposition 11.1.2.   *(i) There is a 1-1 correspondence between fractional ideals in K and elements of I_K.*

(ii) There is a 1-1 correspondence between ideals of R and elements of I_K for which $n_v \geq 0$.

(iii) The element $\displaystyle\sum_{v \in D} n_v v \in I_K$ corresponds to a principal fractional ideal in K if and only if there exists an element $a \in K^\times$ for which $n_v = \mathrm{ord}_{P_v}(a)$ for all discrete v.

Proof. We prove (i) and leave (ii) and (iii) as exercises. By definition, a fractional ideal has the form cJ for $c \in K^\times$ and some ideal J in R. Without loss of generality, we may assume that $c^{-1} \in R$. Let $J = Q_1^{n_1} Q_2^{n_2} \cdots Q_l^{n_l}$ be the unique factorization of J into prime ideals of R, and let $(c^{-1}) = P_1^{m_1} P_2^{m_2} \cdots P_k^{m_k}$ be the unique factorization of the principal ideal (c^{-1}) into prime ideals. For $i = 1, \ldots l$, let v_i be the normalized absolute value corresponding to the prime Q_i, and for $j = 1, \ldots, k$, let ω_j be that for the prime P_j. Define a map $\Psi : \mathcal{F}(R) \to I_K$ by the rule

$$\Psi(cJ) = \sum_{j=1}^{k} (-m_j) \omega_j + \sum_{i=1}^{l} n_i v_i.$$

Next, let $\displaystyle\sum_{v \in D} n_v v$ be an element of I_K. Let $v_1, v_2, \ldots, v_l$ denote the indices corresponding to integers with $n_v < 0$, and let $v_{l+1}, v_{l+2}, \ldots, v_m$ be the indices with $n_v > 0$. For $i = 1, \ldots, m$, let P_{v_i} be the prime ideal corresponding to v_i. Then

$$J = \prod_{j=1}^{m-l} P_{v_{l+j}}^{n_{l+j}}$$

is an ideal of R. Moreover,

$$M = \prod_{i=1}^{l} P_{v_i}^{n_i}$$

is an R-submodule of K that is finitely generated. By Proposition 4.2.2, M is a fractional ideal of R, which necessarily has the form cR for some $c \in K^\times$, $c^{-1} \in R$. Define a map $\Phi : I_K \to \mathcal{F}(R)$ by the rule

$$\Phi\left(\sum_{v \in D} n_v v\right) = cJ.$$

One has $\Psi\Phi = \mathrm{id}_{I_K}$ and $\Phi\Psi = \mathrm{id}_{\mathcal{F}(R)}$, and so Ψ is a bijection. $\square$

Again, we let $\{|\ |_v\}_{v \in S}$ be the set of normalized absolute values on K. For each $v \in S$, let K_v denote the completion of K with respect to the $|\ |_v$-topology on K. Let R_v denote the ring of integers in K_v. For each discrete $|\ |_v$, R_v is a compact subset of K_v. Moreover, R_v is open since $K_v \backslash R_v$ is a closed subset of K_v.

Definition 11.1.2. The **adele ring** V_K of K is the restricted product of the K_v, $v \in S$, with respect to the family $\{R_v : v \in D\}$, together with the restricted product topology on $\prod_{v \in S} K_v$ with respect to the family $\{R_v : v \in D\}$.

Consequently, V_K consists of those vectors $\{a_v\} \in \prod_{v \in S} K_v$ for which $a_v \in R_v$ for all but finitely many v.

A basis for the topology on V_K consists of open sets of the form $\prod_{v \in S} U_v$, where U_v is open in K_v for all v and $U_v = R_v$ for all but finitely many v.

The ring structure of V_K is given componentwise. For $\{a_v\}, \{b_v\} \in V_K$,

$$\{a_v\} + \{b_v\} = \{(a_v + b_v)_v\}, \quad \{a_v\}\{b_v\} = \{(a_v b_v)_v\}.$$

In fact, V_K is a commutative ring.

Proposition 11.1.3. *There is an injection of rings $K \to V_K$.*

Proof. Let $a \in K$, and write $a = b/c$, where $b, c \in R$. Since there are only a finite number of prime divisors of c, $\frac{1}{c} \in R_v$ for all but a finite number of v. Thus $\{a_v\}$, with $a_v = a$ for all $v \in S$, is an adele of K. Moreover, the map $a \to \{a_v\}$ is a ring injection. $\square$

Let $\{1_v\}$ be defined as $1_v = 1$ for all $v \in S$. Then $\{1_v\}$ is in V_K, and so V_K is a commutative ring with unity.

Definition 11.1.3. The multiplicative group of units of V_K is the **idele group of K,** denoted by J_K.

Proposition 11.1.4. *The map $C : J_K \to \mathbb{R}_{>0}$ defined as*

$$C(\{a_v\}) = \prod_{v \in S} |a_v|_v$$

is a homomorphism of multiplicative groups.

Proof. Exercise.   □

The map C is called the **content** of J_K. The **kernel of the content map** is denoted by J_K^1.

Let $\{a_v\} \in V_K$, and let $W_{\{a_v\}}$ be the subset of V_K defined as

$$W_{\{a_v\}} = \{\{x_v\} \in V_K : |x_v|_v \leq |a_v|_v \text{ for all } v \in S\}.$$

Then $W_{\{a_v\}}$ is a compact subset of V_K. An important lemma follows.

Lemma 11.1.1. *For each finite extension $K/\mathbb{Q}$, there exists a positive integer κ such that if $\{a_v\}$ is an idele in J_K with $C(\{a_v\}) > \kappa$, then $W_{\{a_v\}}$ contains an idele $\{\eta_v\}$ in J_K for which $\eta_v = \eta$ for some $\eta \in K$ for all $v \in S$.*

Proof. For a proof, see [CF67, p. 66, Lemma].   □

Our immediate goal is to show that an embedding of rings $K \to V_K$ restricts to an embedding of groups $K^\times \to J_K^1$. We need a lemma.

For each absolute value $|\ |_u$ on $\mathbb{Q}$ listed in (11.2), let $\mathbb{Q}_u$ denote the completion of $\mathbb{Q}$ with respect to $|\ |_u$, and for $v \in S$ let $v|u$ indicate that the absolute value $|\ |_v$ on K is an extension of $|\ |_u$. Let n_v denote the local degree $[K_v : \mathbb{Q}_u]$, defined whenever $v|u$.

We define a norm on $K/\mathbb{Q}$ as follows. We have $K = \mathbb{Q}(\alpha)$ for some $\alpha \in K$ with irreducible polynomial $f(x)$. Let $\lambda_1, \ldots, \lambda_l$ denote the set of embeddings of K into the splitting field of $f(x)$ over $\mathbb{Q}$. We define the **norm of $K/\mathbb{Q}$** to be the map $N_{K/\mathbb{Q}} : K \to \mathbb{Q}, x \mapsto \prod_{i=1}^l \lambda_i(x)$.

The following lemma relates the normalized valuations $|\ |_v$ to the global norm.

Lemma 11.1.2. *Let $b \in K^\times$. Then $|N_{K/\mathbb{Q}}(b)|_u = \prod_{v|u} |b_v|_v$.*

Proof. See [CF67, Theorem, p. 59] for a proof.   □

Proposition 11.1.5. *The injection of K into V_K restricts to an injection of $K^\times$ into J_K^1. In other words, for $b \in K^\times$, the idele $\{b\}$ is in J_K with $C(\{b\}) = 1$.*

Proof. Since $b \neq 0$, $\{b\} \in J_K$. We have

$$\prod_{v \in S} |b|_v = \prod_u \left(\prod_{v|u} |b|_v \right)$$

$$= \prod_u |N_{K/\mathbb{Q}}(b)|_u \quad \text{by Lemma 11.1.2.}$$

Observe that $r = N_{K/\mathbb{Q}}(b) \in \mathbb{Q}^\times$. We claim that $C(\{r\}) = 1$, where $C : J_\mathbb{Q} \to \mathbb{R}_{>0}$ is the content map of the idele group of $\mathbb{Q}$. We have

$$r = \frac{\pm p_1^{e_1} p_2^{e_2} \cdots p_m^{e_m}}{q_1^{f_1} q_2^{f_2} \cdots q_l^{f_l}},$$

where all of the primes p_i, q_j are distinct and the e_i, f_j are positive integers.

For discrete u not corresponding to one of the p_i or q_j, we have $|r|_u = 1$; for discrete u corresponding to a prime p_i, we have $|r|_u = \frac{1}{p_i^{e_i}}$; and for discrete u corresponding to a prime q_j, we obtain $|r|_u = q_j^{f_j}$. Also, $|r|_u = r$ for $u = \infty$, and so

$$C(\{r\}) = \prod_{i=1}^{m} \frac{1}{p_i^{e_i}} \prod_{j=1}^{l} q_j^{f_j} |r|_\infty = 1,$$

which completes the proof of the proposition. $\square$

In view of Proposition 11.1.5, we identify $K^\times$ with a subgroup of J_K^1. Since $J_K^1 \subseteq V_K$, we can endow J_K^1 with the V_K-subspace topology. Through the canonical surjection $s : J_K^1 \to J_K^1/K^\times$, we give $J_K^1/K^\times$ in the quotient topology. Thus a set U is open in $J_K^1/K^\times$ if $s^{-1}(U)$ is open in J_K^1.

Proposition 11.1.6. $J_K^1/K^\times$ *in the quotient topology is compact.*

Proof. Let $\{a_v\}$ be an adele in V_K for which $C(\{a_v\}) > \kappa$ (κ as in Lemma 11.1.1), and let $W_{\{a_v\}}$ be the compact set defined as

$$W_{\{a_v\}} = \{\{x_v\} \in V_K : |x_v|_v \le |a_v|_v \text{ for all } v \in S\}.$$

Now, let $\{b_v\}K^\times$ be an element of $J_K^1/K^\times$. Then

$$C(\{b_v^{-1} a_v\}) = C(\{b_v^{-1}\})C(\{a_v\}) = C(\{a_v\}) > \kappa,$$

and so, by Lemma 11.1.1, there exists an idele $\{\eta\} \in K^\times \subseteq J_K^1$ with $\{\eta\} \in W_{\{b_v^{-1} a_v\}}$. Thus, for all v,

$$|\eta|_v \le |b_v^{-1} a_v|_v = |b_v^{-1}|_v |a_v|_v,$$

and so $|b_v|_v |\eta|_v \le |a_v|_v$. Thus $\{\eta\}\{b_v\} \in W_{\{a_v\}} \cap J_K^1$.

Since $s(\{\eta\}\{b_v\}) = s(\{b_v\})$, the canonical surjection s restricts to a surjection

$$s : W_{\{a_v\}} \cap J_K^1 \to J_K^1/K^\times,$$

which is a continuous map of topological spaces. Now, by [CF67, p. 69, Lemma], J_K^1 is closed in V_K, and therefore $W_{\{a_v\}} \cap J_K^1$ is closed in the compact set $W_{\{a_v\}}$. Hence $W_{\{a_v\}} \cap J_K^1$ is compact by Proposition 2.2.4, and so, by Proposition 2.2.5, $s(W_{\{a_v\}} \cap J_K^1) = J_K^1/K^\times$ is compact. $\square$

Recall that I_K is the ideal group of K. We endow I_K with the discrete topology: an open set is any subset of I_K. A basis for this topology consists of all singleton subsets of I_K.

Proposition 11.1.7. *There is a surjective group homomorphism* $\varrho : J_K^1 \to I_K$ *defined as*

$$\varrho(\{a_v\}) = \sum_{v \in D} \text{ord}_{P_v}(a_v) v,$$

where $r = \mathrm{ord}_{P_v}(a_v)$ *satisfies* $(a_v) \subseteq P_v^r$, $(a_v) \not\subseteq P_v^{r+1}$. *Moreover, the map* ϱ *is a continuous map of topological spaces.*

Proof. For each $v \in D$, let π_v be a uniformizing parameter for R_v such that $\mathrm{ord}_v(\pi_v) = 1$. Let $\sum_{v \in D} n_v v$ be an element of I_K. Let $a_v = \pi_v^{n_v}$ for $v \in D$, and for $v \in S \backslash D$ choose a_v such that $C(\{a_v\}) = 1$. Then $\{a_v\}$ is an idele with $\varrho(\{a_v\}) = \sum_{v \in D} n_v v$, and so ϱ is surjective.

We leave the proof that ϱ is a continuous group homomorphism to the reader.   □

Proposition 11.1.8. *The class group* $C(R)$ *is isomorphic to* $I_K/\varrho(K^\times)$.

Proof. By Proposition 11.1.2 (i), I_K corresponds to the collection of fractional ideals $\mathcal{F}(R)$, and, by Proposition 11.1.2 (iii), $\varrho(K^\times)$ corresponds to the subgroup $\mathcal{PF}(R)$ of principal fractional ideals $\mathcal{F}(R)$.   □

Next, through the canonical surjection $I_K \to I_K/\varrho(K^\times)$, we endow $I_K/\varrho(K^\times)$ with the quotient topology. The quotient topology on $I_K/\varrho(K^\times)$ is equivalent to the discrete topology since I_K has the discrete topology. We can now prove that the class group is finite.

Proposition 11.1.9. *Let K be an algebraic extension of $\mathbb{Q}$. Then the class group $C(R)$ is finite.*

Proof. The surjection $\varrho : J_K^1 \to I_K$ of Proposition 11.1.7 induces a continuous surjective map $J_K^1/K^\times \to I_K/\varrho(K^\times)$. Now, by Proposition 11.1.6, $J_K^1/K^\times$ is compact, and so, by Proposition 2.2.5, $I_K/\varrho(K^\times)$ is compact in the discrete topology.

The collection $\{x\}$, $x \in I_K/\varrho(K^\times)$, is an open covering of $I_K/\varrho(K^\times)$, from which we can extract a finite subcover since $I_K/\varrho(K^\times)$ is compact. Therefore, $C(R) \cong I_K/\varrho(K^\times)$ is finite.   □

The order of the finite group $C(R)$ is the **class number of** K, denoted by $h(K)$. If K^+ is the maximal real subfield of K with ring of integers R^+, then the order of $Cl(R^+)$ is denoted by $h^+(K)$. Let $K = \mathbb{Q}(\zeta_p)$. The prime p satisfies **Vandiver's Conjecture** if p does not divide $h^+(\mathbb{Q}(\zeta_p))$. It is known that Vandiver's Conjecture holds for primes $< 4{,}000{,}000$ [Wa97].

The order of $C(R)$ measures how far R is from being a PID. It is not hard to show that R is a PID if and only if $C(R) = 1$. (Prove this as an exercise.)

Said differently, $C(R)$ is non-trivial if and only if there exists a finitely generated R-submodule of K that is not a free R-module. In fact, since every fractional ideal in R is a projective R-module, the existence of a non-trivial element in $C(R)$ indicates the existence of a locally free R-submodule of K that is not free over R.

There are many number fields K for which $C(R)$ is non-trivial. For example, take $K = \mathbb{Q}(\sqrt{-5})$. Then $R = \mathbb{Z}[\sqrt{-5}]$. It is well-known that R is not a PID, and thus there exists a locally free R-submodule of K that is not free over R. (Can we come up with a specific example?)

11.2   The Class Group of a Hopf Order

Let K be a finite extension of $\mathbb{Q}$ with ring of integers R. The main result in the previous section was to prove that the class group $\mathcal{C}(R)$ is finite. In this section, we generalize $\mathcal{C}(R)$ to Hopf orders in KG, where $G = C_{p^n}$, for a fixed integer $n \geq 0$. We follow the method of A. Fröhlich given in [Fro83, Chapter I, §2]. We know that R can be identified with the R-Hopf order RG in KG in the case $n = 0$, and so $\mathcal{C}(R)$ is the class group of the R-Hopf order $R1$ in $K1$. We want to define the class group for R-Hopf orders in KG, where G is non-trivial.

We have seen that $\mathcal{C}(R) \cong I_K / \varrho(K^\times)$, where I_K is the ideal group and $\varrho(K^\times)$ is the image of $K^\times$ under the surjective map $\varrho : J_K^1 \to I_K$. In fact, there is a surjective map $\varphi : J_K \to I_K$. Let U_K be the kernel of φ such that $I_K = J_K / U_K$. Since $\varrho(K^\times) \cong (K^\times U_K) / U_K$,

$$
\begin{aligned}
\mathcal{C}(R) &\cong I_K / \varrho(K^\times) \\
&\cong (J_K / U_K) / ((K^\times U_K) / U_K) \\
&\cong J_K / (K^\times U_K).
\end{aligned} \tag{11.3}
$$

In this form, $\mathcal{C}(R)$ can be generalized to Hopf orders in KG.

Let n be an integer, and let K be a finite extension of $\mathbb{Q}$. Let S be the set of normalized absolute values on K, and let D be the subset of normalized discrete absolute values on K. Let K_v denote the completion of K with respect to $|\ |_v$. For discrete, v let R_v be the ring of integers in K_v. We adopt the convention that for v Archimedean we define $R_v = K_v$. Let H be an R-Hopf order in KG with G finite and Abelian. For $v \in S$, put $H_v = R_v \otimes_R H$.

For our purposes (and for simplicity), we assume that K contains ζ_{p^n}. By [Se77, Theorem 7], there are p^n irreducible characters of G; they are precisely the homomorphisms

$$
\gamma_m : G \to K,
$$

defined as $\gamma_m(g^l) = \zeta_{p^n}^{ml}$ for $0 \leq m, l \leq p^n - 1$. By linearity, each character γ_m determines a map (also denoted by γ_m)

$$
\gamma_m : K_v G \to K_v,
$$

which restricts to a homomorphism

$$
\gamma_m : U(K_v G) \to K_v^\times,
$$

where $U(K_v G)$ denotes the group of units in $K_v G$. The group of **virtual characters** of G, denoted by C_G, is the free Abelian group on the set of irreducible characters $\{\gamma_m\}_{m=0}^{p^n-1}$. Let $\mathrm{Hom}(C_G, J_K)$ denote the collection of group homomorphisms from C_G to J_K. Then $\mathrm{Hom}(C_G, J_K)$ is a group with group operation defined pointwise: for $f, g \in \text{Hom}(C_G, J_K)$, $(fg)(x) = f(x)g(x)$. Observe that $\text{Hom}(C_G, K^{\times})$ is a subgroup of $\text{Hom}(C_G, J_K)$.

Let M be an H-module. Then M is **locally free and of rank one** if $M_{\nu} = R_{\nu} \otimes_R M$ is a free H_{ν}-module of rank one for all $\nu \in S$ and $M_K = K \otimes_R M$ is a free KG-module of rank one. In what follows, we show that M determines an element of $\text{Hom}(C_G, J_K)$.

Since M is locally free and of rank one, for each $\nu \in S$ there exists an element x_{ν} in M_{ν} for which

$$M_{\nu} = H_{\nu} x_{\nu},$$

and there exists an element x_0 in M_K for which

$$M_K = KG x_0.$$

Consequently, for each $\nu \in S$ there is an element $\lambda_{\nu,M} \in U(K_{\nu}G)$ for which

$$\lambda_{\nu,M} x_0 = x_{\nu}.$$

One has $\lambda_{\nu,M} \in U(H_{\nu})$ for all but a finite number of ν, and so $\{\lambda_{\nu,M}\}_{\nu \in S}$ determines an idele in the idele group

$$J_{KG} = \left\{ \{\lambda_{\nu}\} \in \prod_{\nu \in S} U(K_{\nu}G) : \lambda_{\nu} \in U(H_{\nu}) \text{ almost everywhere} \right\}.$$

Let $\lambda = \{\lambda_{\nu}\} \in J_{KG}$. For each $\lambda_{\nu} \in U(K_{\nu}G)$, one has the map

$$h_{\lambda_{\nu}} : C_G \to K_{\nu}^{\times},$$

defined as

$$h_{\lambda_{\nu}} \left(\sum_{m=0}^{p^n-1} n_m \gamma_m \right) = \prod_{m=0}^{p^n-1} (\gamma_m(\lambda_{\nu}))^{n_m}.$$

One easily checks that $h_{\lambda_{\nu}} \in \text{Hom}(C_G, K_{\nu}^{\times})$ with $h_{\lambda_{\nu}} \in \text{Hom}(C_G, R_{\nu}^{\times})$ for all but a finite number of ν. Passing to the product over all normalized absolute values, we obtain an element of $\text{Hom}(C_G, J_K)$ that we denote as $\mathbf{h}(\lambda)$. If $\lambda \in J_{KG}$ arises from the locally free H-module M (that is, if $\lambda = \{\lambda_{\nu,M}\}_{\nu \in S}$ for some M), we write $\mathbf{h}(\lambda_M)$.

Let

$$U_H = \prod_{\nu \in S} U(H_{\nu}),$$

and let $y = \{y_{\nu}\} \in U_H$. For each $\nu \in S$, $y_{\nu} \in U(K_{\nu}G)$, and so the construction above applies to yield an element $h_{y_{\nu}} \in \text{Hom}(C_G, U(R_{\nu})) \subseteq \text{Hom}(C_G, K_{\nu}^{\times})$. Passing to the product yields an element $\mathbf{h}(y) \in \text{Hom}(C_G, U_K) \subseteq \text{Hom}(C_G, J_K)$, where U_K is the kernel of $\varphi : J_K \to I_K$. Evidently, $\mathbf{h}(U_H)$ is a subgroup of $\text{Hom}(C_G, J_K)$.

We can now give a definition of the class group of H.

Definition 11.2.1. Let $n \geq 1$ be an integer, and assume that $\zeta_{p^n} \in K$. Let H be an R-Hopf order in KC_{p^n}. Then, the **class group of** H is defined as

$$\mathcal{C}(H) = \frac{\mathrm{Hom}(C_G, J_K)}{\mathrm{Hom}(C_G, K^\times)\mathbf{h}(U_H)}.$$

Indeed, recalling the isomorphism (11.3), we see that $\mathcal{C}(H) = \mathcal{C}(R)$ in the case $H = R$, with $\mathrm{Hom}(C_G, J_K)$ playing the role of J_K, $\mathbf{h}(U_H)$ playing the role of U_K, and $\mathrm{Hom}(C_G, K^\times)$ analogous to $K^\times$.

It is helpful to construct an example in which $\mathcal{C}(H)$ is non-trivial. Let $K = \mathbb{Q}(\zeta_{p^2})$ with ring of integers R. Put $P = (1 - \zeta_{p^2})$. Then $pR = P^{p(p-1)}$, and so $e' = p$. Let $G = C_p = \langle g \rangle$, and let i be an integer $0 \leq i < p$. Then $H(i) = R\left[\frac{g-1}{P^i}\right]$ is an R-Hopf order in KG. Let v be the normalized discrete absolute value corresponding to P. Let $H(i)_v = R_v \otimes_R H(i)$. Then $H(i)_v = R_v\left[\frac{g-1}{\pi^i}\right]$, where π is a uniformizing parameter for R_v. Let u be an element of R_v for which $i' > \mathrm{ord}(1 - u) \geq 0$. Then, by Lemma 7.1.6,

$$a_u = \sum_{m=0}^{p-1} u^m e_m \in R_v\left[\frac{g-1}{\pi^{e'}}\right] \backslash H(i)_v,$$

where $\{e_m\}$ is the set of minimal idempotents in $K_v G$. Let

$$\lambda_v = \begin{cases} 1 & \text{if } v \text{ does not correspond to } P \\ a_u & \text{if } v \text{ corresponds to } P. \end{cases}$$

Then $\lambda = \{\lambda_v\} \in J_{KG}$. Define

$$M = H(i)\lambda = KG \cap \left(\bigcap_{v \in S} H(i)_v \lambda_v\right)$$

(see [CR81, §31B, p. 652]). Then $\mathbf{h}(\lambda_M)$ is a non-trivial element in $\mathcal{C}(H(i))$.

A critical observation that we will need presently is the following. Let $\{[M]\}$ be the collection of all isomorphism classes of locally free rank one H-modules M. Let M and N be representatives of classes in $\{[M]\}$. Let $\mathbf{h}(\lambda_M)$ and $\mathbf{h}(\lambda_N)$ be the elements of $\mathrm{Hom}(C_G, J_K)$ determined by M and N, respectively.

Lemma 11.2.1. *If $M \cong N$, then $\mathbf{h}(\lambda_M)$ and $\mathbf{h}(\lambda_N)$ lie in the same class in $\mathcal{C}(H)$.*

Proof. For $v \in S$, there exist elements $x_v \in M_v$ and $y_v \in N_v$ for which

$$M_v = H_v x_v \quad \text{and} \quad N_v = H_v y_v,$$

and elements $x_0 \in M_K$ and $y_0 \in N_K$ for which

$$M_K = KGx_0 \quad \text{and} \quad N_K = KGy_0.$$

Also, there exist unique elements $\lambda_{v,M}, \lambda_{v,N} \in U(K_v G)$, for which

$$\lambda_{v,M}x_0 = x_v \quad \text{and} \quad \lambda_{v,N}y_0 = y_v.$$

Now, since $M_v \cong N_v$ and $M_K \cong N_K$, there exist elements $\beta_v \in U(H_v)$ and $\eta_0 \in U(KG)$, for which $\beta_v x_v = y_v$ and $\eta_0 x_0 = y_0$. Thus

$$(\beta_v^{-1}\lambda_{v,N}\eta_0)x_0 = x_v,$$

and so $\lambda_{v,M} = \beta_v^{-1}\lambda_{v,N}\eta_0 = \lambda_{v,N}\eta_0\beta_v^{-1}$. Passing to the product over all $v \in S$ yields

$$\mathbf{h}(\lambda_M) = \mathbf{h}(\lambda_N)\theta\mathbf{h}(y),$$

where $y \in U_H$ and $\theta \in \text{Hom}(C_G, K^\times)$.   □

Like $\mathcal{C}(R)$, $\mathcal{C}(H)$ is finite and Abelian.

Proposition 11.2.1. *Let $n \geq 1$ be an integer, let K be a finite extension of $\mathbb{Q}$ containing ζ_{p^n}, and let G be the cyclic group of order p^n. Let H be an R-Hopf order in KG. Then the class group $\mathcal{C}(H)$ is a finite Abelian group.*

Proof. Let $\{[M]\}$ denote the collection of isomorphism classes of locally free rank one H-modules M that are finitely generated and projective R-modules satisfying $KM \cong KG$. By the Jordan-Zassenhaus Theorem [CR81, (24.1)], $|\{[M]\}| < \infty$.
   Let $[M], [N] \in \{[M]\}$. If $[M] = [N]$, then

$$\mathbf{h}(\lambda_M) \in \mathbf{h}(\lambda_N)\text{Hom}(C_G, K^\times)\mathbf{h}(U_H)$$

by Lemma 11.2.1. Thus, there is a well-defined map

$$\Omega : \{[M]\} \to \mathcal{C}(H),$$

where $\Omega([M])$ is the class of $\mathbf{h}(\lambda_M)$ in $\mathcal{C}(H)$.
   We show that Ω is surjective. Let $f\text{Hom}(C_G, K^\times)\mathbf{h}(U_H) \in \mathcal{C}(H)$. Then $f \in \text{Hom}(C_G, J_K)$ is determined by its values on the characters γ_m, $m = 0, \ldots p^n - 1$. For each m, $f(\gamma_m) = \{\lambda_{v,m}\}$ is an idele in J_K. There is a subset of ideles $\{\{\lambda_{v,m}\}\}_{m=0}^{p^n-1} \subseteq J_K$.
   For each $v \in S$, let

$$\lambda_v = \lambda_{v,0}e_0 + \lambda_{v,1}e_1 + \cdots + \lambda_{v,p^n-1}e_{p^n-1},$$

where e_m are the idempotents in KC_{p^n}. Then $\lambda = \{\lambda_v\}$ is an idele in J_{KG}. Define

$$M = H\lambda = KG \cap \left(\bigcap_{v \in S} H_v \lambda_v\right).$$

Then M is an H-module that is finitely generated and projective as an R-module. Thus, M is a locally free rank one H-module. We have $\mathbf{h}(\lambda_M) = f$, and so

$$\Omega([M]) = \mathbf{h}(\lambda_M)\mathrm{Hom}(C_G, K^\times)\mathbf{h}(U_H)$$
$$= f\,\mathrm{Hom}(C_G, K^\times)\mathbf{h}(U_H). \qquad \square$$

We consider the class group $\mathcal{C}(H)$ with $H = RG$, where $G = C_{p^n}$. Suppose L/K is a Galois extension with group G, and assume that L/K is tame; that is, assume that each prime Q of R is tamely ramified in S, the ring of integers in L. By Noether's Theorem, S is a locally free rank one RG-module. As such, S gives rise to a class (S) equal to the class of $\mathbf{h}(\lambda_S)$ in the locally free class group $\mathcal{C}(RG)$.

Definition 11.2.2. The set of **realizable classes** $\mathcal{R}(RG)$ in $\mathcal{C}(RG)$ is the collection of classes in $\mathcal{C}(RG)$ of the form (S), where S is the ring of integers in a tame extension L/K with group G.

In fact, $\mathcal{R}(RG)$ is a subgroup of $\mathcal{C}(RG)$. For a proof of this fact, see L. McCulloh's paper [Mc87]. If $\mathcal{R}(RG)$ is non-trivial, then there exists a Galois extension L/K with group G for which S is not free over RG.

We want to find an analog of $\mathcal{R}(RG)$ for Hopf orders H in KG. To do this, we employ the **semilocalization** of R at p, which is defined as

$$R_p = \{r/s : r, s \in R, (s, p) = 1\}.$$

In R_p, there are only a finite number of maximal ideals, namely those prime ideals of R that lie above p. Let $H_p = R_p \otimes_R H$, and let $X^{(p)}$ be a Galois H_p-extension of R_p. Then $X = K \otimes_{R_p} X^{(p)}$ is a Galois KG-extension of K. Let R_X denote the integral closure of R in X. There is an R-order of the form

$$\Theta = X^{(p)} \cap R_X,$$

which is a **semilocal Galois H-extension** of R.

The collection of all semilocal Galois H-extensions of R is denoted by $SGE(H, R)$. By Proposition 10.3.7, H^* is a semilocal Galois H-extension of R.

Proposition 11.2.2. *Each $\Theta \in SGE(H, R)$ is a locally free rank one H-module.*

Proof. (Sketch.) One first shows that Θ is a Galois H-extension. It then follows that $\int_H \Theta = R$. Now, by [CH86, Theorem 5.4], Θ is locally free and of rank one over H. $\qquad\square$

By Proposition 11.2.2, Θ corresponds to a class $(\Theta) \in C(H)$. There exists a map

$$\Psi : SGE(H, R) \to C(H), \quad \Theta \mapsto (\Theta)(H^*)^{-1},$$

which is the **class-invariant map**.

We now have an analog of $\mathcal{R}(RG)$.

Definition 11.2.3. The image of the class-invariant map $\Psi(SGE(H, R))$ is the set of **realizable classes** $\mathcal{R}(H)$ **in** $C(H)$.

By results of N. Byott [By95, Theorem 5.2] and L. McCulloh [Mc83], the image of the class-invariant map $\Psi : SGE(RG, R) \to C(RG)$ is precisely $\mathcal{R}(RG)$, and thus Definition 11.2.3 is a proper generalization of $\mathcal{R}(RG)$.

For the remainder of this section, we assume that $G = C_p$ and $K = \mathbb{Q}(\zeta_{p^2})$. Let R be the ring of integers in K. Then $P = (1 - \zeta_{p^2})$ is the unique prime ideal of R lying above p. One has $\mathrm{ord}_P(p) = e = p(p-1)$, and so $e' = p$. For each j, $0 < j \le p$, there is an R-Hopf order H in KC_p of the form

$$H = H(j) = R\left[\frac{g-1}{P^j}\right].$$

The Hopf order $H(j)$ is a free R-module of rank p with basis

$$\left\{1, \frac{g-1}{(1-\zeta_{p^2})^j}, \left(\frac{g-1}{(1-\zeta_{p^2})^j}\right)^2, \ldots, \left(\frac{g-1}{(1-\zeta_{p^2})^j}\right)^{p-1}\right\}.$$

The linear dual of $H(j)$ is a rank p Hopf order of the form

$$H(j') = R\left[\frac{\gamma-1}{(1-\zeta_{p^2})^{j'}}\right], \quad \langle\gamma\rangle = \hat{C}_p.$$

Proposition 11.2.3. *There exists a semilocal Galois $H(j)$-extension that is integrally closed.*

Proof. Let $w = 1 + (1 - \zeta_{p^2})^{pj'+1}$, and put $L = K(z)$, where $w = z^p$. Let $x = (z-1)/(1-\zeta_{p^2})^j$. Then $R_p[x]$ is a Galois $H(j)_p$-extension of R_p (Proposition 10.4.2). Moreover, $R_p[x]$ is the integral closure of R_p in L (Proposition 10.4.3).

One has $L = K \otimes_{R_p} R_p[x]$. Let S be the ring of integers in L. Then $S = R_p[x] \cap S$, so that S is a semilocal $H(j)$-extension that is integrally closed. $\qquad\square$

Proposition 11.2.4. *Let $K = \mathbb{Q}(\zeta_{p^2})$ with ring of integers R. Put $P = (1 - \zeta_{p^2})$. For an integer $0 < j \le e' = p$, let $H(j) = R\left[\frac{g-1}{P^j}\right]$ be an R-Hopf order in KC_p.*

Then every class in $\mathcal{R}(H(j))$ *is the class of the ring of integers in a field extension* L/K *with group* C_p.

Proof. Note that $H(j)^* = H(j') \in SGE(H(j), R)$. By Proposition 11.2.2, $H(j')$ gives rise to a class $(H(j'))$ in $\mathcal{C}(H(j))$. By Proposition 4.4.5, $\int_{H(j')} = \epsilon_{H(j')}(\int_{H(j')})\iota_0$, where $\iota_0 = \frac{1}{p}\sum_{m=0}^{p-1}\gamma^m$, and, by Proposition 5.3.3, $\epsilon_{H(j')}(\int_{H(j')}) = Rr$, where $r = (1-\zeta_{p^2})^{p(p-1)-(p-1)j'}$. Thus $\Lambda = r\iota_0$ is a generating integral for $H(j')$. By Proposition 4.3.2, $H(j') \cong H(j) \otimes R\Lambda$, and so $H(j')$ is a free rank one $H(j)$-module, which consequently corresponds to the trivial class in $\mathcal{C}(H(j))$. Thus, the class-invariant map $\Psi : SGE(H(j), R) \to \mathcal{C}(H(j))$ reduces to $\Theta \mapsto (\Theta)$, and so the image of the class-invariant map $\mathcal{R}(H(j))$ consists of the classes (M) where $M \in SGE(H(j), R)$.

By Proposition 11.2.3, there exists a class $(S) \in \mathcal{R}(H(j))$ for which S is integrally closed. Let (M) be a class in $\mathcal{R}(H(j))$. Now, by [By95, Theorem 5.6], there exists an element $X \in SGE(H(j), R)$ such that $(X) = (M)$ and for which X is the full ring of integers of some Galois extension of K with group C_p.   □

11.3   The Hopf-Swan Subgroup

Let K be a finite extension of $\mathbb{Q}$ containing ζ_{p^n} with ring of integers R. Let $S = \{v\}$ denote the collection-normalized absolute values on K, and let $D \subseteq S$ denote the subset of discrete normalized absolute values. Let H be an R-Hopf order in KG, $G = C_{p^n}$, with class group $\mathcal{C}(H)$.

In this section, we construct an important subgroup of $\mathcal{C}(H)$. Let r be an element of R that is relatively prime to $\epsilon(\int_H)$. In other words, if $\prod_{i=1}^{l} P_i^{e_i}$ is the factorization of (r) and $\prod_{j=1}^{m} Q_j^{k_j}$ is that for $\epsilon_H(\int_H)$, then $P_i \neq Q_j$ for all i, j.

Definition 11.3.1. The **Hopf-Swan module** is the H-module defined as

$$\langle r, \textstyle\int_H \rangle = rH + \textstyle\int_H,$$

where $r \in R$ is relatively prime to $\epsilon_H(\int_H)$.

Proposition 11.3.1. *The Hopf-Swan module* $\langle r, \int_H \rangle$ *is a locally free rank one H-module.*

Proof. For $v \in D$, let P_v denote the prime of R corresponding to v, and let R_v denote the completion of R with respect to the $|\ |_v$-topology. Put $H_v = R_v \otimes_R H$. By Proposition 4.4.5,

$$\textstyle\int_{H_v} = \epsilon_{H_v}\left(\textstyle\int_{H_v}\right) e_0 = (s)e_0,$$

where e_0 is the idempotent $\frac{1}{p^n}\sum_{g\in G} g$ and $s \in R_v$.

Suppose that $(r) \not\subseteq P_v$. Then $rH_v + (s)e_0 \subseteq H_v$, and since r is a unit in R_v, $H_v \subseteq rH_v + (s)e_0$. Thus $rH_v + \int_{H_v} = H_v$.

Now, suppose $(r) \subseteq P_v$. Note that $(s) \not\subseteq P_v$ since (r) and (s) are relatively prime, and thus s is a unit of R_v. Thus $\int_{H_v} = R_v e_0$. It follows that H_v is the maximal integral order in $K_{P_v}G$. Thus

$$H_v = R_v e_0 \oplus R_v e_1 \oplus \cdots \oplus R_v e_{p^n - 1},$$

where the e_i are the minimal idempotents. Consequently,

$$rH_v + \int_{H_v} = rH_v + R_v e_0$$
$$\cong R_v e_0 \oplus r(R_v e_1 \oplus \cdots \oplus R_v e_{p^n - 1})$$
$$\cong H_v(e_0 + r(1 - e_0)).$$

Let $\alpha = \{a_v\}$ be the idele in J_{KG} defined as $a_v = 1$ for $v \in D$ with $(r) \not\subseteq P_v$, $a_v = e_0 + r(1 - e_0)$, for $v \in D$ with $(r) \subseteq P_v$, and $a_v = 1$ for $v \in S \backslash D$. Then

$$\langle r, \int_H \rangle = H\alpha = KG \cap \left(\bigcap_{v \in S} H_v a_v \right),$$

as required. $\square$

By Proposition 11.3.1, the Hopf-Swan module $\langle r, \int_H \rangle$ gives rise to a class in $\mathcal{C}(H)$ that we denote as $(\langle r, \int_H \rangle)$. The collection of these classes is a subgroup of $\mathcal{C}(H)$ called the **Hopf-Swan subgroup**, which we denote by $\mathcal{T}(H)$. In the case where $H = RG$, $\mathcal{T}(H)$ is the **Swan subgroup**.

We know that $\mathcal{R}(H) \subseteq \mathcal{C}(H)$ and $\mathcal{T}(H) \subseteq \mathcal{C}(H)$. In fact, for the case $H \subseteq KC_p$, D. Replogle and R. Underwood have shown that

$$\mathcal{T}(H)^{(p-1)/2} \subseteq \mathcal{R}(H)$$

(see [RU02, Lemma 2.11]). Moreover, if $\mathcal{T}(H)$ is a group of exponent p, then

$$\mathcal{T}(H) \subseteq \mathcal{R}(H), \tag{11.4}$$

and so, if (11.4) holds, then every element of $\mathcal{T}(H)$ is a realizable class in $\mathcal{C}(H)$. For this reason, it is important to compute $\mathcal{T}(H)$. The structure of $\mathcal{T}(H)$ is the focus of the paper [Un08a], and we review the main results here (largely without proof).

For the remainder of this section, we assume that H is an R-Hopf order in KG, where $G = C_p$. For a ring Y, we let $Y^\times$ denote the multiplicative group of units of Y. Let $q : R \to R/\epsilon(\int_H)$ be the canonical surjection, and put

$$W_{\epsilon(\int_H)} = \left(R/\epsilon \left(\int_H \right) \right)^\times / q(R^\times).$$

Let $\bar{\epsilon} : H/\int_H \to R/\epsilon(\int_H)$ be the map defined as $\bar{\epsilon} \left(h + \int_H \right) = \epsilon(h) + \epsilon \left(\int_H \right)$.

Lemma 11.3.1. *With the notation above,*

$$T(H) \cong W_{\epsilon(\int_H)} / \left(\overline{\epsilon} \left((H/\int_H)^\times \right) / q(R^\times) \right).$$

Proof. One has

$$T(H) \cong (R/\epsilon(\int_H))^\times / \overline{\epsilon} \left((H/\int_H)^\times \right) \quad \text{see [RU02, (2.8)]}$$

$$\cong (R/\epsilon(\int_H))^\times / q(R^\times) / \left(\overline{\epsilon} \left((H/\int_H)^\times \right) / q(R^\times) \right)$$

$$\cong W_{\epsilon(\int_H)} / \left(\overline{\epsilon} \left((H/\int_H)^\times \right) / q(R^\times) \right). \qquad \square$$

By Lemma 11.3.1, there exists a surjection of groups

$$W_{\epsilon(\int_H)} \to T(H).$$

Lemma 11.3.2. *There is a surjective group homomorphism* $T(H) \to W^{p-1}_{\epsilon(\int_H)}$.

Proof. By Lemma 11.3.1,

$$T(H) \cong W_{\epsilon(\int_H)} / \left(\overline{\epsilon} \left((H/\int_H)^\times \right) / q(R^\times) \right).$$

Let $\phi : W_{\epsilon(\int_H)} \to W_{\epsilon(\int_H)}$ denote the homomorphism given as $w \mapsto w^{p-1}$. Since $\phi \left(\overline{\epsilon} \left((H/\int_H)^\times \right) \right) \subseteq q(R^\times)$, we have

$$T(H)^{p-1} \cong W^{p-1}_{\epsilon(\int_H)} / \left(\overline{\epsilon} \left((H/\int_H)^\times \right) / q(R^\times) \right) \cong W^{p-1}_{\epsilon(\int_H)},$$

which proves the lemma. $\square$

So, by Lemma 11.3.2, there are surjections q_1, q_2 for which

$$W_{\epsilon(\int_H)} \xrightarrow{q_1} T(H) \xrightarrow{q_2} W^{p-1}_{\epsilon(\int_H)}. \tag{11.5}$$

We now specialize to the field $K = \mathbb{Q}(\zeta_{p^n})$, where n is a fixed integer $n \geq 1$ and p satisfies Vandiver's Conjecture. For each integer i, $0 \leq i \leq p^{n-1}$, there exists a Larson order $H(i)$ in KC_p (see §5.3). If $i = 0$, then $H(i) = H(0) = RC_p$. Our goal is to compute some Hopf-Swan subgroups $T(H(i))$, including the Swan subgroup $T(RC_p)$. To do this, we need to introduce certain units of R.

Definition 11.3.2. Let K^+ denote the maximal real subfield of K. The *cyclotomic units E^+ of K^+* are the elements of $R^\times$ generated by -1 and quantities of the form

$$c_a = \zeta_{p^n}^{(1-a)/2} \frac{1 - \zeta_{p^n}^a}{1 - \zeta_{p^n}}$$

for $2 \le a \le (p^n - 1)/2$, $(a, p) = 1$. The *cyclotomic units E of K* are the elements of $R^\times$ generated by ζ_{p^n} and E^+.

An initial result is the following.

Lemma 11.3.3. *For each i, $0 \le i \le p^{n-1}$, $T(H(i)) \cong W_{\epsilon_{H(i)}}\left(\mathcal{J}_{H(i)}\right)$.*

Proof. In view of the sequence (11.5), it suffices to show that the exponent of

$$W_{\epsilon_{H(i)}}\left(\mathcal{J}_{H(i)}\right) = \left(R/\epsilon_{H(i)}\left(\mathcal{J}_{H(i)}\right)\right)^\times / q(R^\times)$$

is relatively prime to $p - 1$. Let $\lambda = \zeta_{p^n} - 1$. By Proposition 5.3.3, $\epsilon_{H(i)}\left(\mathcal{J}_{H(i)}\right) = (\lambda)^{(p^{n-1}-i)(p-1)}$. Since

$$R = \mathbb{Z} \oplus \lambda\mathbb{Z} \oplus \lambda^2\mathbb{Z} \oplus \cdots \oplus \lambda^{p^{n-1}(p-1)-1}\mathbb{Z},$$

$(R/(\lambda)^{(p^{n-1}-i)(p-1)})^\times$ has order $(p - 1)p^{(p^{n-1}-i)(p-1)-1}$ as a multiplicative group. Thus $(R/(\lambda)^{(p^{n-1}-i)(p-1)})^\times$ has the form

$$C_{p^n}^{q_0} \times C_{p^{n-1}}^{q_1} \times C_{p^{n-2}}^{q_2} \times \cdots \times C_p^{q_{n-1}} \times C_{p-1}$$

for some integers $q_i \ge 0$, $i = 0, \ldots, n - 1$. The factor C_{p-1} is identified with the group of units $\mathbb{F}_p^\times$.

Let $\langle a \rangle = \mathbb{F}_p^\times$, and let c_a be the corresponding cyclotomic generator. Since $c_a^{p^n} \equiv a \mod p$, $q(R^\times)$ contains the factor C_{p-1}, and consequently the exponent of $(R/(\lambda)^{(p^{n-1}-i)(p-1)})^\times / q(R^\times)$ is relatively prime to $p - 1$. $\square$

Proposition 11.3.2. *For each i, $0 \le i \le p^{n-1}$,*

$$T(H(i)) \cong (R/(\lambda)^{(p^{n-1}-i)(p-1)})^\times / q(E).$$

Proof. In view of Lemma 11.3.3, it suffices to show that $q(R^\times) = q(E)$. First note that the quotient $R^\times/E$ has order $h^+(R)$ by [Wa97, Theorem 8.2]. Moreover, there exists a surjection $R^\times/E \to Q$ with $Q = q(R^\times)/q(E)$. Note that both $q(E)$ and $q(R^\times)$ contain the factor C_{p-1} in their cyclic decompositions. Consequently, if $q(E)$ is a proper subgroup of $q(R^\times)$, then p divides the order of the quotient Q, and hence p divides $[R^\times : E] = h^+(K)$. By [Wa97, Corollary 10.6], p then divides $h^+(\mathbb{Q}(\zeta_p))$, which is a contradiction since p satisfies Vandiver's Conjecture. It follows that $q(R^\times) = q(E)$. $\square$

Proposition 11.3.2 is a useful mechanism for computation. It says that $T(H(i))$ can be calculated by finding the cyclic decompositions of

$$(R/(\lambda)^{(p^{n-1}-i)(p-1)})^\times$$

and $q(E)$ and then taking the quotient.

To illustrate this, we set $n = 1$, $i = 0$, and consider the R-Hopf order $H(0) = RC_p$. Then $\epsilon_{RC_p}\left(\int_{RC_p}\right) = (p)$. It is not hard to show that

$$(R/(p))^\times \cong C_p^{p-2} \times C_{p-1}.$$

(Prove this as an exercise.) Moreover, the structure of $q(E)$ has been determined by D. Replogle [Re01], who has proved the following (though not stated exactly in this manner). First, we define the **index of irregularity** of the prime p is defined as the number s of numerators in the set $B_0, B_2, B_4, \ldots, B_{p-3}$ that are divisible by p.

Proposition 11.3.3. *(D. Replogle) Suppose the prime $l \geq 3$ satisfies Vandiver's Conjecture. Then $q(E) \cong C_p^{(p-1)/2-s} \times C_{p-1}$, where s is the index of irregularity of p.*

Proof. For a proof, see [Re01, Theorem 1]. $\qquad\qquad\qquad\qquad\qquad\qquad\qquad\square$

Consequently, the Swan subgroup is computed as

$$T(RC_p) \cong (C_p^{p-2} \times C_{p-1})/(C_p^{(p-1)/2+s} \times C_{p-1})$$
$$\cong C_p^{(p-3)/2+s}.$$

For example, take $p = 5$. Then it is known that the index of irregularity is $s = 0$, and thus $T(RC_5) = C_5$.

We next consider the case $n = 2$, $i = 0$, such that $H(i) = H(0) = RC_p$ and $\epsilon_{RC_p}\left(\int_{RC_p}\right) = (p)$. We seek to compute the Swan subgroup $T(RC_p)$. In this case, $T(RC_p) \cong (R/(p))^\times/q(E)$. We first consider the numerator of this quotient.

Proposition 11.3.4. $(R/(p))^\times \cong C_{p^2}^{p-2} \times C_p^{p^2-3p+3} \times C_{p-2}$.

Proof. With $\lambda = \zeta_{p^2} - 1$, observe that

$$R = \mathbb{Z} \oplus \lambda\mathbb{Z} \oplus \lambda^2\mathbb{Z} \oplus \cdots \oplus \lambda^{p(p-1)-1}\mathbb{Z},$$

and thus $R/(\lambda)^{p(p-1)} = R/(p)$ is isomorphic to $C_p^{p(p-1)}$ as additive groups. Consequently, there are $(p-1)p^{p(p-1)-1}$ elements in $(R/(p))^\times$. The elements

$$1 + \lambda, 1 + \lambda^2, 1 + \lambda^3, \ldots, 1 + \lambda^{p-2}$$

have order p^2 in $(R/(p))^\times$. Moreover,

$$1 + \lambda^{p-1}, 1 + \lambda^p, 1 + \lambda^{p+1}, \ldots, 1 + \lambda^{p(p-1)-1}$$

have order p in $(R/(p))^\times$.
   For $r = 1, 2, \ldots, p-2$,

$$(1 + \lambda^r)^p \equiv 1 + \lambda^{pr} \mod (p),$$

and so $(R/(p))^\times$ is generated by C_{p-1} together with the elements $1 + \lambda^r$ for $r = 1, 2, \ldots, p - 2$ and the elements $1 + \lambda^s$ for

$$p - 1 \leq s \leq p(p - 1) - 1, \quad \text{with } (s, p) = 1.$$

Since there are $p - 2$ values of s with $(s, p) \neq 1$, $(R/(p))^\times$ contains

$$p(p - 1) - (p - 1) - (p - 2) = p^2 - 3p + 3$$

copies of C_p in its decomposition. It follows that

$$(R/(p))^\times \cong C_{p-1} \times C_p^{p^2-3p+3} \times C_{p^2}^{p-2}. \qquad \square$$

The computation of $q(E)$ is much more complicated, as we shall see. We begin with the following lemmas. Let $\{c_a\}$ denote the collection of cyclotomic generators of E.

Lemma 11.3.4. *Modulo p, the expansions of c_a in powers of $\lambda = \zeta_{p^2} - 1$ are as follows:*

(i) If $a = pm + 1$, then

$$c_a \equiv 1 + m\lambda^{p-1} + \frac{1}{2}m\lambda^p + t_{2(p-1)+1}\lambda^{2(p-1)+1} + \cdots + t_{p(p-1)-1}\lambda^{p(p-1)-1}.$$

(ii) If $a = pm - 1$, then

$$c_a \equiv -1 + m\lambda^{p-1} + \frac{1}{2}m\lambda^p + t_{2(p-1)+1}\lambda^{2(p-1)+1} + \ldots t_{p(p-1)-1}\lambda^{p(p-1)-1}.$$

(iii) If $a = pm + r$, $r \neq \pm 1$, then

$$c_a \equiv r + t_2\lambda^2 + \cdots + t_{p(p-1)-1}\lambda^{p(p-1)-1}$$

with $t_2 \not\equiv 0 \mod p$.

Proof. Note that

$$c_a = \zeta_{p^2}^{(a-1)(p^2-1)/2} + \zeta_{p^2}^{(a-1)(p^2-1)/2+1} + \cdots + \zeta_{p^2}^{(a-1)(p^2-1)/2+(a-1)},$$

and write the polynomial

$$f_a(x) = x^{(a-1)(p^2-1)/2} + x^{(a-1)(p^2-1)/2+1} + \cdots + x^{(a-1)(p^2-1)/2+(a-1)},$$

so that $f_a(\zeta_{p^2}) = c_a$.

To establish the lemma, we compute the Taylor series expansion of $f_a(x)$ in the indeterminate x about the point 1 and reduce modulo p. We seek the coefficients t_k such that

$$f_a(x) = t_0 + t_1(x-1) + t_2(x-1)^2 + \dots.$$

Let $a = pm + r, 0 \leq r \leq p-1$, and put $h = (a-1)(p^2-1)/2$. We calculate

$$t_k = \frac{1}{k!} \sum_{i=h}^{h+(a-1)} i(i-1)(i-2)\dots(i-(k-1))$$

$$= \sum_{i=h}^{h+a-1} \binom{i}{k},$$

$k \geq 0$. Now, using the identity

$$\sum_{i=k}^{n} \binom{i}{k} = \binom{n+1}{k+1},$$

we obtain

$$t_k = \binom{h+a}{k+1} - \binom{h}{k+1},$$

which yields modulo p

$$t_k \equiv \begin{cases} r & \text{if } k = 0 \\ 0 & \text{if } k = 1 \\ \frac{1}{24}(a(a-1)(a+1)) & \text{if } k = 2. \end{cases}$$

Now, if $a \not\equiv \pm 1 \mod p$, then $p > 3$ and $t_2 \not\equiv 0 \mod p$. Thus the cyclotomic generators c_a with $a \not\equiv \pm 1$ have the expansions claimed.

We next consider the generators c_a, $a \equiv \pm 1 \mod p$. Let X and Y be integers, $X \geq Y$, whose p-adic expansions $X = \sum x_i p^i$ and $Y = \sum y_i p^i$, $0 \leq x_i, y_i \leq p-1$, satisfy $x_i \geq y_i$ for all i. Then a classical result of E. Lucas states that

$$\binom{X}{Y} \equiv \prod_{x_i, y_i} \binom{x_i}{y_i} \mod p.$$

An induction argument using Lucas's formula yields

$$\binom{px + x_0}{py + y_0} \equiv \binom{x}{y}\binom{x_0}{y_0} \mod p, \tag{11.6}$$

which is valid for $0 \leq y_0 \leq x_0 \leq p - 1$ and $x \geq y \geq 1$. Moreover, if $x \geq y \geq 1$ and $0 \leq x_0 < y_0 \leq p - 1$, then

$$
\binom{px + x_0}{py + y_0} \equiv 0 \mod p. \tag{11.7}
$$

Now, with $a = pm + 1$,

$$
\binom{h + a}{k + 1} - \binom{h}{k + 1} = \binom{p(m(p^2 - 1)/2 + m) + 1}{k + 1} - \binom{p(m(p^2 - 1)/2)}{k + 1},
$$

and, for $a = pm - 1$,

$$
\binom{h + a}{k + 1} - \binom{h}{k + 1} = \binom{p(m(p^2 - 1)/2 - p + m)}{k + 1} - \binom{p(m(p^2 - 1)/2 - p) + 1}{k + 1}.
$$

Thus, by (11.6) and (11.7),

$$
t_k \equiv \begin{cases} 0 & \text{if } 2 \leq k \leq p - 2 \\ m & \text{if } k = p - 1 \\ \frac{1}{2}m & \text{if } k = p \\ 0 & \text{if } p + 1 \leq k \leq 2(p - 1). \end{cases}
$$

Thus, the cyclotomic generators c_a have the expansions claimed.   □

Using Lemma 11.3.4, we can compute the orders of the generators of $q(E)$.

Lemma 11.3.5.   (i) $q(\zeta_{p^2})$ has order p^2.
(ii) If $a = pm + 1$, then $q(c_a)$ has order p.
(iii) If $a = pm - 1$, then $q(c_a)$ has order $2p$.
(iv) Suppose $a \not\equiv \pm 1$, and let the order of a in C_{p-1} be w. Then $w > 2$ and $q(c_a)$ has order $p^2 w$.

Proof. For (i), observe that $\zeta_{p^2} = 1 + \lambda$. The statements (ii), (iii), and (iv) follow from the expansions of Lemma 11.3.4.   □

We now proceed with the calculation of $q(E)$. The group of cyclotomic units E is generated by the set $\{-1, \zeta_{p^2}\} \cup A \cup B$, where

$$
A = \{c_a : 2 \leq a \leq (p^2 - 1)/2, (a, p) = 1, a \not\equiv \pm 1\}
$$

and

$$
B = \{c_a : 2 \leq a \leq (p^2 - 1)/2, (a, p) = 1, a \equiv \pm 1\}.
$$

Let $\langle -1, \zeta_{p^2}, A \rangle$ denote the subgroup of E generated by $\{-1, \zeta_{p^2}\} \cup A$, and let $\langle B \rangle$ denote the subgroup of E generated by B.

Lemma 11.3.6. $q(\langle -1, \zeta_{p^2}, A \rangle) \cong C_{p^2}^{(p-1)/2-s} \times C_p^b \times C_{p-1}$, where s is the index of irregularity of the prime $p \geq 3$ and b is an integer with

$$0 \leq b \leq (p^2 - 4p + 3)/2 + s.$$

Proof. We have $q(E)^p \cong \tilde{q}(E)$, where $\tilde{q} : R \rightarrow R/(\lambda)^{p-1}$ is the canonical surjection. Moreover,

$$q(E)^p = q(\langle -1, \zeta_{p^2}, A \rangle)^p$$

since all elements of $q(B)$ have order either p or $2p$ by Lemma 11.3.5 (ii), (iii). Hence

$$q(\langle -1, \zeta_{p^2}, A \rangle)^p \cong \tilde{q}(E).$$

Now $\tilde{q}(E) \cong q'(E')$, where E' is the group of cyclotomic units in $\mathbb{Z}[\zeta_p]$ and $q' : \mathbb{Z}[\zeta_p] \rightarrow \mathbb{Z}[\zeta_p]/(p)$ is the canonical surjection. Thus, by Proposition 11.3.3,

$$\tilde{q}(E) \cong C_p^{(p-1)/2-s} \times C_{p-1},$$

and so

$$q(\langle -1, \zeta_{p^2}, A \rangle)^p \cong C_p^{(p-1)/2-s} \times C_{p-1}.$$

Now, $q(\langle -1, \zeta_{p^2}, A \rangle) \cong C_{p^2}^{(p-1)/2-s} \times C_p^b \times C_{p-1}$ for some integer b. But since $|A| = p(p-3)/2$ and $q(\zeta_{p^2})$ has order p^2,

$$b + \frac{p-1}{2} - s \leq \frac{p(p-3)}{2} + 1,$$

and hence

$$0 \leq b \leq \frac{p^2 - 4p + 3}{2} + s. \qquad \square$$

Lemma 11.3.7. $q(\langle B \rangle) \cong C_p^d \times C_2$, where d is an integer with $1 \leq d \leq p - 1$.

Proof. We have $|B| = p - 1$. By Lemma 11.3.5 (ii), (iii), each element in $q(B)$ has order either p or $2p$, and thus

$$q(\langle B \rangle) \cong C_p^d \times C_2$$

with $1 \leq d \leq p - 1$. $\qquad \square$

We now have the computation of $q(E)$.

Proposition 11.3.5. $q(E) \cong C_{p^2}^{(p-1)/2-s} \times C_p^r \times C_{p-1}$, where s is the index of irregularity of the prime p and where r is an integer $1 \leq r \leq (p-1)^2/2 + s$.

Proof. By Lemma 11.3.6, $q(\langle -1, \zeta_{p^2}, A \rangle) \cong C_{p^2}^{(p-1)/2-s} \times C_p^b \times C_{p-1}$, and, by Lemma 11.3.7, $q(\langle B \rangle) \cong C_p^d \times C_2$. Thus

$$q(E) \cong C_{p^2}^{(p-1)/2-s} \times C_p^r \times C_{p-1},$$

where r is an integer with

$$1 \le r \le \frac{p^2 - 4p + 3}{2} + s + p - 1 = (p-1)^2/2 + s. \qquad \square$$

Finally, we compute $T(RC_p)$.

Proposition 11.3.6. *Let $p \ge 3$ be a prime that satisfies Vandiver's Conjecture, and let $K = \mathbb{Q}(\zeta_{p^2})$. Then*

$$T(RC_p) \cong C_{p^2}^{(p-3)/2+s-t} \times C_p^{p^2-3p+3-r+2t},$$

where s is the index of irregularity of p and where t satisfies $0 \le t \le \min\{r, p-2\}$ and is determined by

$$(R/(p))^{\times p} \cap q(E) \cong C_p^{(p-1)/2-s+t} \times C_{p-1}.$$

Proof. By Proposition 11.3.2, $T(RC_p) \cong (R/(p))^\times / q(E)$. By Proposition 11.3.4, $(R/(p))^\times \cong C_{p^2}^{p-2} \times C_p^{p^2-3p+3} \times C_{p-1}$, and, by Proposition 11.3.5, $q(E) \cong C_{p^2}^{(p-1)/2-s} \times C_p^r \times C_{p-1}$. Hence

$$T(RC_p) \cong (C_{p^2}^{p-2} \times C_p^{p^2-3p+3} \times C_{p-1}) / \left(C_{p^2}^{(p-1)/2-s} \times C_p^r \times C_{p-1} \right)$$

$$\cong C_{p^2}^{p-2-(p-1)/2+s-t} \times C_p^{p^2-3p+3-(r-t)+t}$$

$$\cong C_{p^2}^{(p-3)/2+s-t} \times C_p^{p^2-3p+3-r+2t},$$

where t is as claimed. $\qquad \square$

Observe that $T(RC_p)$ is a group of exponent p, and hence (11.4) holds; that is, $T(RC_p)$ is a subgroup of the realizable classes $R(RC_p)$.

Example 11.3.1. One can use a computer algebra system (such as GAP) to compute $T(RC_p)$ for certain primes. For $p = 3$ and $p = 5$, several computations using GAP yield Table 11.1 below (note that the index of irregularity for these primes is $s = 0$). Observe that, for $p = 5$, $q(E) = C_{25}^2 \times C_5^7 \times C_4$, and so $r = 7$. Consequently, $t = 0$ since $T(RC_5) = C_{25} \times C_5^6$.

Table 11.1 Computation of the Hopf-Swan subgroup

p	$q(\langle -1, \zeta, A \rangle)$	$q(\langle B \rangle)$	$q(E)$	$T(RC_p)$
3	$C_9 \times C_2$	$C_3^2 \times C_2$	$C_9 \times C_3^2 \times C_2$	C_3
5	$C_{25}^2 \times C_5^4 \times C_4$	$C_5^4 \times C_2$	$C_{25}^2 \times C_5^7 \times C_4$	$C_{25} \times C_5^6$

So far in this section, we have computed some Swan subgroups. We next compute some Hopf-Swan subgroups. Let $p \geq 3$ be a prime that satisfies Vandiver's Conjecture, and let $K = \mathbb{Q}(\zeta_{p^2})$. Put $\lambda = \zeta_{p^2} - 1$, so that $(p) = (\lambda)^{p(p-1)}$. For each integer i, $0 \leq i \leq p$, there exists a Larson order $H(i)$ in KC_p, given as $H(i) = R\left[\frac{g-1}{\lambda^i}\right]$. In this section, we compute the Hopf-Swan subgroups $T(H(p))$, $T(H(p-1))$, and $T(H(p-2))$.

We begin with some observations. for $i = p, p-1, p-2, \ldots, 2, 1, 0$, we have $\epsilon_{H(i)}\left(\int_{H(i)}\right) = (\lambda)^{(p-i)(p-1)}$. Let $q_i : R \to R/(\lambda)^{(p-i)(p-1)}$ be the canonical surjections, each playing the role of q in the computation of the Swan subgroup.

Using (11.5), it is not difficult to prove the following.

Proposition 11.3.7. $T(H(p)) = 1$.

Proof. Exercise.   □

We shall compute $T(H(p-1))$ and $T(H(p-2))$ by using the formula

$$T(H(i)) \cong (R/(\lambda^{(p-i)(p-1)})^\times / q_i(E)$$

for $i = p-1, p-2$, which follows from Proposition 11.3.2. We have the following lemma.

Lemma 11.3.8. *For* $i = p-1, p-2, p-3, \ldots, 2, 1$,

$$(R/(\lambda)^{(p-i)(p-1)})^\times \cong C_{p^2}^{p-i-1} \times C_p^{(p-2)+(p-i-1)(p-3)} \times C_{p-1}.$$

Proof. We have

$$R = \mathbb{Z} \oplus \lambda\mathbb{Z} \oplus \lambda^2\mathbb{Z} \oplus \cdots \oplus \lambda^{p(p-1)-1}\mathbb{Z}.$$

For $p-2 \leq i \leq p-1$, $(R/(\lambda)^{(p-i)(p-1)})^\times$ has order $(p-1)p^{(p-i)(p-1)-1}$ as a multiplicative group.

There are $p - i - 1$ copies of C_{p^2} in the cyclic decomposition of $(R/(\lambda)^{(p-i)(p-1)})^\times$. Let N be the number of copies of C_p that occur in the cyclic decomposition of $(R/(\lambda)^{(p-i)(p-1)})^\times$. Then $N + 2(p-i-1) = (p-i)(p-1)-1$, so that

$$N = (p-i)(p-1) - 1 - 2(p-i-1)$$
$$= (p-2) + (p-i-1)(p-3).$$

Thus $(R/(\lambda)^{(p-i)(p-1)})^{\times} \cong C_{p^2}^{p-i-1} \times C_p^{(p-2)+(p-i-1)(p-3)} \times C_{p-1}.$ $\qquad\square$

Proposition 11.3.8. *Suppose* $p \geq 3$ *satisfies Vandiver's Conjecture. Then* $T(H(p-1)) \cong C_p^{(p-3)/2+s}$, *where s is the index of irregularity of p.*

Proof. Exercise. $\qquad\square$

Finally, we compute $T(H(p-2))$. We already know the numerator of $T(H(p-2))$ by Lemma 11.3.8, so it is a matter of computing $q_{p-2}(E)$. Again we use the fact that E is generated by the set $\{-1, \zeta_{p^2}\} \cup A \cup B$.

Lemma 11.3.9. $q_{p-2}(\langle -1, \zeta_{p^2}, A \rangle) \cong C_{p^2} \times C_p^{\eta} \times C_{p-1}$, *where* $\eta \geq (p-3)/2 - s.$

Proof. Since $\zeta_{p^2} = 1 + \lambda$, $q_{p-2}(\zeta_{p^2})$ has order p^2. By Lemma 11.3.4 (iii), $q_{p-2}(c_a)$, $a \not\equiv \pm 1$, has order pw with $2p < pw \leq p(p-1)$. Thus

$$q_{p-2}(\langle -1, \zeta_{p^2}, A \rangle) \cong C_{p^2} \times C_p^{\eta} \times C_{p-1}$$

for some $\eta \geq 0$.

Observe that $q_{p-1}(\langle -1, \zeta_{p^2}, A \rangle) = q_{p-1}(E)$ since $c_{pm\pm 1} \equiv \pm 1 \mod (\lambda)^{p-1}$ by Lemma 11.3.4 (i), (ii). Moreover, by Proposition 11.3.8,

$$q_{p-1}(E) \cong C_p^{(p-1)/2-s} \times C_{p-1},$$

and hence

$$q_{p-1}(\langle -1, \zeta_{p^2}, A \rangle) \cong C_p^{(p-1)/2-s} \times C_{p-1}.$$

But this says that $(p-1)/2 - s \leq \eta + 1$, and thus $\eta \geq (p-3)/2 - s.$ $\qquad\square$

Lemma 11.3.10. $q_{p-2}(\langle B \rangle) \cong C_p \times C_2.$

Proof. We observe that, for integers m, n, $1 \leq m, n \leq (p-1)/2$,

$c_{pm+1}c_{pn+1} - c_{p(m+n)+1}$

$$= \left(\zeta_{p^2}^{-pm/2} \frac{1 - \zeta_{p^2}^{pm+1}}{1 - \zeta_{p^2}} \right) \left(\zeta_{p^2}^{-pn/2} \frac{1 - \zeta_{p^2}^{pn+1}}{1 - \zeta_{p^2}} \right) - \zeta_{p^2}^{-p(m+n)/2} \frac{1 - \zeta_{p^2}^{pm+pn+1}}{1 - \zeta_{p^2}}$$

$$= \zeta_{p^2}^{-p(m+n)/2}(1 - \zeta_{p^2})^{-2} \left(\left(1 - \zeta_{p^2}^{pn+1} - \zeta_{p^2}^{pm+1} + \zeta_{p^2}^{pm+pn+2} \right) \right.$$

$$\left. - (1 - \zeta_{p^2}) \left(1 - \zeta_{p^2}^{pm+pn+1} \right) \right)$$

$$= \zeta_{p^2}^{-p(m+n)/2}(1 - \zeta_{p^2})^{-2}$$

$$\times \left(1 - \zeta_{p^2}^{pn+1} - \zeta_{p^2}^{pm+1} + \zeta_{p^2}^{pm+pn+2} - 1 + \zeta_{p^2}^{pm+pn+1} + \zeta_{p^2} - \zeta_{p^2}^{pm+pn+2} \right)$$

$$= \zeta_{p^2}^{-p(m+n)/2}(1 - \zeta_{p^2})^{-2}\zeta_{p^2}(1 - \zeta_{p^2}^{mp})(1 - \zeta_{p^2}^{np})$$

$$\equiv 0 \mod (\lambda)^{2(p-1)}.$$

Thus

$$c_{p+1}^i \equiv c_{pi+1} \mod (\lambda)^{2(p-1)}$$

for $i = 1, \ldots, p - 1$. Moreover,

$$c_{pi+1} \equiv -c_{p(p-i)-1} \mod (\lambda)^{2(p-1)},$$

and thus, for $i = (p - 1)/2 + 1, \ldots, p - 1$,

$$-c_{p+1}^i \equiv c_{p(p-i)-1} \mod (\lambda)^{2(p-1)}.$$

It follows that $q_{p-2}(\langle B \rangle)$ is generated by the two elements $q_{p-2}(c_{p+1})$ and $q_{p-2}(-1)$. Since these elements have order p and 2, respectively (use Lemma 11.3.4), we conclude that

$$q_{p-2}(\langle B \rangle) \cong C_p \times C_2.   \qquad \square$$

Lemma 11.3.11. $q_{p-2}(E) \cong C_{p^2} \times C_p^\eta \times C_{p-1}$, where $\eta \geq (p - 3)/2 - s$.

Proof. By Lemma 11.3.9, $q_{p-2}(\langle -1, \zeta_{p^2}, A \rangle) \cong C_{p^2} \times C_p^\eta \times C_{p-1}$, with $\eta \geq (p - 3)/2 - s$, and, by Lemma 11.3.10, $q_{p-2}(\langle B \rangle) \cong C_p \times C_2$. Thus $q_{p-2}(E) \cong C_{p^2} \times C_p^\eta \times C_{p-1}$, where $\eta \geq (p - 3)/2 - s$. $\square$

Proposition 11.3.9. *Suppose $p \geq 3$ satisfies Vandiver's Conjecture. Then*

$$\mathcal{T}(H(p - 2)) \cong C_p^\tau,$$

with $(p - 3)/2 + s \leq \tau \leq (3p - 7)/2 + s$, where s is the index of irregularity of p.

Proof. We have $\mathcal{T}(H(p-2)) \cong (R/(\lambda)^{2(p-1)})^\times / q_{p-2}(E)$. Now $(R/(\lambda)^{2(p-1)})^\times \cong C_{l^2} \times C_l^{2l-5} \times C_{l-1}$ by Lemma 11.3.8, and $q_{p-2}(E) \cong C_{p^2} \times C_p^\eta \times C_{p-1}$ by Lemma 11.3.11. Hence

$$\begin{aligned}
\mathcal{T}(H(p - 2)) &\cong (C_{p^2} \times C_p^{2p-5} \times C_{p-1})/(C_{p^2} \times C_p^\eta \times C_{p-1}) \\
&\cong C_p^{2p-5-\eta} \\
&\cong C_p^\tau,
\end{aligned}$$

with $0 \leq \tau \leq (3p - 7)/2 + s$. Also, in view of the surjection $\mathcal{T}(H(p - 2)) \rightarrow \mathcal{T}(H(p - 1))$, τ is bounded below by $(p - 3)/2 + s$. $\square$

For $p \geq 5$, $j = 1, 2$, $\mathcal{T}(H(p - j))$ is a non-trivial p-group. Thus (11.4) holds and there is a non-trivial class (M) in $\mathcal{R}(H(p - j))$. By Proposition 11.2.4, (M) is the class of a ring of integers S in a field extension L/K with group C_p. Moreover, since (S) is non-trivial, S is a semilocal Galois $H(p - j)$-extension that is not free over $H(p - j)$.

11.4   Chapter Exercises

Exercises for §11.1

1. Show that there exists an adele $\{a_v\}$ in $V_{\mathbb{Q}}$ for which $C(\{a_v\}) > 1$.
2. Let $\{a_v\}$ be an idele in $V_{\mathbb{Q}}$ for which $C(\{a_v\}) > 1$. Prove that $|a_v|_v = 1$ for almost all v.
3. Prove Proposition 11.1.2, parts (ii) and (iii).
4. Prove Proposition 11.1.4.
5. Let R be the ring of integers in an algebraic number field. Prove that R is a PID if and only if $\mathcal{C}(R) = 1$.
6. Find an example of a locally free $\mathbb{Z}[\sqrt{-5}]$-submodule of $\mathbb{Q}(\sqrt{-5})$ that is not free over $\mathbb{Z}[\sqrt{-5}]$.

Exercises for §11.2

7. Let $n \geq 1$ be an integer, and let H be an R-Hopf order in KC_{p^n}, $\zeta_{p^n} \in K$.

   (a) Let M be a locally free H-module of rank k. Prove that M is a free H-module of rank k if and only if the class of $\mathbf{h}(\lambda_M)$ is trivial in $\mathcal{C}(H)$.
   (b) Let M and N be locally free H-modules that satisfy

   $$M \oplus H^m \cong N \oplus H^n$$

   for some m, n. Show that $(\mathbf{h}(\lambda_M)) = (\mathbf{h}(\lambda_N))$ in $\mathcal{C}(H)$.

Chapter 12
Open Questions and Research Problems

In the chapters of this book, we have developed some of the central themes in the study of Hopf algebras and Hopf orders. As one might expect, there are many outstanding problems in this field that have not been solved. We now revisit some of the topics in this book and give an account of some open questions and research problems.

12.1 The Spectrum of a Ring

Problem 1. In the manner illustrated in Figures 1.1, 1.2, and 1.3, find the structure of the associated map $\operatorname{Spec} \mathbb{Z}V \to \operatorname{Spec} \mathbb{Z}$ induced from the ring inclusion $\mathbb{Z} \to \mathbb{Z}V$, where V denotes the Klein 4-group.

Problem 2. Find the structure of the associated map $\operatorname{Spec} \mathbb{Z}C_{p^n} \to \operatorname{Spec} \mathbb{Z}$ induced from the ring inclusion $\mathbb{Z} \to \mathbb{Z}C_{p^n}$.

12.2 Hopf Algebras

Problem 3. A major open problem is the following. Let p be a prime, and let $n \geq 1$ be an integer. Let K be a finite extension of $\mathbb{Q}_p$, and let C_{p^n} denote the cyclic group of order p^n. Give a classification of all of the Hopf orders in KG (More on this is presented below).

Problem 4. Let K be a finite extension of $\mathbb{Q}_p$. It has been conjectured that every Hopf order in KC_{p^n} can be determined by n valuation parameters and $n(n-1)/2$ unit parameters. Prove or disprove this conjecture. (Certainly, the conjecture is true for the cases $n = 1, 2$. For $n = 3$, no Hopf order in KC_{p^3} has yet been constructed that requires more than six parameters.)

R.G. Underwood, *An Introduction to Hopf Algebras*, DOI 10.1007/978-0-387-72766-0_12,   261
© Springer Science+Business Media, LLC 2011

12.3   Valuations and Larson Orders

Problem 5. There are various problems regarding the counting of order bounded and p-adic group valuations on finite groups (see for instance the paper of A. Koch and A. Malagon [KM07]). Let $K = \mathbb{Q}(\zeta_{p^n})$. Compute the number of p-adic obgvs on C_{p^n} and C_p^n. More generally, compute the number of p-adic obgvs on a p-group of order p^n.

Problem 6. R. Larson [Lar76] has given the following generalization of group valuations. Let $\psi : \mathbb{Z}_{>0} \to \mathbb{Z}$ be a function defined as

$$\psi(n) = \begin{cases} p & \text{if } n = p^a, \text{ for some } a \\ 1 & \text{otherwise.} \end{cases}$$

Let G be a finite Abelian group, let K be a finite extension of $\mathbb{Q}$, and let I_K denote the group of fractional ideals in K. A **global group valuation** is a function $\eta : G \to I_K \cup \{0\}$ that satisfies

(i) $\eta(g) \subseteq R$, $\eta(g) = \{0\}$, if and only if $g = 1$;
(ii) $\eta(gh) \subseteq \eta(g) + \eta(h)$.

A global group valuation η is **order bounded** if

$$\psi(|g|)R \subseteq \eta(g)^{\phi(|g|)}.$$

An order-bounded global group valuation (obggv) η is p**-adic** if

$$\eta(g^{\psi(|g|)}) \subseteq \eta(g)^{\psi(|g|)}.$$

Let H be an R-Hopf order in KC_p. Let P be a prime ideal of R. By Proposition 7.1.2, the R_v-Hopf order H_{R_v} in $K_P C_p$ is a Larson order and is given by a p-adic obgv on C_p. Is H determined by a p-adic obggv on C_p?

12.4   Hopf Orders in KC_{p^2}

Problem 7. Let p be prime and let K be a finite extension of $\mathbb{Q}_p$. In §8.4, we showed that every Hopf order in KC_{p^2} for $p = 2, 3$ can be written in the form

$$H = R\left[\frac{g^p - 1}{\pi^i}, \frac{gu - 1}{\pi^j}\right]$$

for some $u = u_0 e_0 + u_1 e_1 + \cdots + u_{p-1} e_{p-1}$, $u_m \in K$, where e_m are the idempotents in $K\langle g^p \rangle$. It would be of interest to extend this result to primes $p > 3$ for then one could give an alternate (perhaps shorter) proof of the valuation condition for $n = 2$.

12.5   Hopf Orders in KC_{p^3}

Problem 8. As we indicated above, a major problem is to complete the classification of R-Hopf orders in KC_{p^3}. Can every Hopf order in KC_{p^3} be written as a circulant matrix Hopf order or the linear dual of such a Hopf order?

Problem 9. Does every circulant matrix Hopf order satisfy the valuation condition for $n = 3$? Does an arbitrary Hopf order in KC_{p^3} satisfy the valuation condition for $n = 3$?

These two questions are important for the following reason. If the Hopf order H in KC_{p^3}, $\langle g \rangle = C_{p^3}$, satisfied the valuation condition for $n = 3$, and induced the short exact sequence

$$E : R \rightarrow A(i, j, u) \rightarrow H \rightarrow H(k) \rightarrow R,$$

with $pk \leq \Xi(H)(g^p)$, then there would be a short exact sequence of Hopf orders

$$E_0 : R \rightarrow A(i, j, u) \rightarrow H_0 \rightarrow H(k) \rightarrow R,$$

where H_0 is an R-Hopf order in $K(C_{p^2} \times C_p)$ of the form

$$H_0 = R \left[\frac{g^{p^2} - 1}{\pi^i}, \frac{g^p a_u - 1}{\pi^j}, \frac{g - 1}{\pi^k} \right].$$

The Baer product $E^* E_0^{-1} = E'$ would then yield an extension

$$E' : R \rightarrow A(i, j, u) \rightarrow H' \rightarrow H(k) \rightarrow R,$$

which over K would appear as

$$K \rightarrow KC_{p^2} \rightarrow KC_{p^3} \rightarrow KC_p \rightarrow K.$$

Thus the classification of Hopf orders in KC_{p^3} would reduce to the case $K(C_{p^2} \times C_p)$, presumably a simpler problem.

Of course, all of this is predicated on giving a precise diagramwise definition of the Baer product of two extensions of $A(i, j, u)$ by $H(k)$ similar to the way we defined the Baer product in §8.3.

Problem 10. The construction of formal group Hopf orders in KC_{p^n} (Chapter 6) when applied to the case $n = 3$ yields a collection of R-Hopf orders in KC_{p^3}. What is the structure of the linear duals of these Hopf orders? Also, what is the precise relationship between the formal group Hopf orders in KC_{p^3}, the duality Hopf orders of §9.1, and the circulant matrix Hopf orders of §9.2? Also, it would be of interest to compute the Larson order $A(\Xi(H))$, where H is one of these Hopf orders in KC_{p^3}.

Problem 11. Another approach to the classification of Hopf orders in KC_{p^3} would be to extend the method of §8.4 to show that every Hopf order in KC_{p^3} could be written in the form

$$R\left[\frac{g^{p^2}-1}{\pi^i}, \frac{g^p a_v - 1}{\pi^j}, \frac{gu-1}{\pi^k}\right]$$

for some $u \in KC_{p^2}$.

12.6 Hopf Orders and Galois Module Theory

Problem 12. Let $p \geq 5$ be prime, and let $K^{(p)}$ be the splitting field of the polynomial $h_p(x) = f_p(x)/(x+1)$ defined in §10.1. By Proposition 10.1.10, the Galois group of $K^{(p)}$ is C_{p-1}. What are the equivalence classes of extensions of $|\ |_\infty$ to $K^{(p)}$? For q prime, what are the equivalence classes of extensions of $|\ |_q$ to $K^{(p)}$?

Problem 13. Compute the ring of integers $R^{(p)}$ of $K^{(p)}$ for primes $p \geq 5$. Compute $\mathrm{disc}(R^{(p)}/\mathbb{Z})$. Which primes $q \in \mathbb{Z}$ ramify in $K^{(p)}$?

Problem 14. Let $R^{(p)}$ be as in Problem 14. Let $\mathbb{Z} \to R^{(p)}$ be the ring inclusion. Illustrate the spectral diagram for the associated map $\psi : \mathrm{Spec}\ R^{(p)} \to \mathrm{Spec}\ \mathbb{Z}$.

Problem 15. Let K be a Galois extension of $\mathbb{Q}$ with Abelian Galois group G and ring of integers R. Since R is Dedekind, there exists a unique factorization

$$\mathrm{disc}(R/\mathbb{Z}) = (p_1)^{e_1} p_2^{e_2} \cdots (p_k)^{e_k}$$

for primes $p_i \in \mathbb{Z}$. By Proposition 10.2.14, these are precisely the primes that ramify in R. Let q be a prime of $\mathbb{Z}$ that is unramified in R, let Q be a prime of R lying above q, and let τ_Q be the Frobenius element at Q. For a prime Q' lying above q, one has $\tau_{Q'} = \tau_Q$. Thus we may define a function

$$F_{K/\mathbb{Z}} : \mathrm{Spec}\ \mathbb{Z}\backslash\{p_1, p_2, \ldots, p_k, \omega\} \to G$$

by the rule $F_{K/\mathbb{Z}}(q) = \tau_Q$, where Q is a prime lying above q.

Proposition 12.6.1. *Let K be a Galois extension of $\mathbb{Q}$ with Abelian Galois group G and ring of integers R. For each element $\tau \in G$, the primes $q \in \mathbb{Z}$ with $F_{K/\mathbb{Z}}(q) = \tau$ have density $\frac{1}{|G|}$.*

Proposition A is known as the Tchebotarev Density Theorem for Abelian Extensions (TDT). The TDT implies that the density of primes of $\mathbb{Z}$ that completely split in $\mathbb{Z}[i]$ is $1/2$. Use a computer algebra/graphics program to extend the spectral diagram in Figure 1.1 to the first 1000 primes of $\mathbb{Z}$. Show that this empirical data is consistent with the Tchebotarev density of $1/2$.

Let $K^{(p)}$ be the number field as in Problem 13, and let $\mathbb{Z} \to R^{(p)}$ be the ring inclusion. The TDT implies that the density of primes of $\mathbb{Z}$ that completely split in $R^{(p)}$ is $\frac{1}{p-1}$. Construct the spectral diagram for the associated map ψ : Spec $R^{(p)} \to$ Spec $\mathbb{Z}$. Include at least 1000 primes of $\mathbb{Z}$. Show that this empirical data is consistent with the Tchebotarev density of $\frac{1}{p-1}$.

Problem 16. Let Spec $\mathbb{Z}C_3 \to$ Spec $\mathbb{Z}$ be the map illustrated in §1.3. Use a computer algebra/graphics program to extend the spectral diagram in Figure 1.3 to the first 1000 primes of $\mathbb{Z}$. What does this suggest about the ratio of primes in $\mathbb{Z}$ that factor as $P_1 P_2$ (as does 11) to primes in $\mathbb{Z}$ that factor as $P_1 P_2 P_3$ (like 13)? Is there an analog of the TDT for rings that are not rings of integers in number fields?

12.7 The Class Group of a Hopf Order

Problem 17. Let $K = \mathbb{Q}(\sqrt{d})$ be a real quadratic extension with $d > 0$ and square-free, let $S = \{v\}$ be the set of normalized absolute values on K, and let J_K denote the group of ideles over K. Find a constant κ such that for an adele $\{a_v\} \in J_K$ there exists an element $b \in K^+, b \neq 0$, for which $v(b) \leq v(a_v), \forall v \in S$.

Problem 18. The group of units in an algebraic number field can be determined using the following result of Dirichlet.

Dirichlet's Unit Theorem. *Let K be a finite extension of $\mathbb{Q}$ of degree n. Then*

$$U(R) \cong \mathbb{Z}^{r+s-1} \times W,$$

where r is the number of embeddings of K into $\mathbb{R}$, s is the number of pairs of conjugate embeddings of K into $\mathbb{C}$, and W is the subgroup of $R^\times$ generated by the roots of unity in R. Moreover, $n = r + 2s$.

Let $p \geq 5$ be prime, and let $K^{(p)}$ be the splitting field of the polynomial $h_p(x) = f_p(x)/(x+1)$. Since every root of $h_p(x)$ is real, one has

$$U(R^{(p)}) \cong \langle -1 \rangle \times \underbrace{\mathbb{Z} \times \mathbb{Z} \times \cdots \mathbb{Z}}_{p-2}.$$

Find a system of fundamental units for $R^{(p)}$. (We were very close to giving a proof of Dirichlet's Unit Theorem using the theory of adeles developed in Chapter 11. See [CF67, Chapter II, §18].)

Problem 19. For the prime $p = 7$, use a computer algebra system (like GAP) to compute the Hopf-Swan subgroup for each Larson order $H(i)$ in $K = \mathbb{Q}(\zeta_{49})$.

Problem 20. Let $n \geq 1$ be an integer and let $K = \mathbb{Q}(\zeta_{p^n})$. Compute the structure of $T(H)$ for a Larson order H in KC_p. Let $K = \mathbb{Q}(\zeta_{p^2})$ and let H be an R-Hopf order in KC_{p^2}. Compute $T(H)$.

Bibliography

[Br82]   K. Brown, Cohomology of Groups, Springer-Verlag, New York, 1982.

[By93a] N. Byott, Cleft extensions of Hopf algebras I, *J. Algebrax*, **157**, (1993), 405–429.

[By93b] N. Byott, Cleft extensions of Hopf algebras II, *Proc. London Math. Soc.*, **67**, (1993), 227–307.

[By95]   N. P. Byott, Hopf orders and a generalisation of a theorem of L. R. McCulloh, *J. Algebra*, **177**, (1995), 409–433.

[By02]   N. Byott, Integral Hopf-Galois structures in degree p^2 extensions of p-adic fields, *J. Algebra*, **248**, (2002), 334–365.

[By04]   N. Byott, Monogenic Hopf orders and associated orders of valuation rings, *J. Algebra*, **275**, (2004), 575–599.

[CF67]   J. W. S. Cassels, A. Fröhlich (eds.), Algebraic Number Theory, Academic Press, London, 1967.

[Ch79]   L. N. Childs, A Concrete Introduction to Higher Algebra, Springer-Verlag, New York, 1979.

[CH86]   L. N. Childs, S. Hurley, Tameness and local normal bases for objects of finite Hopf algebras, *Trans. Amer. Math. Soc.*, **298**, (1986), 763–778.

[Ch87]   L. N. Childs, Taming wild extensions with Hopf algebras, *Trans. Amer. Math. Soc.*, **304**, (1987), 111–140.

[CZ94]   L. N. Childs, K. Zimmermann, Congruence-torsion subgroups of dimension one formal groups, *J. Algebra*, **170**, (1994), 929–955.

[Ch96]   L. N. Childs, Hopf Galois structures on degree p^2 cyclic extensions of local fields, *New York J. Math.*, **2**, (1996), 86–102.

[Ch98]   L. N. Childs, Introduction to polynomial formal groups and Hopf algebras, in Hopf Algebras, Polynomial Formal Groups and Raynaud Orders, *Mem. Amer. Math. Soc.*, **136**(651), (1998), 1–10.

[CHR65] S. Chase, D. Harrison, A. Rosenberg, Galois theory and Galois cohomology of commutative rings, *Mem. Amer. Math. Soc.*, 52, (1965), 15–33.

[CMS98] L. N. Childs, D. J. Moss, J. Sauerberg, Dimension one polynomial formal groups, in Hopf Algebras, Polynomial Formal Groups and Raynaud Orders, *Mem. Amer. Math. Soc.*, **136**(651), (1998), 11–19.

[CMSZ98] L. N. Childs, D. J. Moss, J. Sauerberg, K. Zimmermann, Dimension two polynomial formal groups, in Hopf Algebras, Polynomial Formal Groups and Raynaud Orders, *Mem. Amer. Math. Soc.*, **136**(651), (1998), 21–50.

[CS98]   L. N. Childs, J. Sauerberg, Degree two formal groups and Hopf algebras, in Hopf Algebras, Polynomial Formal Groups and Raynaud Orders, *Mem. Amer. Math. Soc.*, **136**(651), (1998), 55–89.

R.G. Underwood, *An Introduction to Hopf Algebras*, DOI 10.1007/978-0-387-72766-0,
© Springer Science+Business Media, LLC 2011

[Ch00]   L. N. Childs, Taming wild extensions: Hopf algebras and local Galois module theory, *Amer. Math. Soc., Math. Surveys Monographs*, **80**, (2000).

[CU03]   L. N. Childs, R. G. Underwood, Cyclic Hopf orders defined by isogenies of formal groups, *Amer. J. Math.*, **125**, (2003), 1295–1334.

[CU04]   L. N. Childs and R. G. Underwood, Duals of formal group Hopf orders in cyclic groups, *Ill. J. Math.*, **48**(3), (2004), 923–940.

[CR81]   C. Curtis, I. Reiner, Methods of Representation Theory, vol. 1, John Wiley & Sons, New York, 1981.

[DG70]   M. Demazure, P. Gabriel, Groupes algebriques, Masson, Paris, 1970.

[Ei95]   D. Eisenbud, Commutative Algebra, Springer-Verlag, New York, 1995.

[FLSU08]   M. Filaseta, F. Luca, P. Stanica, R. G. Underwood, Galois groups of polynomials arising from circulant matrices, *J. Num. Theory*, **128**(1), (2008), 59–70.

[Fr03]   J. Fraleigh, A First Course in Abstract Algebra, 7th ed., Addison Wesley, Boston, 2003.

[Fro68]   A. Fröhlich, Formal Groups, Lecture Notes in Mathematics., **74**, Springer-Verlag, New York, 1968.

[Fro83]   A. Fröhlich, Galois Module Structure of Algebraic Integers, Ergebnisse. der Mathematisches "mathematik", 3 Folge, Band 1, Springer-Verlag, Berlin, 1983.

[FT91]   A. Fröhlich, M. J. Taylor, Algebraic Number Theory, Cambridge University. Press, Cambridge, 1991.

[Gr92]   C. Greither, Extensions of finite group schemes and Hopf Galois theory over a complete discrete valuation ring, *Math. Z.*, **210**, (1992), 37–67.

[GC98]   C. Greither, L. N. Childs, *p*-elementary group–schemes constructions, and Raynaud's theory, in Hopf Algebras, Polynomial Formal Groups and Raynaud Orders, *Mem. Amer. Math. Soc.*, **136**(651), (1998), 91–118.

[GRRS99]   C. Greither, D. R. Replogle, K. Rubin, A. Srivastav, Swan modules and Hilbert-Speiser number fields, *J. Number. Theory*, **79**, (1999), 164–173.

[HS71]   P. J. Hilton, U. Stammbach, A Course in Homological Algebra, Springer-Verlag, New York, 1971.

[HK71]   K. Hoffman, R. Kunze, Linear Algebra, 2nd ed., Prentice-Hall, Englewood Cliffs, New Jersey, 1971.

[Ho64]   F. E. Hohn, Elementary Matrix Algebra, 2nd ed., Macmillan, New York, 1964.

[IR90]   K. Ireland, M. Rosen, A Classical Introduction to Modern Number Theory, 2nd ed., Springer-Verlag, New York, 1990.

[Ja51]   N. Jacobson, Lectures in Abstract Algebra, D. Van Nostrand, Princeton, New Jersey, 1951.

[KM07]   A. Koch, A. Malagon, *p*-adic order bounded group valuations on abelian groups, *Glasgow Math. J.*, **49**(2), (2007), 269–279.

[La84]   S. Lang, Algebra, 2nd ed., Addison-Wesley, Reading, Massachusetts, 1984.

[Lar67]   R. G. Larson, Group rings over Dedekind domains, *J. Algebra*, **5**, (1967), 358–361.

[LS69]   R. G. Larson, M. E. Sweedler, An associative orthogonal bilinear form for Hopf algebras, *Amer. J. Math.*, **91**, (1969), 75–93.

[Lar71]   R. G. Larson, Characters of Hopf algebras, *J. Algebra*, **17**, (1972), 352–368.

[Lar72]   R. G. Larson, Orders in Hopf algebras, *J. Algebra*, **22**, (1972), 201–210.

[Lar76]   R. G. Larson, Hopf algebra orders determined by group valuations, *J. Algebra*, **38**, (1976), 414–452.

[Lar88]   R. G. Larson, Hopf algebras via symbolic algebra, in Computers in Algebra M. C. Tangora (ed), Lecture Notes in Pure and Applied Mathematics, **111**, Marcel Dekker, New York, 1988, 91–97.

[Lu79]   J. Lubin, Canonical subgroups of formal groups, *Trans. Amer. Math. Soc.*, **251**, (1979), 103–127.

[Mc83]   L. R. McCulloh, Galois module structure of elementary abelian extensions, *J. Algebra*, **82**, (1983), 102–134.

[Mc87]   L. R. McCulloh, Galois module structure of abelian extensions, *J. Reine Angew. Math.*, **375/376**, (1987), 259–306.

[Mi96]   J. S. Milne, Class Field Theory, unpublished class notes, 1996.

[Mo93]   S. Montgomery, Hopf Algebras and Their Actions on Rings, American Mathematical Society, CBMS Regional Conference Series in Mathematics, American Mathematical Society, Providence Rhode Island, **82**, (1993).

[Mu75]   J. Munkres, Topology, Prentice-Hall, Englewood Cliffs, New Jersey, 1975.

[Ne99]   J. Neukirch, Algebraic Number Theory, Grundlehren der mathematischen Wissenschaften, **322**, Springer-Verlag, Berlin, 1999.

[Ra74]   M. Raynaud, Schemas en groupes de type $(p, ..., p)$, *Bull. Soc. Math. France*, **102**, (1974), 241–280.

[RUl74]   I. Reiner, S. V. Ullom, A Mayer-Vietoris sequence for class groups, *J. Algebra*, **31**, (1974), 305–342.

[Re01]   D. Replogle, Cyclotomic Swan subgroups and irregular indices, *Rocky Mtn. J. Math.*, **31**(2), (2001), 611–618.

[RU02]   D. Replogle, R. Underwood, Nontrivial tame extensions over Hopf orders, *Acta Arith.*, **104**(1), (2002), 67–84.

[Ro73]   L. Roberts, Comparison of algebraic and topological K-theory, algebraic K-theory, II: "Classical" algebraic K-theory and connections with arithmetic, Proceedings of a Conference, Battelle Memorial Institute, Seattle, Washington, 1972, Lecture Notes in Mathematics, **342**, Springer-Verlag, Berlin, 1973, 74–78.

[Rot02]   J. Rotman, Advanced Modern Algebra, Pearson, Upper Saddle River, New Jersey, 2002.

[Ru76]   W. Rudin, Principles of Mathematical Analysis, 3rd ed., McGraw-Hill, New York, 1976.

[SS94]   T. Sekiguchi, N. Suwa, Theories de Kummer-Artin-Schreier-Witt theory, *Comptes Rendus l'Acad. Sci.*, **319**, ser. I, (1994), 105–110.

[Se77]   J.-P. Serre, Linear Representations of Finite Groups, Springer-Verlag, New York, 1977.

[Se79]   J.-P. Serre, Local Fields, Springer-Verlag, New York, 1979.

[Sh74]   I. R. Shafarevich, Basic Algebraic Geometry, Springer-Verlag, New York, 1974.

[Sm97]   H. Smith, Constructing Hopf orders in elementary abelian group rings, doctoral dissertation, SUNY Albany, 1997.

[Sn94]   V. Snaith, Galois Module Structure, American. Mathematical Society, Providence, Rhode Island, Fields Institute Monograph, **2**, 1994.

[Sw68]   R. G. Swan, Algebraic K-Theory, Lecture Notes in Mathematics, **76**, Springer-Verlag, New York, 1968.

[Swe69]   M. E. Sweedler, Hopf Algebras, W. A. Benjamin, New York, 1969.

[TO70]   J. Tate, F. Oort, Group schemes of prime order, *Ann. Sci. Ec. Norm. Sup.*, **3**, (1970), 1–21.

[TB92]   M. J. Taylor, N. P. Byott, Hopf orders and Galois module structure, DMV Seminar **18**, Birkhauser Verlag, Basel, 1992, 154–210.

[Ul76]   S. V. Ullom, Nontrivial lower bounds for class groups of integral group rings, *Ill. J. Math.*, **20**, (1976), 361–371.

[Un94]   R. Underwood, R-Hopf algebra orders in KC_{p^2}, *J. Algebra*, **169**, (1994), 418–440.

[Un96]   R. Underwood, The valuative condition and R-Hopf algebra orders in KC_{p^3}, *Amer. J. Math.*, **118**, (1996), 701–743.

[Un98]   R. Underwood, The Structure and realizability of R-Hopf orders in KC_{p^3}, *Comm. Alg.*, **26**(11), (1998), 3447–3462.

[Un99]   R. Underwood, Isogenies of polynomial formal groups, *J. Algebra*, **212**, (1999), 428–459.

[Un03]   R. Underwood, Galois module theory over a discrete valuation ring, in Recent Research on Pure and Applied Algebra, O. Pordavi (ed). Nova Science Publishers, Hauppauge, New York, 2003.

[UC05]   R. Underwood, L. N. Childs, Duality for Hopf orders, *Trans. Amer. Math. Soc.*, **358**(3), (2006), 1117–1163.

[Un06]   R. Underwood, Realizable Hopf orders in KC_8, *Inter. Math. Forum*, **1**(17–20), (2006), 833–851.

[Un08a]   R. Underwood, Sequences of Hopf-Swan subgroups, *J. Number Theory*, **128**(7), (2008), 1900–1915.

[Un08b]   R. Underwood, Realizable Hopf orders in KC_{p^3}, *J. Algebra*, **319**(11), (2008), 4426–4455.

[Wa97]   L. C. Washington, Introduction to Cyclotomic Fields, Springer-Verlag, New York, 1997.

[Wat79]   W. Waterhouse, Introduction to Affine Group Schemes, Springer-Verlag, New York, 1979.

[We63]   E. Weiss, Algebraic Number Theory, Chelsea, New York, 1963.

Index

R.G. Underwood, *An Introduction to Hopf Algebras*, DOI 10.1007/978-0-387-72766-0,
© Springer Science+Business Media, LLC 2011